Thermal Engineering Volume 2

Shiv Kumar

Thermal Engineering
Volume 2

Ane Books
Pvt. Ltd.

Springer

Shiv Kumar
Department of Mechanical and Automation
Engineering
Guru Gobind Singh Indraprastha University
Delhi, Delhi, India

ISBN 978-3-030-89218-0   ISBN 978-3-030-89216-6   (eBook)
https://doi.org/10.1007/978-3-030-89216-6

Jointly published with ANE Books Pvt. Ltd.
In addition to this printed edition, there is a local printed edition of this work available via Ane Books in
South Asia (India, Pakistan, Sri Lanka, Bangladesh, Nepal and Bhutan) and Africa (all countries in the
African subcontinent).
ISBN of the Co-Publisher's edition: 978-9-3854-6232-0

This Springer imprint is published by the registered company Springer Nature Switzerland AG
The registered company address is: Gewerbestrasse 11, 6330 Cham, Switzerland

Dedicated to

My Parents

My Wife Dr. Kusum Lata and My Son Tanishq

Preface

This second volume on Thermal Engineering is intended to cover the advanced concepts of thermodynamics and its applications in Vapor Power Cycle, IC Engines, Gas Turbines, and Jet Propulsions. The book provides a conceptual background that would facilitate students to understand the advanced subjects like Internal Combustion Engines, Power Plant Engineering, Refrigeration, and Air-Conditioning of engineering disciplines, better.

Thermal Engineering is often seen as the most challenging core subject encountered by engineering students. This text overcomes this difficulty by introducing the concepts through examples before moving on to the more involved mathematics. The various laws of thermodynamics correlated with real phenomena and combined with innumerable figures, help the reader place the subject in context.

Writing this book made me think a lot more than the material it covers. The methods I used in this book are primarily the ones that worked best for my students. The suggestions from the teachers and students for the further improvement of the text are welcome and will be implemented in the next edition. The readers are requested to bring out the error to the notice, which will be gratefully acknowledged.

Shiv Kumar

Acknowledgements

First of all, I would like to express my deep gratitude to God for giving me the strength and health for completing this book. I am very thankful to my colleagues in the mechanical engineering department for their highly appreciable help and my students for their valuable suggestions.

I am also thankful to my publishers Shri. Sunil Saxena and Shri. Jai Raj Kapoor of Ane Books Pvt. Ltd. and the editorial group for their help and assistance.

A special thanks goes to my wife Dr. Kusum Lata for her help, support, and strength to complete the book.

Shiv Kumar

Contents

About the Author

Shiv Kumar is former Head of Department of Mechanical and Automation Engineering and Associate Professor at Guru Premsukh Memorial College of Engineering affiliated to Guru Gobind Singh Indraprastha University, Delhi, India. He obtained his Bachelor's degree in Mechanical Machine Design and Automation Engineering from R. E. C. (presently NIT), Jalandhar, and Master's degree in Thermal Engineering from Delhi College of Engineering, India. He has over 18 years of teaching experience. He has published several books in the field of Mechanical Engineering.

Chapter 1
Properties of Pure Substance

Nomenclature

p_v	kPa	Vapour pressure of a liquid
T_{sat}	°C or K	Saturated temperature
T_{sup}	°C or K	Superheated temperature
h_f	kJ/kg	Specific enthalpy at saturated liquid state
h_g	kJ/kg	Specific enthalpy at saturated vapour state
h_{fg}	kJ/kg	Latent heat of evaporization
c_w	kJ/kgK	Specific heat of the water
v	m³/kg	Specific volume
CP	—	Critical point
v_f	m³/kg	Specific volume of saturated liquid state
v_g	m³/kg	Specific volume of saturated vapour state
TP	—	Triple point
m_f	kg	Mass of liquid
m_g	kg	Mass of vapour
x	—	Dryness fraction
s_f	kJ/kgK	Specific entropy at saturated liquid state
s_g	kJ/kgK	Specific entropy at saturated vapour state
u_f	kJ/kg	Specific internal energy at saturated liquid line
u_g	kJ/kg	Specific internal energy at saturated vapour line
q	kJ/kg	Heat transfer per unit mass
w	kJ/kg	Work transfer per unit mass
p_{atm}	kPa	Atmospheric pressure
p	kPa	Absolute pressure
p_g	kPa	Gauge pressure

(continued)

© The Author(s) 2022
S. Kumar, *Thermal Engineering Volume 2*,
https://doi.org/10.1007/978-3-030-89216-6_1

(continued)

c_{ps}	kJ/kgK	Specific heat of superheated steam
V	Volt	Voltage
V	m/s	Velocity
V	m^3	Volume
I	A	Current

1.1 Introduction

A substance that has a fixed chemical composition throughout each phase during a thermodynamics process is called a **pure substance**, for example, water, carbon dioxide, nitrogen, etc. Water is one of the pure substances which exists in three different phases, *i.e.*, in the solid phase as ice, in the liquid phase as water and in the gaseous phase as steam, and in all the three phases it retains the same chemical composition, *i.e.*, H_2O.

A mixture of two or more phases is also called pure substance, if the chemical composition of all phases is the same. For example, a mixture of liquid water and steam (or ice and liquid water) is a pure substance because both phases have the same chemical composition. A mixture of gaseous air and liquid air is not a pure substance because the composition of gaseous air is different from the composition of liquid air.

1.2 Vapourization, Evaporation, and Boiling

Vapourization is a process of phase change from the liquid phase to vapour phase. It is a more general term and applies to both evaporation and boiling.

Evaporization is a process of phase change from the liquid to vapour that occurs below the boiling point where particles of the liquid move from the free surface of liquid to the vapour state. It occurs at any temperature when the vapour pressure of a liquid is less than the surrounding pressure (or pressure on the liquid surface) at given temperature.

Condition for evaporation,

$$p_v < p$$

where
$$p_v = \text{vapour pressure of the liquid,}$$
$$p = \text{pressure of the liqiud or surronding pressure.}$$

$p_v < p$.

where p_v = vapour pressure of the liquid,

p = pressure of the liquid or surrounding pressure.

Boiling is a process of phase change from the liquid to vapour at fixed temperature and at given pressure. Boiling takes place when the vapour pressure p_v becomes equal to or greater than the pressure on liquid free surface at given temperature.

Condition for boiling,

$$p_v \geq p \text{ at } T = C$$

In boiling, vapourization can occur anywhere within the liquid, not just at the surface.

Difference between evaporation and boiling

S. No	Evaporation	Boiling
1	Evaporation takes place slowly and proceeds more rapidly as the temperature increases	Boiling occurs at a constant rate for a particular temperature
2	It takes place at any temperature. For water, at 1 atm pressure, evaporation takes place at 0 °C to less than 100 °C	It occurs at a definite temperature which is called boiling temperature. For water, at 1 atm pressure, boiling temperature is 100 °C
3	The temperature decreases during evaporation	The temperature remains constant during boiling
4	It takes place only on the surface of the liquid	Boiling takes place at all regions of the liquid, *i.e.*, on the surface of liquid, within the liquid and the bottom of the liquid
5	The rate of evaporation depends on the area of the free surface of the liquid	The rate of boiling does not depend on the area of the free surface of the liquid

1.3 Formation of Steam at Constant Pressure

Consider the heating of water at constant pressure. If various properties are to be measured, an experiment can be set up where water is heated in a vertical cylinder closed by a frictionless piston on which there is a weight. The weight acting down under gravity on a piston of fixed size ensures that the fluid in the cylinder is always subjected to constant pressure. Initially, the cylinder contains 1 kg of water at 25 °C. As this is heated the water changes into steam and certain characteristics may be noted.

1. At state *A*, water is at 25 °C and atmospheric pressure (101.325 kPa), water exists in the liquid phase, and it is called a subcooled liquid, or a compressed

liquid, meaning that it is not about to vapourize. Heat is now supplied to the water until its temperature rises, say, 60 °C. As the temperature rises, the liquid water expands slightly, and its specific volume increases. The pressure in the cylinder remains constant, *i.e.*, 101.325 kPa because the piston can move up slightly. Water is still a compressed liquid at this state since vapourization has not started yet. As more heat is supplied, the temperature of water rises and continues to rise until it reaches 100 °C (boiling point) at state B at constant atmospheric pressure

[**Note:** The boiling temperature of water depends on the pressure acting on it].

At state B, water is still a liquid, but further small amount of heat addition will cause some of the liquid to vapourize. That is, a phase-change process from liquid to vapour is about to take place. The state of liquid at which a vapour just begins to form (*i.e.*, liquid about to vapourize) is called a saturated liquid. Therefore, state B is called saturated liquid state.

Some terms are related to process A–B and state B.

(i)   **Saturated liquid state:** The state of liquid at which a vapour just begins to form is called **saturated liquid state**. It is denoted by point B in Fig. 1.1b. For water, saturated liquid state lies at 1 atm (101.325 kPa) pressure and 100 °C temperature.

(ii) **Sensible heat:** The amount of heat required to raise the temperature of the substance at constant pressure when phase change does not occur is called **sensible heat**. It is denoted by Q_s.

For present case, the water is heated from 25 to 100 °C.
The sensible heat required per unit mass:

$$q_s = c_w(T_B - T_A)$$

where
$c_w = 4.18$ kJ/kgK, specific heat of the water,
$\therefore$

$$q_s = 4.18(100 - 25) = 313.5 \text{ kJ/kg}.$$

If water is heated from 0 °C to 100 °C at atmospheric pressure, the sensible heat required per unit mass:

$$q_s = c_w(T_B - 0)$$
$$= 4.18(100 - 0) = 418 \text{ kJ/kg}$$

(iii) **Saturated temperature:** T_{sat}. It is the temperature at which the liquid starts boiling at a given pressure. It is denoted by T_{sat}. For water, at pressure of 101.325 kPa, $T_{sat} = 100$ °C.

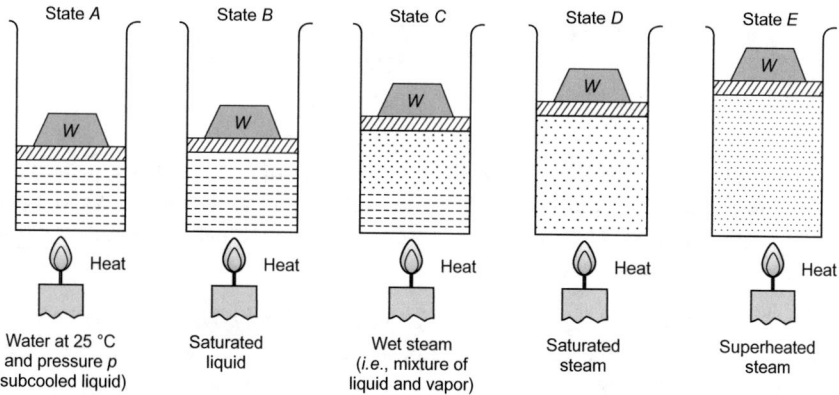

(a) Generation of steam at constant pressure

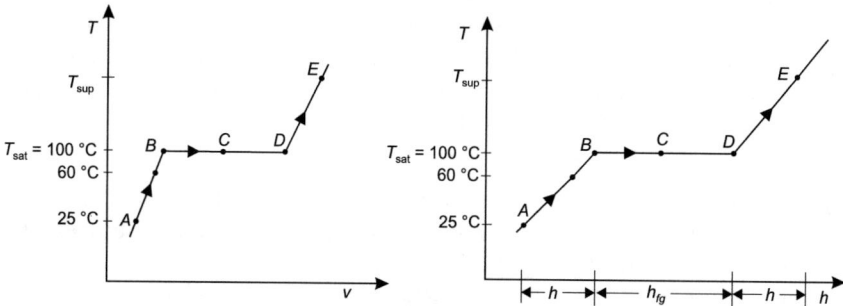

(b) T–v and T–h diagrams for generation of steam at atmospheric pressure i.e., $p = 101.325$ kPa

Fig. 1.1 Formation of steam at constant pressure

(iv)   **Saturated pressure:** p_{sat}. It is the pressure at which the liquid starts boiling at a given temperature. It is denoted by p_{sat}. For water, at temperature of 100 °C, $p_{sat} = 101.325$kPa.

Both the saturated temperature T_{sat} and saturated pressure p_{sat} remain constant during phase change from liquid to vapour.

(v)   **Compressed liquid or subcooled liquid:** The state of liquid at which it is not about to vapourize is called **compressed liquid or subcooled liquid**. In other words, when the temperature of liquid is less than the saturated temperature corresponding to given pressure, the liquid is called compressed liquid or subcooled liquid.

At pressure of 101.325 kPa, the compressed liquid exists at temperature of 0 °C to less than 100 °C.

The difference between the temperature of subcooled liquid and saturated temperature at given pressure is called as **degree of subcooling**.

Degree of subcooling at state A $= (T_{\text{sat}} - T_A)$
$$= (100 - 25) \quad \text{at} \quad p = 101.325\text{kPa}$$
$$= 75^{\circ}\text{C or K}$$

2. At state B, the condition of water is saturated liquid. As heat is transferred at state B, the temperature stops rising until the liquid is completely vapourized, *i.e.*, the temperature will remain constant during the phase change from liquid to vapour at given constant pressure.

During the phase-change process (*i.e.*, boiling), the only change we will observe is a large increase in the volume and slowly the liquid is converted into vapours. At state D, the liquid is completely converted into vapour on continuous supply of heat. The vapour at state D is called a saturated vapour (or also called dry saturated steam). Any heat loss from this vapour will cause some of the vapour to condense. A vapour that is about to condense is called a saturated vapour. Therefore, state D is a saturated vapour state. A substance at states between B and D is called a saturated liquid-vapour mixture (*i.e.*, wet steam) since the liquid and vapour phases coexist in equilibrium at these states. At state C, the midway of the vapourization line BD, the cylinder contains equal amount of liquid and vapour.

Some terms are related to process B–D and state D.

(i) **Saturated vapour state:** The state of vapour at which a condensation just begins on loss of heat from the vapour is called **saturated vapour state.** It is denoted by point D in Fig. 1.1b.

(ii) **Latent heat:** The amount of heat required to change the phase from liquid to vapour at constant temperature and pressure is called the **latent heat** or **latent heat of vapourization**. It is denoted by h_{fg}.

(iii) **Wet steam:** The mixture of water-vapour is called wet steam. This steam lies between saturated liquid state B and saturated vapour state D.

(iv) **Dry saturated steam:** When all the liquid water in wet steam is just to vapourize is called dry saturated steam. This steam lies at saturated vapour state D.

3. At saturated vapour state D, the vapour is called dry saturated steam. If the dry saturated steam is further heated beyond the state D to any state say E, this heating process is called superheating. The steam obtained is called the superheated steam which follows the gas laws.

Some terms are related to process D–E and state E.

(i) **Superheated steam:** The steam obtained by heating the dry saturated steam above the saturated temperature is called superheated steam. The heating process of the dry saturated steam from the state D to any state say E is called superheating.

The difference between the temperature of superheated steam and saturated temperature of steam at given pressure is called **degree of superheating**.

Degree of superheating $=$ temperature of superheated steam

$-$ saturated temperature of steam

$= T_E - T_D$

$= T_{sup} - T_{sat}$

1.4 Graphical Representation of T-V Diagram for Generation of Superheated Steam from −20 °C of ICE

Let us consider 1 kg of ice of −20 °C below the freezing point. The initial state of the substance had been denoted by state *A*.

Process A–B: When heat is supplied, temperature of ice rises from −20 to 0 °C with increase in volume. The heat supplied during process *A–B* is called sensible heat.

Process B–C: At state *B* is ice at 0 °C. On heating further beyond state *B*, at constant 0 °C temperature, the phase changes from ice (solid) to water (liquid) with decrease in volume. This transformation of phase from solid to liquid is called the melting or fusion of ice, and the amount of heat required for complete transformation of ice into water at constant temperature of 0 °C is called the latent heat of fusion. At 1 atm (101.325 kPa) pressure, the latent heat of ice is 80 kcal/kg or 334.96 kJ/kg Fig. 1.2.

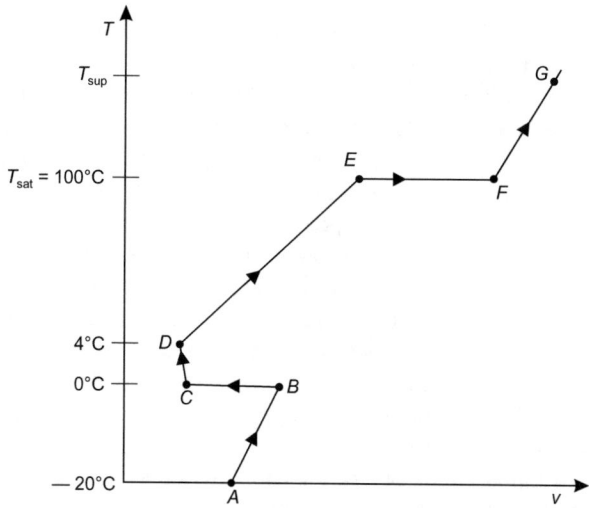

Fig. 1.2 Generation of superheated steam from −20 °C of ice at constant 1 atm pressure

Process C–D: At state C is water at 0 °C. On heating further beyond state C, the volume of water decreases when the temperature rises from 0 to 4 °C. The maximum density of water at 4 °C is 1000 kg/m³.

Process D–E: Further addition of heat at state D, temperature of water rises from 4 to 100 °C (*i.e.*, boiling point E) with increase in volume. The amount of heat required to raise the temperature of the liquid water from 0 to 100 °C liquid water at constant 1 atm pressure is called the sensible heat.

Process E–F: As more heat is added at the saturated liquid state E, boiling of water starts at constant temperature and pressure with increase in volume. During this phase change from state E to state F, there exists a two-phase mixture of water and its vapour. At state F, all the water has vapourized the state of steam at this point is called dry saturated steam. The amount of heat required during the phase change that occurs at constant temperature and pressure is called the latent heat of vapourization. For water boiling at 100 °C under 1 atm pressure (*i.e.*, 101.325 kPa), the latent heat = 2257 kJ/kg.

Process F–G: As more heat is added at the saturated vapour state F, temperature of steam rises at given constant pressure with increase in volume. The steam obtained by heating the dry saturated steam is called superheated steam. The heating process of the dry saturated steam from the state F to any state say G is called superheating.

1.5 Property Diagrams for Phase-Change Processes

The variation of properties during phase-change processes are studied with the help of property diagrams. We will develop and discuss the T-s, p–v and h–s diagrams for pure substances.

T-s diagram: The T-s (temperature-specific entropy) diagram shows the phase-change process of water at different constant pressure. As the pressure increases, the specific volume of water decreases and boiling point increases. By experiment, if the water is heated at constant pressure p_1, we get saturated liquid state (*i.e.*, boiling point) and saturated vapour state. If the experiment is repeated a number of times using different constant pressure, like p_2, p_3, p_4, and so on, we get a number of saturated liquid states and saturated vapour states as shown in Fig. 1.3. The horizontal line between the saturated liquid state and saturated vapour state is called saturated line. Thus, saturated line represents the latent heat of vapourization of the water. It is clear from the Fig. 1.3 that the saturated liquid states and the saturated vapour states comes close to each other and saturated lines become shorter with increase in pressures. As pressure is increased further, a state is reached at which saturated liquid state and saturated vapour state are identical and saturated line is disappeared. That state of water is called critical point.

The saturated liquid line is obtained by joining the saturated liquid states. The saturated vapour line is obtained by joining the saturated vapour states. The point at which the saturated liquid line and the saturated vapour line meet with increase in temperature or pressure is called critical point.

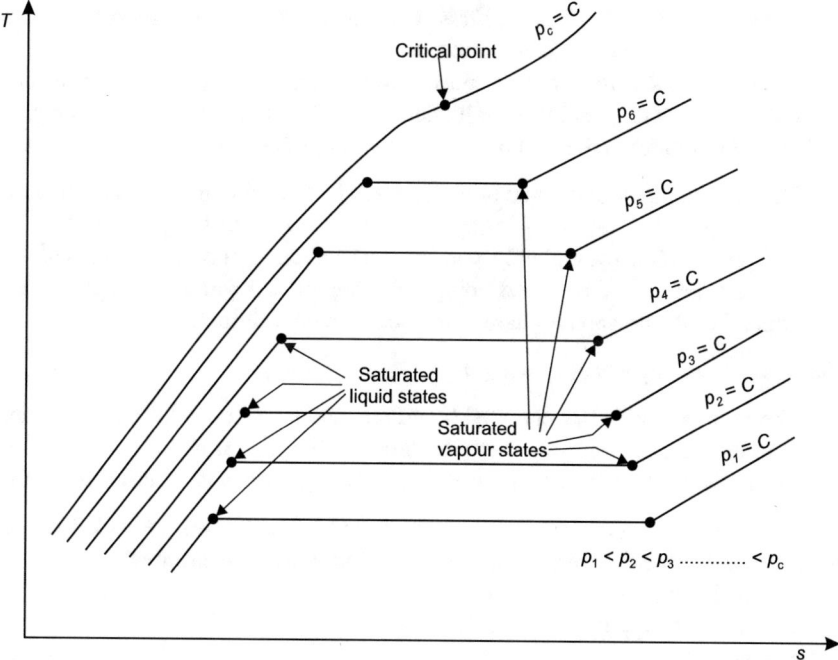

Fig. 1.3 T-s diagram of constant-pressure phase-change process of a pure substance at various pressures

Now, we can get the shape of saturated dome in *T-s* diagram as shown in Fig. 1.4. The shape of the saturated dome is identical for all the liquid. The saturated dome

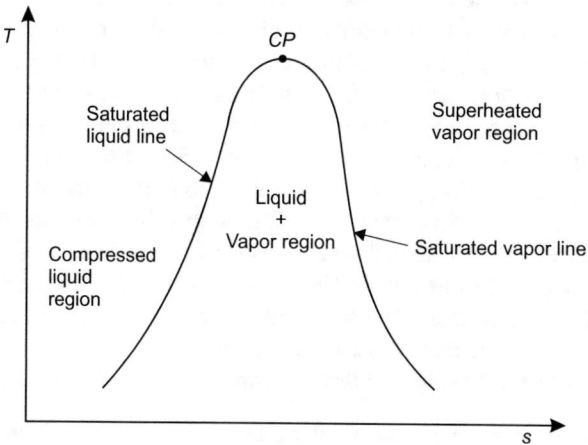

Fig. 1.4 Saturated dome in *T-s* diagram for a pure substance

may be narrow or broad or lies at different ranges of pressures and temperatures but the shape of the dome must be the same.

All the states that involve both liquid and vapour phases in equilibrium are located under the dome which is called the saturated liquid–vapour region, or the wet region.

The critical point is defined in different ways as following:

1. The critical point is the state at which there is no distinction between the saturated liquid and saturated vapour, *i.e.*, at which all of the properties of the two phases (density, enthalpy, entropy, internal energy) become identical. For a pure substance, the critical point is the highest temperature and pressure at which liquid and vapour phases can coexist in equilibrium.

That is, $h_f = h_g, v_f = v_g, s_f = sg, u_f = u_g$.

2. The point at which the saturated liquid line and the saturated vapour line meet with increase in temperature or pressure is called critical point.
3. The point at which the latent heat of vapourization is zero is called critical point.

The temperature, pressure, and volume at the critical point are called critical temperature T_c, critical pressure p_c, and critical volume v_c, respectively.

At critical point for pure substance:

Latent heat of vapourization is zero.

For water,

$$T_c = 374.15\,^{\circ}\text{C},$$
$$p_c = 22.12\,\text{MPa} = 221.2\,\text{bar},$$
$$v_c = 0.003155\,\text{m}^3/\text{kg}.$$

Latent heat of vapourization: $h_{fg} = 0$ always for all pure substance

The gas can never be liquefied above the critical point.

p–v **diagram:** The *p–v* (pressure-specific volume) diagram shows the phase change process of water at different constant pressure. In similar way as *T-s* diagram, we can mark a number of saturated liquid states, saturated vapour states as shown in Fig. 1.5. And we can get the saturated liquid line, the saturated vapour line, critical point, and shape of the saturated dome as shown in Figs. 1.6 and 1.7.

The *h–s* (specific enthalpy-specific entropy) diagram shows the phase-change process of water at different constant pressure. In this diagram, specific enthalpy is laid along the axis of ordinate and specific entropy along the axis of abscissa. This diagram was suggested by Mollier in 1904 and that is why *h–s* diagram is also called the Mollier diagram or chart. The saturated liquid line, the saturated vapour line, critical point, and the shape of the saturated dome are also obtained as similar way as *T-s* diagram. The following observations can be made from *h–s* diagram.

1. Constant pressure lines and constant temperature lines coincide in the wet region. The pressure lines diverge from one another due to the increase in saturation temperature with increase in pressure.

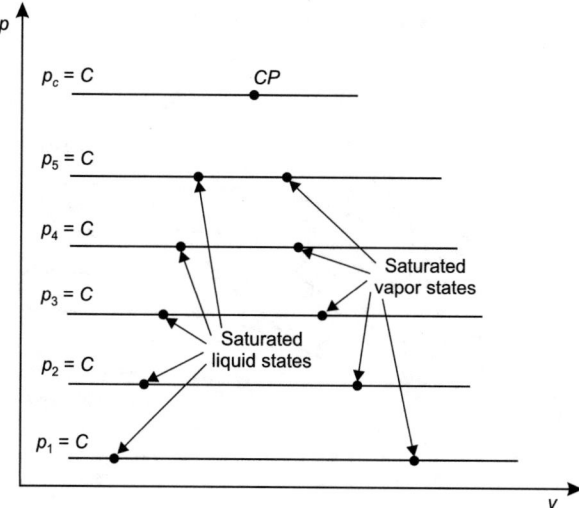

Fig. 1.5 *p–v* diagram of constant-pressure phase-change processes of a pure substance at various pressures

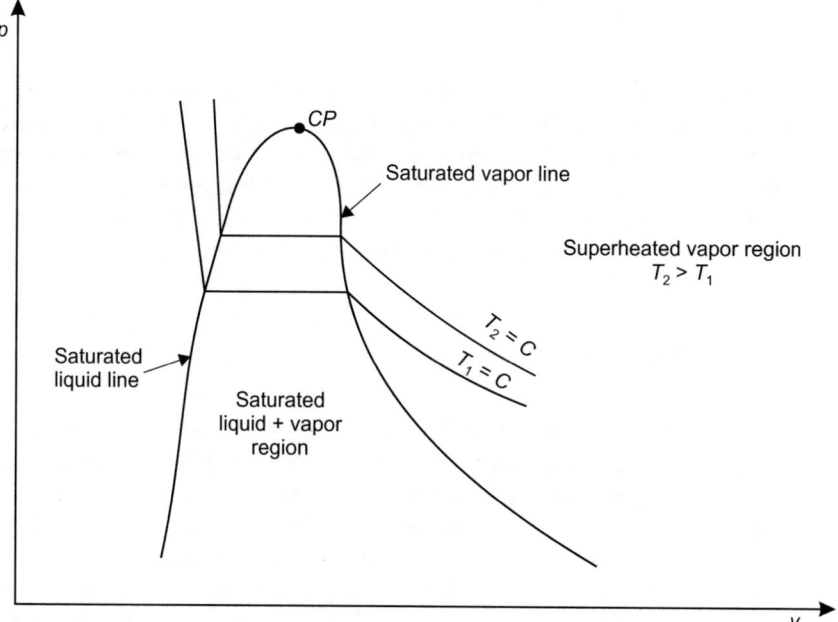

Fig. 1.6 Saturated dome in *p–v* diagram for a pure substance

$h - s$ diagram or Mollier diagram:

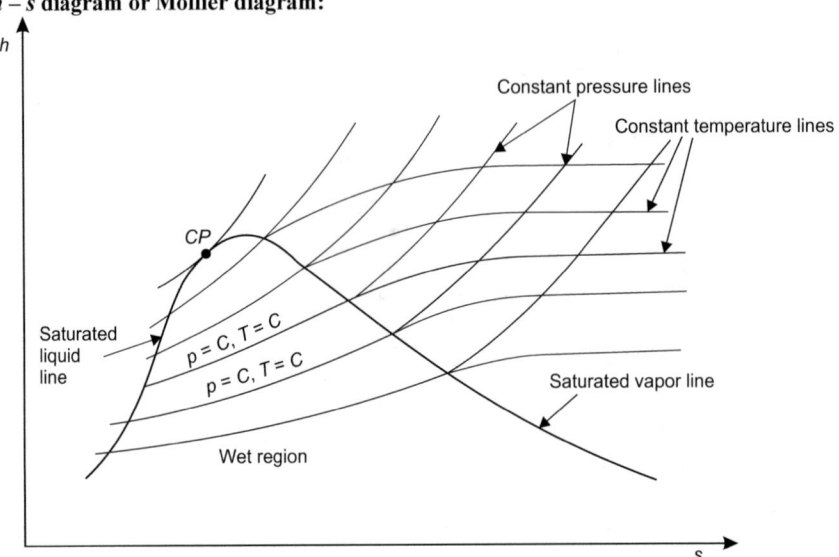

Fig. 1.7   h–s diagram or Mollier diagram

2.   In the region of superheated vapour, the isobaric lines have a positive slope and turn upwards, and the constant temperature lines tend towards a horizontal straight line. This may be attributed to the fact that at high temperature, the behaviour of superheated vapour approaches as on ideal gas, *i.e.*, $h = f(T)$.
3.   The throttling process, which is a constant enthalpy process, plots as a horizontal line. A reversible adiabatic process plots as a vertical line.

1.6   *P–v-T* Surface

All the information contained on both the *p–v* and the *p–T* diagram can be shown on a single diagram if the three coordinates *p*, *v*, and *T* are plotted in three-dimensional space. The three dimensional space among *p*, *v*, and *T* is called a *p–v-T* surface. Such a surface is shown in Fig. 1.8 for a substance like carbon dioxide which expands on melting and in Fig. 1.9 for a substance like water which contracts on melting.

If the *p–v-T* surface is projected on the *p–v* plane, a *p–v* diagram, and projected on the *p–T* plane, the surface generates a *p–T* diagram as shown in Fig. 1.8. Any point on *p–v-T* surface represent an equilibrium state of the substance. The whole solid-vapour region projects to the vapourization curve, the whole liquid–vapour region projected to the vapourization curve, and the whole solid–liquid region projected to the fusion curve, and the triple line projected to the triple point to the *p–T* plane.

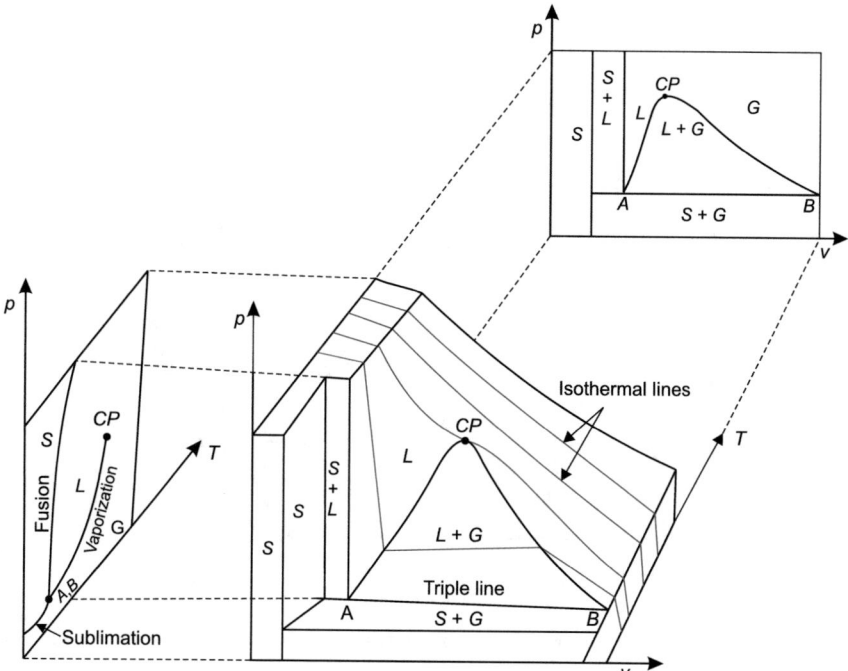

Fig. 1.8 *p–v-T* surface for a substance which expands on melting or contracts on freezing (like carbon dioxide)

1.7 *p–T* DIAGRAM

The *p–T* diagram or phase diagram for pure substance like water is shown in Fig. 1.10. Some terms are related to *p–T* diagram:

1. Sublimation line,
2. Fusion line,
3. Vapourization line,
4. Triple point,
5. Critical point

1. **Sublimation line (0–TP):** The sublimation line separates the solid and vapour regions. The solid and vapour coexist in equilibrium on sublimation line 0–TP. The sublimation refers to the heating process in which solid is directly transformed into vapour phase without melting into liquid. The solid carbon dioxide follows the sublimation process in which solid CO_2 (dry ice) is directly transformed into vapour phase at atmospheric conditions. In the particular case of water, the sublimation line is called the frost line.

2. **Fusion line (TP–1):** The fusion line separates the solid and liquid phases. The solid and liquid coexist in equilibrium on fusion line *TP*–1. The fusion refers

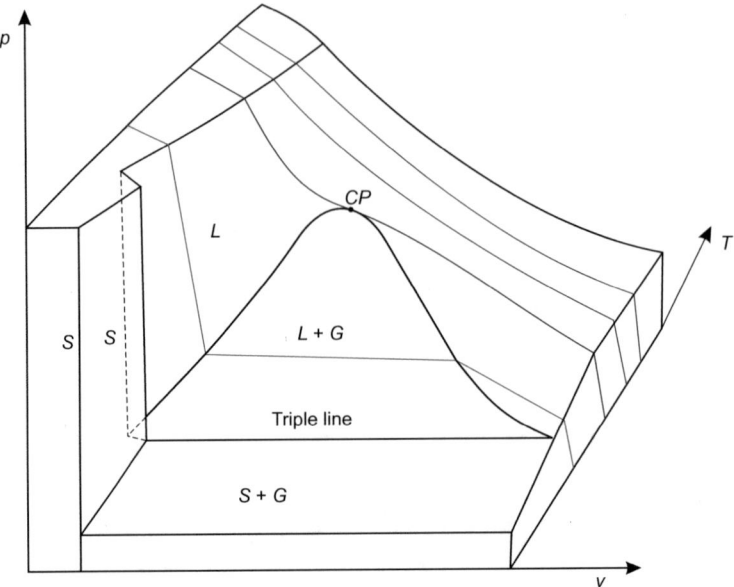

Fig. 1.9 p–v-T surface for a substance which contracts on melting or expands on freezing (like water)

Fig. 1.10 p-T diagram for water

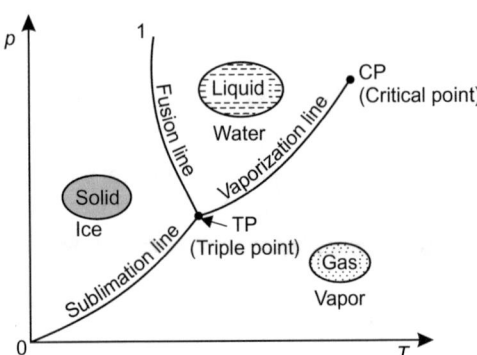

to the heating process in which solid phase is transformed into liquid phase. In the particular case of water, the fusion line is called ice line.

3. **Vapourization line (TP-CP):** The vapourization line separates the liquid phase and vapour phase. The liquid and vapour coexist in equilibrium on vapourization line TP-CP. The vapourization refers to the heating process which liquid phase is converted into vapour phase. In the particular case of water, the vapourization line is called the steam line.

4. **Triple point (TP):** The triple point is a state where all three phases like solid, liquid, and vapour coexist in equilibrium. The triple point may be defined as

the point at which the sublimation fusion and vapourization lines meet. The triple point exists at a definite pressure and temperature. For water, the values of temperature and pressure at triple point are as follows:

$$T_{TP} = 0.01°C = (273.15 + 0.01)K = 273.16 \text{ K}$$

and

$$p_{TP} = 611.2 \text{ Pa} = 0.6112\text{kPa} = 0.006112 \text{ bar}$$

For CO_2,

$$T_{TP} = -56.60°C$$
$$p_{TP} = 5.178\text{bar}$$

5. **Critical point (*CP*):** The point at which the vapourization line end is called the critical point. There is no distinction between liquid and vapour phases above the critical point.

1.8 Types of Steam

The steam is classified into the following three categories:

1. Wet steam or saturated liquid–vapour mixture,
2. Dry saturated steam, and
3. Superheated steam.

1.8.1 Wet Steam

The wet steam is the mixture of saturated liquid and saturated vapour. This steam lies between saturated liquid line and saturated vapour line at saturated temperature. Wet steam can exist with different qualities, *i.e.*, having different proportions of water particles and dry steam. Therefore, it will be necessary to express the quality of the wet steam. The quality of the wet steam is indicated by dryness fraction.

Dryness fraction. The dryness fraction of the wet steam is defined as the ratio of mass of vapour to the total mass of the wet steam. It is denoted by x.

$$\text{Dryness fraction: } x = \frac{\text{mass of vapor}}{\text{total mass of the wet steam}}$$
$$x = \frac{m_{\text{vapor}}}{m}$$

where

$$m = m_{\text{liquid}} + m_{\text{vapor}} = m_f + m_g$$

$\therefore$

$$x = \frac{m_g}{m_f + m_g}$$

[Note that a liquid was referred to as a fluid and a vapour as a gas, hence the subscripts f and g for liquid and vapour, respectively.]

The quality x (or dryness fraction) has significance for wet steam only. It has no meaning in the compressed liquid or superheated steam. Its value is between 0 and 1. The quality of saturated liquid is 0, and the quality of saturated vapour is 1.

1.8.2 Dry Saturated Steam

The steam that has no water particles at its saturated temperature is called dry saturated steam or simply dry steam. This steam exists at saturated vapour line. When the heat is abstracted from dry saturated steam at the constant pressure its saturation temperature will remain same but only some portion of the steam will condense and the dry steam thus becomes wet steam. Since the dry saturated steam does not contain water particles, the dryness fraction of the dry saturated steam is one (Fig. 1.11).

Fig. 1.11 Types of steam

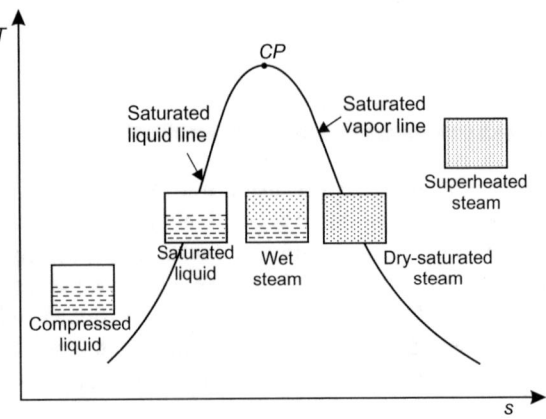

1.8.3 Superheated Steam

When the dry saturated steam is heated further at constant pressure, its temperature
will raise from the saturation temperature and the steam becomes superheated steam.
This steam lies in the region right to the saturated vapour line. The superheated steam
at a given pressure can exist at any temperature higher than the saturated temperature
corresponding to given pressure is called a superheated temperature. It is denoted as
T_{sup}. The superheated steam behaves as a perfect gas.

1.9 Properties of Wet Steam

At any state 1 between saturated liquid state f and saturated vapour state g, shown
in Fig. 1.12, liquid and vapour exist as a mixture in equilibrium.

Let v_f = specific volume of the saturated liquid,

v_g = specific volume of the saturated vapour,

m = total mass of the wet steam,

m_f = mass of the liquid phase,

m_g = mass of the vapour phase.

At state 1, the total volume of the wet steam (*i.e.*, mixture of the liquid and vapour
phases) is the sum of the volume occupied by the liquid and that occupied by the
vapour.

That is,

$$V = V_f + V_g \tag{1.1}$$

By definition of the specific volume, $v = \frac{V}{m} =$ for wet steam.

or

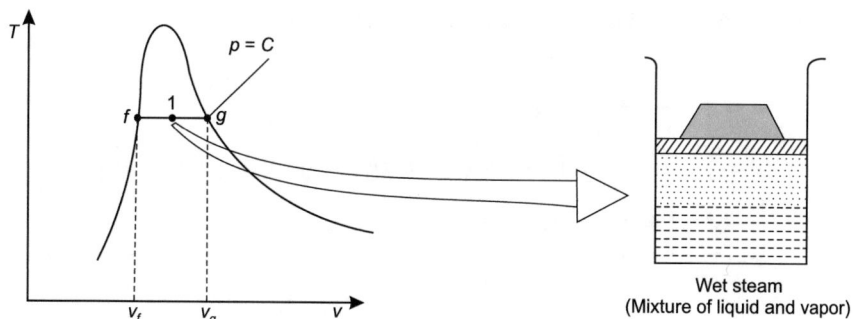

Fig. 1.12 T-v diagram showing the wet steam at state 1

$$V = mv.$$

$$v_f = \frac{V_f}{m_f} = \text{ for liquid.}$$

or

$$V_f = m_f V_f.$$

and

$$v_g = \frac{V_f}{m_f} = \text{ for vapour.}$$

or

$$V_g = m_g v_g$$

Substituting the values of V, V_f, and V_g in Eq. (1.1), we get

$$mv = m_f v_f + m_g v_g$$

or

$$v = \frac{m_f}{m} v_f + \frac{m_g}{m} v_g \tag{1.2}$$

The ratio of the mass of saturated vapour to the total mass of the wet steam is called the quality of the wet steam.

That is, $x = \frac{m_g}{m}$.
   and total mass: $m = m_f + m_g$.
   or $m_f = m - m_g$.
Substituting the values of x and m_f in Eq. (1.2), we get

$$v = \frac{(m - m_g)}{m} v_f + x v_g$$

$$v = \left(1 - \frac{m_g}{m}\right) v_f + x v_g = (1 - x) v_f + x v_g = v_f - x v_f + x v_g$$

$$v = v_f + x\left(v_g - v_f\right)$$

$$v = v_f + x v_{fg}$$

where $v_{fg} = v_g - v_f$.
   Note that the percentage liquid in a wet steam is $100 (1 - x)$ and the percentage vapour is $100\, x$.

The specific volume of the wet steam at state 1,

$$v = v_f + x v_{fg}$$

where

$$V_f = \text{pecific volume at saturated liquid state,}$$

$$x = \text{dryness fraction (or quality) at state 1,}$$

$$v_{fg} = v_g - v_f,$$

specific volumes difference between saturated vapour state and saturated liquid state.

Similarly, the specific enthalpy of the wet steam at state 1,

$$h = h_f + x(h_g - h_f)$$
$$h = h_f + x h_{fg}$$

where

$$h_f = \text{specific enthalpy at saturated liquid state,}$$

$$x = \text{dryness fraction at state 1,}$$

$$h_{fg} = h_g - h_f$$

specific enthalpies difference between saturated vapour state and saturated liquid state Fig. 1.13.

The specific entropy of the wet steam at state 1,

$$s = s_f + x(s_g - s_f)$$
$$s = s_f + x s_{fg}$$

where

$$S_f = \text{specific entropy at saturated liquid state,}$$

$$x = \text{ryness fraction at state 1,}$$

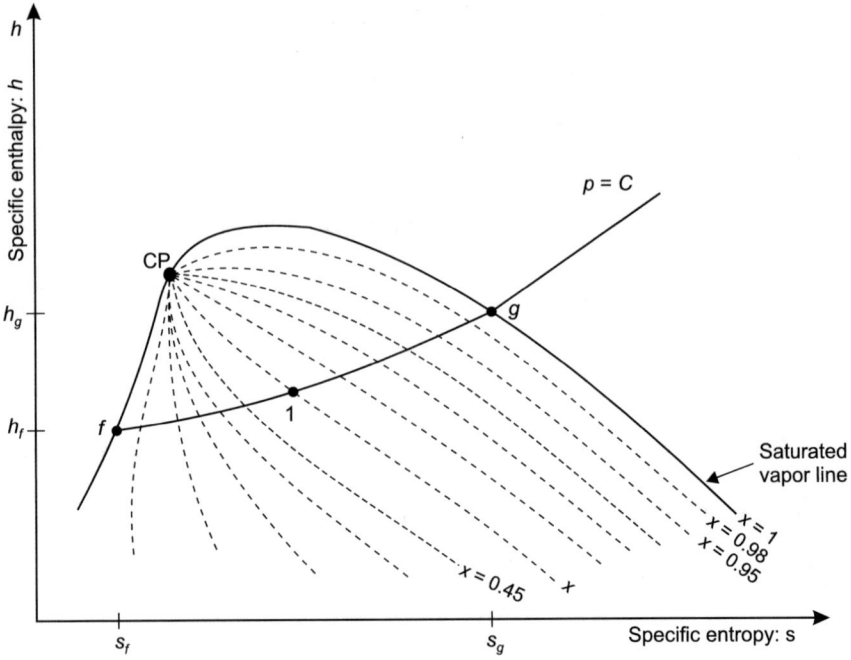

Fig. 1.13 h–s diagram showing the wet steam at state 1

$$S_g = S_f - S_{fg}$$

specific entropies difference between saturated vapour state and saturated liquid state.

The specific internal energy of the wet steam at state 1,

$$u = u_f + x\left(u_g - u_f\right)$$
$$u = u_f + xu_{fg}$$

where

$$u_f = \text{specific internal energy at saturated liquid state,}$$

$$u_{fg} = u_g - u_f,$$

specific internal energies difference between saturated vapour state and saturated liquid state.

1.10 Steam Tables

The laws of ideal gas do not apply to the wet steam and dry saturated steam. Thus, the variations between their properties can be experimentally determined and recorded in the form of tables. The preparation of steam tables has been made possible by studying experimentally the various properties of steam at different temperature and pressure. The experimental data and equations are used to obtain the steam properties. The steam tables are tabular presentation of the steam properties such as specific volume, specific enthalpy, specific entropy and specific internal energy for saturated liquid and saturated vapour of different saturation temperatures and pressure. Table 1.1 gives the properties of saturated steam on absolute pressure based, and Table 1.2 gives the properties of saturated steam on temperature based.

Similarly, the properties of superheated steam are also estimated and tabulated separately. The values of specific volume (v), specific enthalpy (h), and specific entropy (s) are not depended on pressure only but also dependent on temperature or degree of superheat. Sample of superheated steam table is given in Table 1.3.

Table 1.1 Saturated steam on pressure based

Absolute pressure	Saturated temperature	Specific volume			Specific internal energy		Specific enthalpy			Specific entropy		
p kP a	T_{sat} °C	v_f v_{fg} v_g m³/kg			u_f u_g kJ/kg		h_f h_{fg} h_g kJ/kg			s_f s_{fg} s_g kJ/kgK		

Table 1.2 Saturated steam on temperature based

Saturated temperature	Absolute pressure	Specific volume			Specific internal energy		Specific enthalpy			Specific entropy		
T_{sat} °C	P kPa	v_f v_{fg} v_g m³/kg			u_f u_g kJ/kg		h_f h_{fg} h_g kJ/kg			s_f s_{fg} s_g kJ/kgK		

Table 1.3 Superheated steam table

Absolute pressure: p kPa	Saturated temperature: T_{sat} °C	Specific volume: v Specific entropy: h Specific entropy: s	Temperature in °C				
			200	250	300	350	400
500	151.9	v: m³/kg	0.4249	0.4749	0.5226	0.5701	0.6173
		h: kJ/kg	2855.4	2960.7	3064.2	3167.7	3271.9
		s: kJ/kgK	7.059	7.271	7.460	7.633	7.945

1.11   Thermodynamics Processes for the Steam

Thermodynamics processes are used to study the behaviour of change in states of the steam. Following are the main processes.

1.   Isothermal process,
2.   Isobaric process,
3.   Isochoric process,
4.   Adiabatic process,
5.   Polytropic process,
6.   Hyperbolic process, and
7.   Throttling process

1.   **Isothermal process:** Isothermal process takes place at constant temperature. This process is discussed in the following conditions of the steam process.

   (i)   **For wet-wet steam process:** The wet-wet steam process means both the states of the process are wet as shown in Fig. 1.14.

Let

$$x_1, x_2 = \text{dryness fractions at states 1 and 2, respectively,}$$

$$p_1 = p_1, \text{ pressures at states 1 and 2,}$$

$$v_f = \text{specific volume at saturated liquid state corresponding}$$
$$\text{temperature } T = T_1 = T_2,$$

$$v_g = \text{specific volume at saturated vapour state}$$
$$\text{corresponding temperature } T = T_1 = T_2.$$

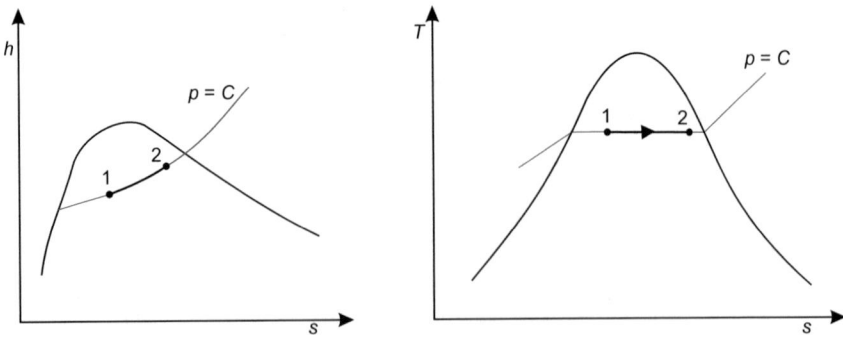

Fig. 1.14 Wet-wet steam isothermal process

Specific volume at state 1,

$$v_1 = v_f + x_1(v_g - v_f).$$

As

$$v_g >> v_f.$$

$\therefore$

$$v_g >> v_f.$$

$$v_1 = x_1 v_g.$$

and at state, 2, $v_2 = x_2 v_g$.
Specific enthalpies at states 1 and 2,

$$h_1 = h_f + x_1 h_{fg}$$
$$\text{and} \quad h_2 = h_f + x_2 h_{fg}$$

Work done:

$$\underset{1-2}{W} = \int_1^w p\,dv \quad \text{for non - flow process}$$

$$= p \int_1^2 dv$$

$$= p(v_2 - v_1)$$

$$= p(x_2 v_g - x_1 v_g)$$

$$\underset{1-2}{w} = p v_g (x_2 - x_1)$$

Heat transfer:

$$\underset{1-2}{q} = h_2 - h_1 = h_f + x_2 h_{fg} - h_f - x_1 h_{fg}$$

$$= (x_2 - x_1) h_{fg}$$

Change in specific internal energy:

$$u_2 - u_1 = (h_2 - h_1) - p v_g (x_2 - x_1)$$

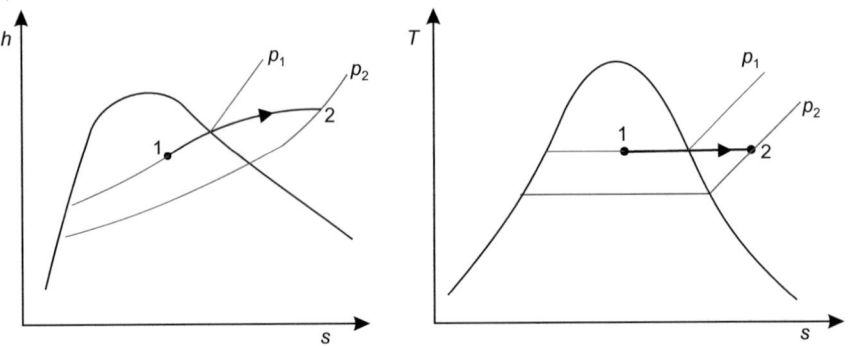

Fig. 1.15 Wet-superheated isothermal process

(ii)   **For wet-superheated steam process:** The wet-superheated steam process
means initial state of the steam is wet and final state is superheated as shown
in Fig. 1.15.

Let

$$x_1 = \text{dryness fraction at state 1},$$

$$p_1 = \text{pressure at state 1},$$

$$v_{f1} = \text{specific volume at saturated liquid state, corresponding pressure} \, p_1,$$

$$h_{f1} = \text{specific enthalpy at saturated liquid state, corresponding pressure} \, p_1,$$

$$v_{g1} = \text{specific volume at saturated vapour state, corresponding pressure} \, p_1,$$

$$h_{g1} = \text{specific enthalpy at saturated vapour state, corresponding pressure} \, p_1.$$

$\therefore$ Specific volume at state 1, $v_1 = v_{f1} + x_1(v_{g1} - v_{f1})$.
As $v_g \gg v_f$.
$\therefore v_1 = x_1 v_{g1}$.
and specific enthalpy:

$$h_1 = h_{f1} + x_1(h_{g1} - h_{f1})$$
$$= h_{f1} + x_1 h_{fg1}$$

From superheated steam table,
at p_2 and T_2

we get v_2 and h_2.

Heat transfer: $\underset{1-2}{q} = h_2 - h_1$.

Change in specific internal energy:

$$u_2 - u_1 = (h_2 - p_2 v_2) - (h_1 - p_1 v_1)u_2 - u_1$$
$$= (h_2 - h_1) - p_2 v_2 + p_1 v_1$$
$$= (h_2 - h_1) - \left(p_2 v_2 - p_1 x_1 v_{g1}\right)$$

According to the first law of thermodynamic for a process,

$$\underset{1-2}{q} = (u_2 - u_1) + \underset{1-2}{w}$$
$$h_2 - h_1 = (h_2 - h_1) - p\left(v_2 - x_1 v_g\right) + \underset{1-2}{w}$$

or

$$\underset{1-2}{w} = p_2 v_2 - p_1 x_1 v_{g1}$$

(iii) **For superheated-superheated steam process.** The superheated-superheated steam process means both the states of the process are superheated as shown in Fig. 1.16.

From superheated steam table,
corresponding p_1, T_1 and p_2, $T_2 = T_1$,
we get h_1, v_1 and h_2, v_2.
Heat transfer:

$$\underset{1-2}{q} = h_2 - h_1$$

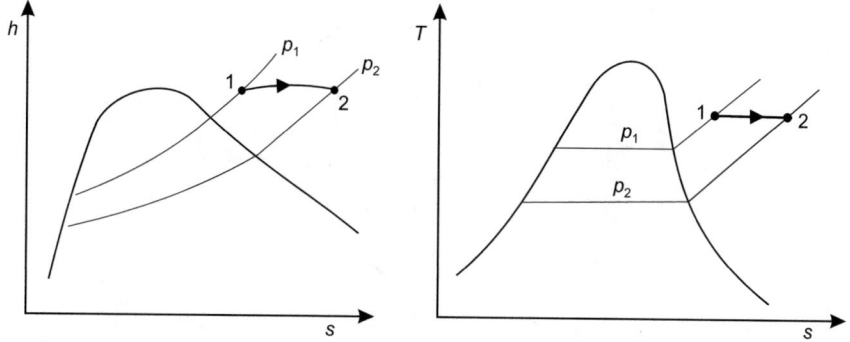

Fig. 1.16 Superheated-superheated isothermal process

Change in specific internal energy:

$$u_2 - u_1 = (h_2 - p_2v_2) - (h_1 - p_1v_1)$$
$$= (h_2 - h_1) - (p_2v_2 - p_1v_1)$$

According to the first law of thermodynamic for a process,

$$\underset{1-2}{q} = (u_2 - u_1) + \underset{1-2}{w}$$
$$h_2 - h_1 = (h_2 - h_1) - (p_1v_2 - p_1v_1) + \underset{1-2}{w}$$

or

$$\underset{1-2}{w} = \boldsymbol{p_2v_2 - p_1v_1}$$

2. **Isobaric process:** Isobaric process takes place at constant pressure. This process discussed the following condition of the steam process.

    (i)   **For wet-wet steam process.**

The isobaric and isothermal processes are identical for wet-wet steam process (Fig. 1.17).
Heat transfer:

$$\underset{1-2}{q} = h_2 - h_1$$

where

$$h_1 = h_f + x_1 h_{fg} h_2$$

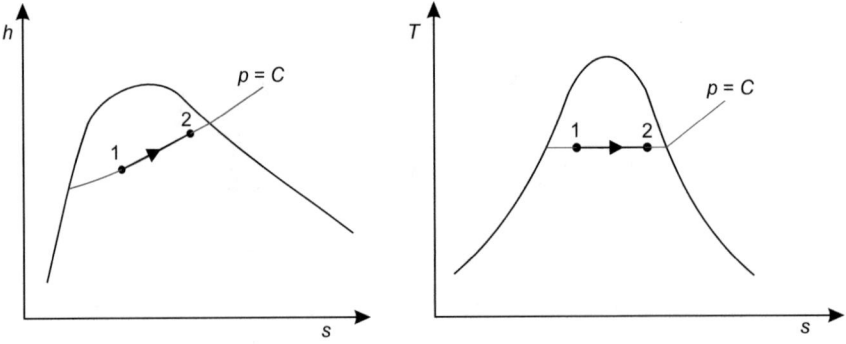

Fig. 1.17 Wet-wet isobaric process

$$h_2 = h_f + x_2 h_{fg}$$
$$\underset{1-2}{q} = h_f - x_2 h_{fg} - h_f - x_1 h_{fg}$$
$$= (x_2 - x)h_{fg}$$

Work done:

$$\underset{1-2}{w} = p(v_2 - v_1)$$

$$\underset{1-2}{w} = p\left(x_2 v_g - x_1 v_g\right)$$

$$\underset{1-2}{w} = \boldsymbol{p v_g (x_2 - x_1)}$$

Change in specific internal energy:

$$u_2 - u_1 = (h_2 - h_1) - p v_g (x_2 - x_1)$$

(ii)   For wet-superheated steam process:

Heat transfer:

$$\underset{1-2}{q} = h_2 - h_1$$

where

$$h_1 = h_f + x_1 h_{fg}$$
$$h_2 = h_g + c_{ps}(T_2 - T_{sat})$$

or

h_2 is found out from superheated steam table, corresponding p_2 and T_2 (Fig. 1.18).

$\therefore$

$$
\begin{aligned}
q_{1-2} &= h_g + c_{ps}(T_2 - T_{sat}) - h_f - x_1 h_{fg} \\
&= c_{ps}(T_2 - T_{sat}) + h_g - h_f - x_1 h_{fg} \\
&= c_{ps}(T_2 - T_{sat}) + h_{fg} - x_1 h_{fg} \\
&= c_{ps}(T_2 - T_{sat}) + (1 - x_1)h_{fg}
\end{aligned}
$$

Change in specific internal energy:

$$u_2 - u_1 = (h_2 - h_1) - p\left(v_2 - x_1 v_g\right)$$

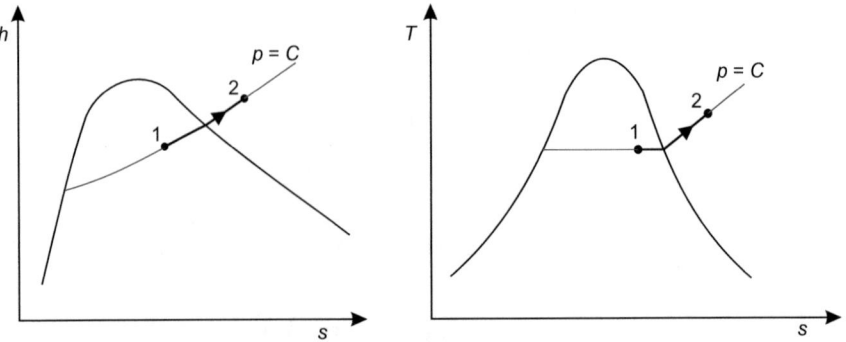

Fig. 1.18 Wet-superheated isobaric process

According to the first law of thermodynamic for a process,

$$\underset{1-2}{q} = (u_2 - u_1) + \underset{1-2}{w}$$
$$h_2 - h_1 = (h_2 - h_1) - p(v_2 - x_1 v_g) + \underset{1-2}{w}$$

or

$$\underset{1-2}{w} = \boldsymbol{p(v_2 - x_1 v_g)}$$

(iii)   For superheated-superheated steam process:

From superheated steam table,
corresponding p, T_1 and p, T_2
we get h_1, v_1 and h_2, v_2 (Fig. 1.19).
Heat transfer:

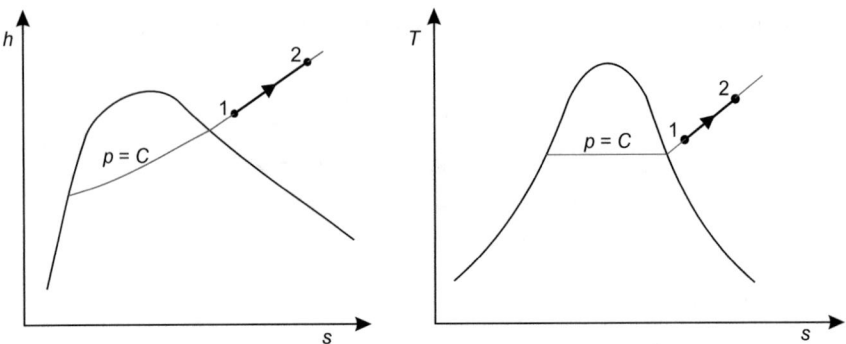

Fig. 1.19 Superheated-superheated steam isobaric process

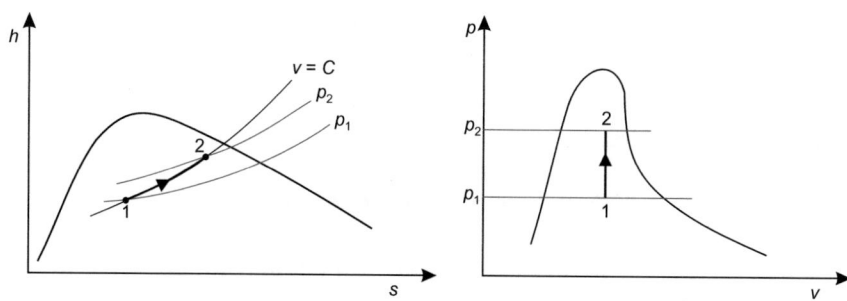

Fig. 1.20 Wet-wet steam isochoric process

$$q_{1-2} = h_2 - h_1$$

Work done:

$$w_{1-2} = p(v_2 - v_1)$$

Change in specific internal energy:

$$u_2 - u_1 = (h_2 - h_1) - (p_2 v_2 - p_1 v_1)$$
$$= (h_2 - h_1) - p(v_2 - v_1) \qquad \therefore p_1 = p_2 = p$$

3.   Isochoric process:

Isochoric process takes place at constant volume. This process is discussed in the following conditions of the steam process.

(i)   For wet-wet steam process.

Specific volume at state 1 (Fig. 1.20),

$$v_1 = x_1 v_{g1}$$

Specific volume at state 2,

$$v_2 = x_2 v_{g2}$$

Since the volume remains constant,

$$v_1 = v_2$$
$$x_1 v_{g1} = x_2 v_{g2}$$

or

$$x_2 = \frac{x_1 v_{g1}}{v_{g2}}$$

Thus, x_2 can be calculated.

Work done:

$w = 0 \therefore v_1 = v_2$ Change is specific internal energy:

$$
\begin{aligned}
u_2 - u_1 &= (h_2 - p_2 v_2) - (h_1 - p_1 v_1) \\
&= (h_2 - h_1) - p_2 v + p_1 v \\
&= (h_2 - h_1) - v(p_2 - p_1)
\end{aligned}
$$

According to the first law of thermodynamic for a process,

$$\underset{1-2}{q} = (u_2 - u_1) + \underset{1-2}{w}$$

$$\underset{1-2}{q} = u_2 - u_1 \quad \therefore \underset{1-2}{w} = 0$$

$$= (h_2 - h_1) - v(p_2 - p_1)$$

where

$$h_1 = h_{f1} + x_1 h_{fg1}$$
$$h_2 = h_{f2} + x_2 h_{fg2}$$

(ii) For wet-superheated steam process:

Specific volume at state 1,

$$v_1 = x_1 v_{g1}$$

.

Specific volume at state 2 is found out from the superheated steam table, corresponding p_2 and T_2 (Fig. 1.21).

Since the volume remains constant,

$$v_1 = v_2$$
$$x_1 v_{g1} = v_2$$

or

$$x_1 = \frac{v_2}{v_{g1}}$$

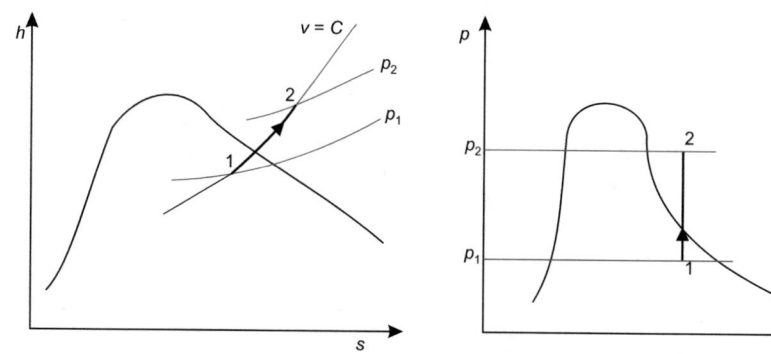

Fig. 1.21 Wet-superheated isochoric process

Work done: $\underset{1-2}{w} = 0 \qquad \because v_1 = v_2.$

Change in specific internal energy:

$$u_2 - u_1 = (h_2 - p_2 v_2) - (h_1 - p_1 v_1)$$
$$= (h_2 - h_1) - p_2 v + p_1 v$$
$$= (h_2 - h_1) - v(p_2 - p_1)$$

where

$$h_1 = h_{f1} + x_1 h_{fg1}$$

h_2 is found from superheated steam table, corresponding p_2, T_2.

According to the first law of thermodynamic for a process,

$$\underset{1-2}{q} = (u_2 - u_1) + \underset{1-2}{w}$$
$$\underset{1-2}{w} = u_2 - u_1 \qquad \because \quad \underset{1-2}{w} = 0$$
$$= (h_2 - h_1) - v(p_2 - p_1)$$

(iii)   For superheated-superheated steam process:

At constant volume process,

$$\frac{T_2}{p_2} = \frac{T_1}{p_1}$$

From superheated steam table, corresponding p_1, T_1 and p_2 T_2, we get h_1, v_1 and h_2, $v_2 = v_1$ (Fig. 1.22).

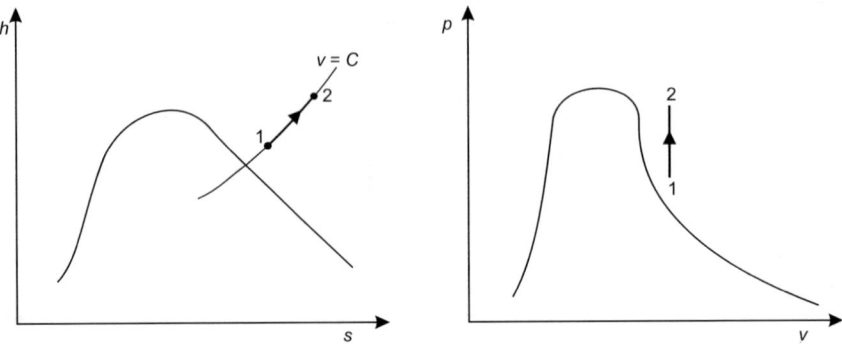

Fig. 1.22 Superheated-superheated steam isochoric process

Work done:

$$w_{1-2} = 0$$

Change in specific internal energy:

$$u_2 - u_1 = (h_2 - p_2 v_2) - (h_1 - p_1 v_1)$$
$$= (h_2 - h_1) - v(p_2 - p_1)$$

Heat transfer:

$$q_{1-2} = u_2 - u_1 = (h_2 - h_1) - v(p_2 - p_1)$$

4.   Adiabatic process:

When there is no heat transfer between the system and surroundings during a process, it is known as adiabatic process. This process is discussed in the following conditions of the steam process.

(i)   **For wet-wet steam process:**

Specific volume at state 1, $v_1 = x_1 v_{g1}$.

Specific volume at state 2, $v_2 = x_2 v_{g2}$.

Specific enthalpy at state 1, $h_1 = h_{f1} + x_1 h_{fg1}$ (Fig. 1.23).

Specific enthalpy at state 2, $h_2 = h_{f2} + x_2 h_{fg2}$.

Heat transfer: $q_{1-2} = 0$.

   Change in specific internal energy:

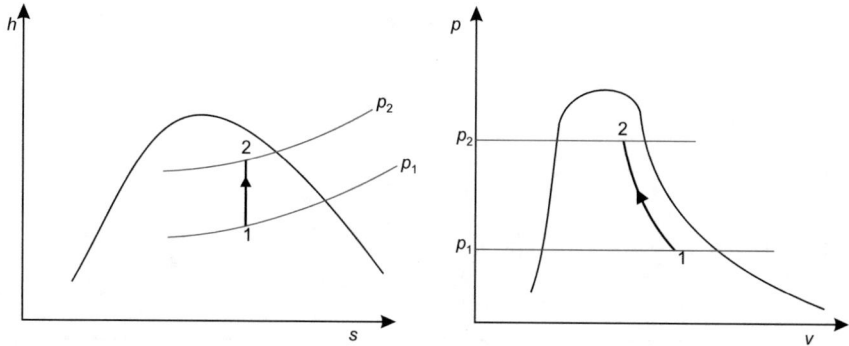

Fig. 1.23 Wet-wet steam adiabatic process

$$u_2 - u_1 = (h_2 - p_2v_2) - (h_1 - p_1v_1)$$
$$= (h_2 - h_1) - \left(p_2 x_2 v_{g2} - p_1 x_1 v_{g1}\right)$$

According to the first law of thermodynamic for a process,

$$\underset{1-2}{q} = (u_2 - u_1) + \underset{1-2}{w}$$
$$0 = (u_2 - u_1) + \underset{1-2}{w}$$

or

$$\underset{1-2}{w} = -(u_2 - u_1) = u_1 - u_2$$

(ii) For wet-superheated steam process:

Heat transfer:

$$\underset{1-2}{w} = 0$$

Change in specific internal energy (Fig. 1.24):

$$u_2 - u_1 = (h_2 - h_1) - \left(p_2 v_2 - p_1 x_1 v_{g1}\right)$$

Work done: $\underset{1-2}{q} = u_1 - u_2.$

(iii) For superheated-superheated steam process:

Heat transfer:

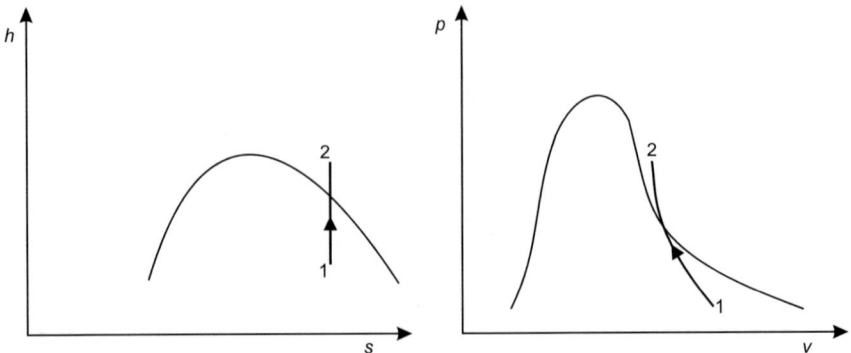

Fig. 1.24 Wet-superheated steam adiabatic process

$$q \atop 1-2 = 0$$

½ergy:

$$u_2 - u_1 = (h_2 - h_1) - (p_2 v_2 - p_1 v_1)$$

Work done:

$$w \atop 1-2 = u_1 - u_2$$

5. Polytropic Process: The polytropic process followed the law of $pv^n = C$, where n is polytropic index. This process is discussed in the following conditions of the steam process (Fig. 1.25).

   (i) For wet-wet steam process.

Specific volume at state 1, $v_1 = x_1 v_{g1}$.

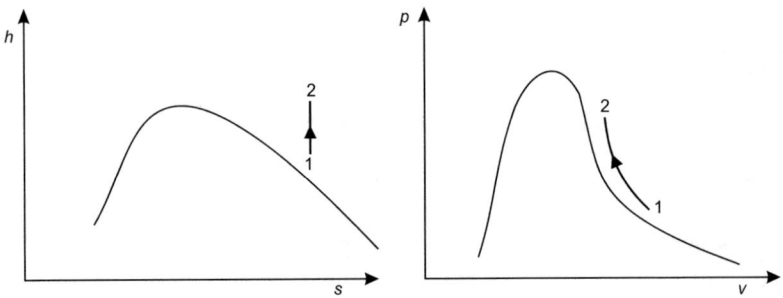

Fig. 1.25 Superheated-superheated steam adiabatic process

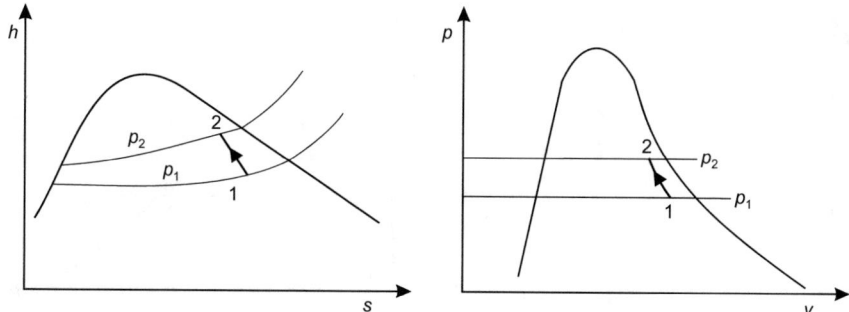

Fig. 1.26 Wet-wet steam polytropic process

Specific volume at state 2, $v_2 = x_2 v_{g2}$.

Polytropic law for process 1–2, $p_1 v_1^n = p_2 v_2^n$ (Fig. 1.26).

$$p_1 \left(x_1 v_{g1} \right)^n = p_2 \left(x_2 v_{g2} \right)^n$$

or

$$x_2 = \left(\frac{p_1}{p_2} \right)^{\frac{1}{n}} \times \frac{v_{g1}}{v_{g2}} x_1$$

Work done:

$$\underset{1-2}{w} = \frac{p_1 v_1 - p_2 v_2}{n - 1} = \frac{p_1 x_1 v_{g_1} - p_2 x_2 v_{g_2}}{n - 1}$$

Change in specific internal energy:

$$
\begin{aligned}
u_2 - u_1 &= (h_2 - p_2 v_2) - (h_1 - p_1 v_1) \\
&= (h_2 - h_1) - (p_2 v_2 - p_1 v_1) \\
&= (h_2 - h_1) - \left(p_2 x_2 v_{g2} - p_1 x_1 v_{g1} \right)
\end{aligned}
$$

According to the first law of thermodynamic for a process (Fig. 1.27),

$$
\begin{aligned}
\underset{1-2}{q} &= du + \underset{1-2}{w} \\
&= (u_2 - u_1) + \frac{p_1 x_1 v_{g1} - p_2 x_2 v_{g2}}{n - 1}
\end{aligned}
$$

(ii) For wet-superheated steam process:

Polytropic law for process 1-2,

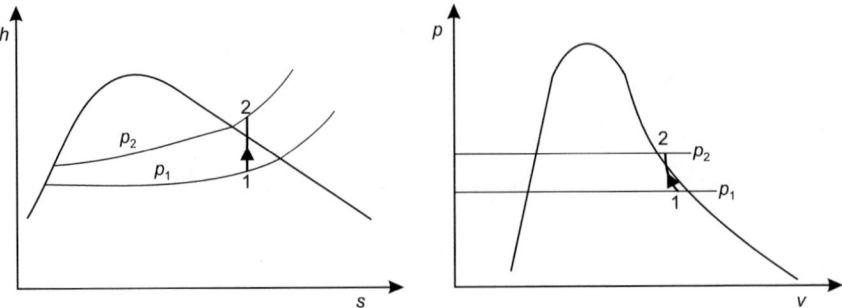

Fig. 1.27 Wet-superheated steam polytropic process

$$p_1 v_1^n = p_2 v_2^n$$
$$p_1 (x_1 v_{g1})^n = p_2 v_2^n$$

Work done:

$$\underset{1-2}{w} = \frac{p_1 v_1 - p_2 v_2}{n - 1} = \frac{p_1 x_1 v_{g1} - p_2 v_2}{n - 1}$$

Change in specific internal energy:

$$u_2 - u_1 = (h_2 - p_2 v_2) - (h_1 - p_1 v_1)$$
$$= (h_2 - h_1) - (p_2 v_2 - p_1 v_1)$$
$$= (h_2 - h_1) - (p_2 v_2 - p_1 x_1 v_{g1})$$

According to the first law of thermodynamic for a process,

$$\underset{1-2}{q} = du + \underset{1-2}{w}$$
$$= (u_2 - u_1) + \frac{p_1 x_1 v_{g1} - p_2 v_2}{n - 1}$$

(iii)   For superheated-superheated steam process:

Polytropic law for process 1-2,

$$p_1 v_1^n = p_2 v_2^n$$

Work done:

$$\underset{1-2}{w} = \frac{p_1 v_1 - p_2 v_2}{n - 1}$$

Change in specific internal energy:

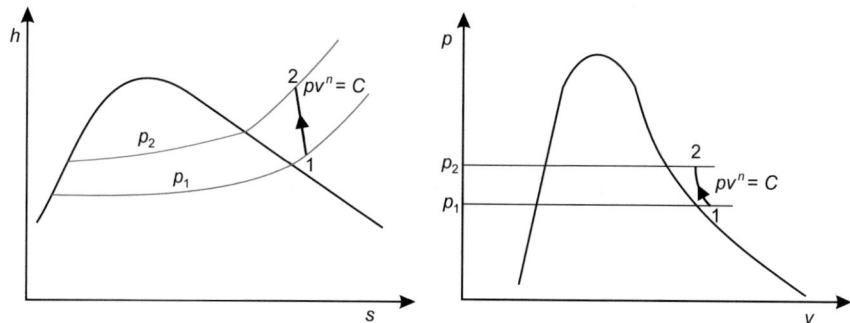

Fig. 1.28 Superheated-superheated steam polytropic process

$$u_2 - u_1 = (h_2 - p_2 v_2) - (h_1 - p_1 v_1)$$
$$= (h_2 - h_1) - (p_2 v_2 - p_1 v_1)$$

According to the first law of thermodynamic for a process,

$$\underset{1-2}{q} = du + \underset{1-2}{w}$$
$$= (u_2 - u_1) + \frac{p_1 v_1 - p_2 v_2}{n - 1}$$

6.   **Hyperbolic process:**

When the product of pressure and volume remains constant during the process of compression or expansion, the process is called the hyperbolic process because such a process on p-v diagram forms a rectangular hyperbolic. Hence, the hyperbolic process follows the law $pv = C$. This process is discussed in the following conditions of the steam process (Fig. 1.28).

(i)   **For wet-wet steam process.**

Initial specific volume of steam: $v_1 = x_1 v_{g1}$.
    Final specific volume of steam: $v_2 = x_2 v_{g2}$.

$$\text{As } pv = c$$

$$\therefore p_1 v_1 = p_2 v_2$$

$$p_1 x_1 v_{g1} = p_2 x_2 v_{g2}$$

Specific enthalpies at initial and final states,

$$h_1 = h_{f1} + x_1 h_{fg1}$$
$$h_2 = h_{f2} + x_2 h_{fg2}$$

Change in specific internal energy :

$$u_2 - u_1 = (h_2 - p_2 v_2) - (h_1 - p_1 v_1)$$
$$= (h_2 - h_1) - (p_2 x_2 v_{g2} - p_1 x_1 v_{g1})$$
$$u_2 - u_1 = h_2 - h_1$$
$$\therefore p_1 v_1 = p_2 v_2$$

$$\therefore p_1 v_1 = p_2 v_2$$

Hence, the change in specific internal energy is equal to the change in specific enthalpy.

(ii)   For wet-superheated steam process:

$$p_1 v_1 = p_2 v_2$$
$$p_1 x_1 v_{g1} = p_2 v_2$$

Specific enthalpy at states 1: $h_1 = h_{f1} + x_1 h_{fg1}$.
Change in specific internal energy:

$$u_2 - u_1 = (h_2 - p_2 v_2) - (h_1 - p_1 v_1)$$
$$= h_2 - h_1$$
$$= h_2 - (h_{f1} + x_1 h_{fg1})$$
$$= h_2 - h_{f1} - x_1 h_{fg1}$$

(iii)   **For superheated-superheated steam process:**

For superheated steam, both the hyperbolic and isothermal processes are identical (Fig. 1.29).
   Let

$$v_1 = \text{specific volume at state 1,}$$

$$v_2 = \text{specific volume at state 2,}$$

$$p_1 = \text{pressure at state 1,}$$

$$p_2 = \text{pressure at state 2.}$$

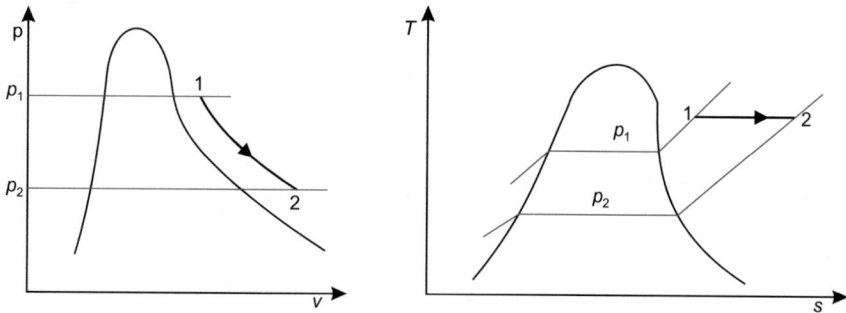

Fig. 1.29 Superheated-superheated steam hyperbolic process

For process 1-2,

$$p_1 v_1 = p_2 v_2$$

Work done: $\underset{1-2}{w} = \int_1^2 p\,dv$ for non-flow process.

Work done: $= \int_1^2 \frac{c}{v} dv \quad \therefore pv = c \text{ or } p = \frac{c}{v}$

$$= c \int_1^2 \frac{dv}{v}$$
$$= c \log_e \frac{v_2}{v_1}$$
$$\underset{1-2}{w} = p_1 v_1 \log_e \frac{v_2}{v_1}$$

Change in specific internal energy:

$$u_2 - u_1 = (h_2 - p_2 v_2) - (h_1 - p_1 v_1)$$
$$u_2 - u_1 = h_2 - h_2 \quad \therefore \quad p_1 v_1 = p_2 v_2$$

$$\therefore p_1 v_1 = p_2 v_2$$

Hence, the change in specific internal energy is equal to the change in specific enthalpy.

According to the first law of thermodynamic for a process,

$$\underset{1-2}{q} = (u_2 - u_1) + \underset{1-2}{w}$$
$$= (u_2 - u_1) + p_1 v_1 \log_e \frac{v_2}{v_1}$$
$$\underset{1-2}{q} = p_1 v_1 \log_e \frac{v_2}{v_1} + (u_2 - u_1)$$

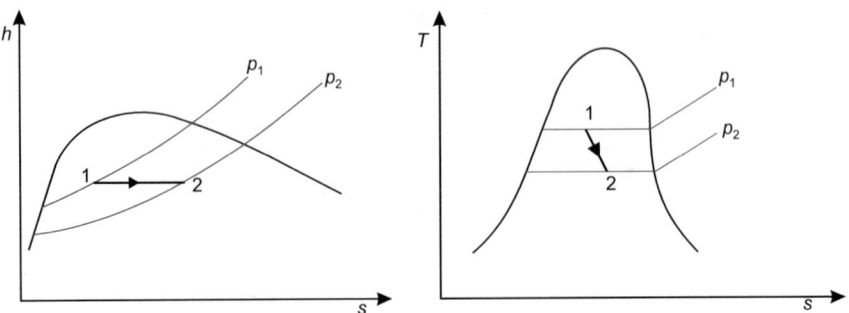

Fig. 1.30 Wet-wet steam throttling process

7. **Throttling process:**

The throttling of steam is an expansion process in which the steam is passed through an orifice or restricted passage and the steam will expand from high pressure to low pressure adiabatically without doing any work. Thus, the enthalpy of the steam will remain same before and after throttling, *i.e.*, $h_1 = h_2$ (Fig. 1.30).

This process will be discussed in the following conditions of the steam process.

(i)   **For wet-wet steam process:**

Specific enthalpy at state 1, $h_1 = h_{f1} + x_1 h_{fg1}$.

where

$\quad h_{f1} =$ specific enthalpy at saturate liquid state, corresponding pressure p_1

$$x_1 = \text{dryness fraction at state 1}$$

$\quad h_{fg1} = h_{g1} - h_{f1,}$, corresponding pressure p_1.
Specific enthalpy at state 2, $h_2 = h_{f2} + x_2 h_{fg2}$.

where

$\quad h_{f2} =$ specific enthalpy at saturated liquid state corresponding pressure p_2,

$h_{fg2} = h_{g2} - h_{f2,}$, corresponding pressure p_2.
Since the enthalpy during throttling process remains constant

$$h_1 = h_2$$
$$h_{f1} + x_1 h_{fg1} = h_{f2} + x_2 h_{fg2}$$

(ii)   **For wet-superheated steam process:**

Specific enthalpy before throttling at state 1,

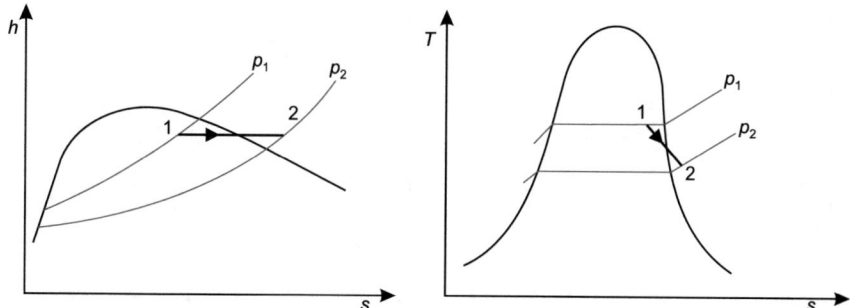

Fig. 1.31 Wet-superheated steam throttling process

$$h_1 = h_{f1} + x_1 h_{fg1}$$

Specific enthalpy after throttling at state 2 (Fig. 1.31),

$$h_2 = h_{g2} + c_{ps}(T_2 - T_{sat2})$$

where

h_{g2} = specific enthalpy at saturated vapour state, corresponding pressure p_2,

c_{ps} = specific heat of the superheated steam,

T_{sat2} = saturated temperature of the steam, corresponding pressure p_2, ,

T_2 = temperature of the steam after throttling at state 2

Since the enthalpy during throttling process remains constant

$$h_1 = h_2$$
$$h_{f1} + x_1 h_{fg1} = h_{g2} + c_{ps}(T_2 - T_{sat2})$$

(iii)   **For superheated-superheated steam process:**

Specific enthalpy before throttling at state 1,

$$h_1 = h_{g1} + c_{ps}(T_1 - T_{sat\,1})$$

where

h_{g1} = specific enthalpy at saturated vapour state, corresponding pressure p_1,

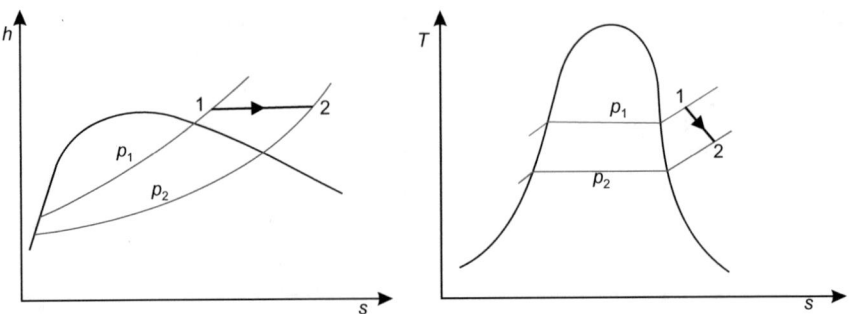

Fig. 1.32 Superheated-superheated steam throttling process

$$c_{ps} = \text{specific heat of the superheated steam,}$$

$$T_1 = \text{temperature of the steam before throttling process at state 1.}$$

$$T_{sat1} = \text{saturated temperature of the steam, corresponding pressure } p_1$$

(Fig. 1.32).

Specific enthalpy after throttling process at state 2, $h_2 = h_{g2} + c_{ps}(T_2 - T_{sat2})$.
Since the enthalpy during throttling process remains constant

$$h_1 = h_2$$
$$h_{g1} + c_{ps}(T_1 - T_{sat1}) = h_{g2} + c_{ps}(T_2 - T_{sat2})$$

1.12 Experimental Methods of Determination of Dryness Fraction

The dryness fraction of wet steam is determined by the following types of calorimeters.

1. Separating calorimeter,
2. Throttling calorimeter,
3. Combined separating and throttling calorimeter, and
4. Electrical calorimeter.

1. **Separating Calorimeter:**

This calorimeter is used to find the dryness fraction of the wet steam when the steam is very wet. The known mass of wet steam is passed through the sampling tube and is passed further through control valve which is completely kept open. The wet steam coming out of the sampling tube in the inner chamber strikes the baffle plates and follows the zig-zig path inside the baffle plates. When the wet steam, containing the water particles, strikes the baffle plates the particles of water having greater density and hence greater inertia, clings to the baffle plates, gets separated from the steam, and gets collected in the inner chamber. The quantity of separated water particles is measured by the calibrated gauge glass tube. The dry steam after it is being separated from the water particles moves up from the inner chamber to the outer chamber. The dry steam then passes into the condenser in which it is condensed. The liquid thus separated out is collected in a collecting tank and is measured. Thus by knowing the values of $m_1 = m_f$ and $m_2 = m_g$, the dryness fraction can be calculated as follows (Fig. 1.33):

Dryness fraction:

$$x = \frac{\text{mass of vapor} : m_g}{\text{total mass of the wet steam} : m}$$
$$= \frac{m_g}{m} = \frac{m_g}{m_f + m_g}$$
$$= \frac{m_g}{m_f + m_g}$$

In practice, it is not possible to separate all the water particles completely by this type of mechanical separation, the dryness fraction obtained by this calorimeter will not be accurate and it will always be higher than the actual value. The advantage of

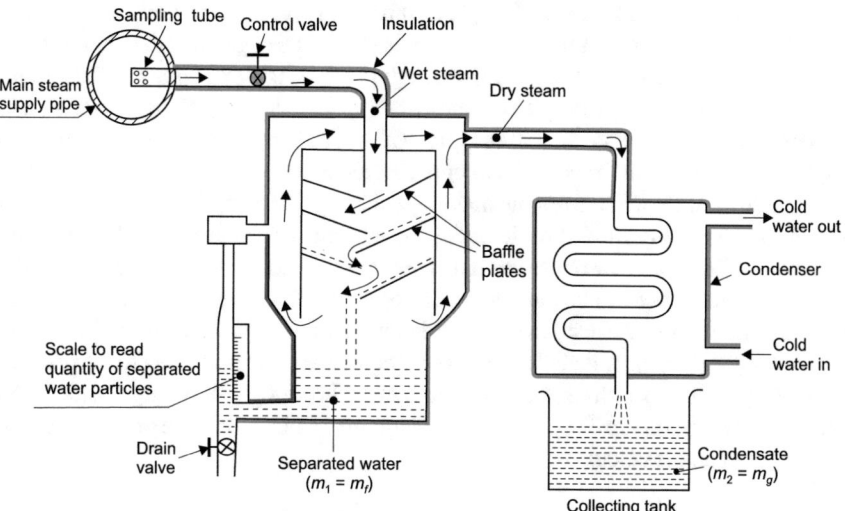

Fig. 1.33 Separating calorimeter

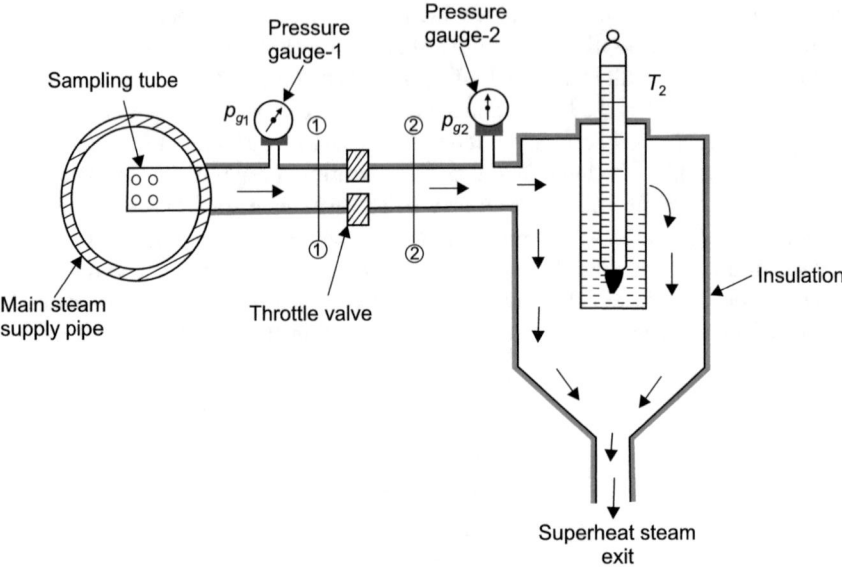

Fig. 1.34 Throttling calorimeter

this calorimeter is only the quick determination of the dryness fraction of very wet steam Fig. (1.34).

2. **Throttling Calorimeter:**

This calorimeter works on the principle of throttling process. The throttling of steam is an expansion process in which the steam is passed through an orifice or restricted passage and the steam will expand from high pressure to low pressure adiabatically and without doing any work. Thus, the enthalpy (*i.e.*, total heat) of the steam will remain same before and after throttling. After throttling, the steam will become superheated due to fall in pressure at constant enthalpy.

Sample steam is taken from the main pipe through the sampling tube. The sample steam passes through the throttle valve, which will be subjected to the throttling expansion. The gauge pressures of the steam before and after throttling is recorded by the pressure gauges. The temperature after the throttled steam is measured by the mercury thermometer placed in pocket filled with oil. If the temperature of the steam after throttling is greater than the saturated temperature corresponding to the pressure of the steam after throttling, then the steam after throttling will be superheated. The process is shown on the *T-s, p-h,* and *h-s* diagrams in Fig. 1.35. The dryness fraction of the wet steam is calculated as follows:

Let

p_{g1} = gauge pressure of the steam before throttling read by pressure gauge-1,

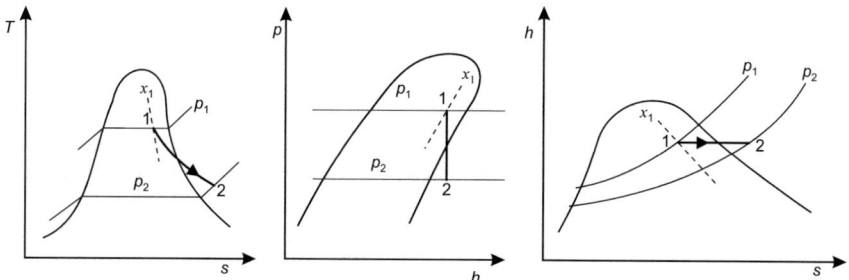

Fig. 1.35 Throttling process on T-s, p–h, and h–s diagrams

p_{g2} = gauge pressure of the steam before throttling read by pressure gauge-2,

p_{atm} = atmospheric pressure,

x_1 = dryness fraction of the sample steam,

T_2 = temperature of the steam after throttling read by the mercury thermometer in °C.

The properties of the steam are given based upon absolute pressure. Let

p_1 = absolute pressure of the steam before throttling

= $p_{g1} + p_{atm}$ *and* p_2 = absolute pressure of the steam after throttling

= $p_{g1} + p_{atm}.$

From pressure-based saturated steam table, corresponding to the absolute pressure p_1, we can find properties h_f, h_g, and h_{fg}.
∴ The specific enthalpy before throttling at state 1,

$$h_1 = h_f + x_1 h_{fg}$$

From the superheated steam table, corresponding to the absolute pressure p_2 and temperature T_2 at state 2, the specific enthalpy h_2 at state 2 is determined.

[**Note:** The specific enthalpy at superheated steam at state 2 is also calculated by the following equation],

$$h_2 = h_{g,p2} + c_{ps}(T_2 - T_{sat})$$

where

$h_{g,p2}$ = specific enthalpy at saturated vapour state corresponding to the pressure p_2,

c_{ps} = specific heat of superheated steam,

T_{sat} = saturated temperature of the dry saturated steam corresponding to the pressure p_2,

T_2 = temperature of the steam after throttling read by the mercury thermometer.

Since the enthalpy or specific enthalpy of the steam remains constant before and after throttling process,
    i.e.,

$$h_1 = h_2$$
$$h_f + x_1 h_{fg} = h_2$$

or

$$x_1 = \frac{h_3 - h_f}{h_{fg}}$$

The dryness fraction found out from the above equation will be accurate provided the wet steam becomes superheated after throttling. This can happen only if the dryness fraction of the wet steam before throttling is quite high, i.e., $x \geq 0.9$. Therefore, this calorimeter cannot be used to determine when the dryness fraction of the steam is low because the steam remains wet after the throttling.

3.   **Combined Separating and Throttling Calorimeter:**

In a separating calorimeter, it does not possible to separate all the water particles with mechanical separation, and hence the dryness faction of the wet steam determined by the separating calorimeter is always higher than the actual value.

In a throttling calorimeter, it is not possible to determine the dryness fraction of the steam, if the steam remains wet after throttling. The throttling calorimeter can be used only if the steam has high dryness fraction (i.e., $x \geq 0.9$) before throttling so that it must be superheated after throttling.

As discussed above, the separating and throttling calorimeters cannot be used individually to accurately determine the dryness fraction of the wet steam at any quality. However, by combining both the separating and throttling calorimeters as shown in Fig. 1.36, the dryness fraction of very wet steam can be very determined accurately.

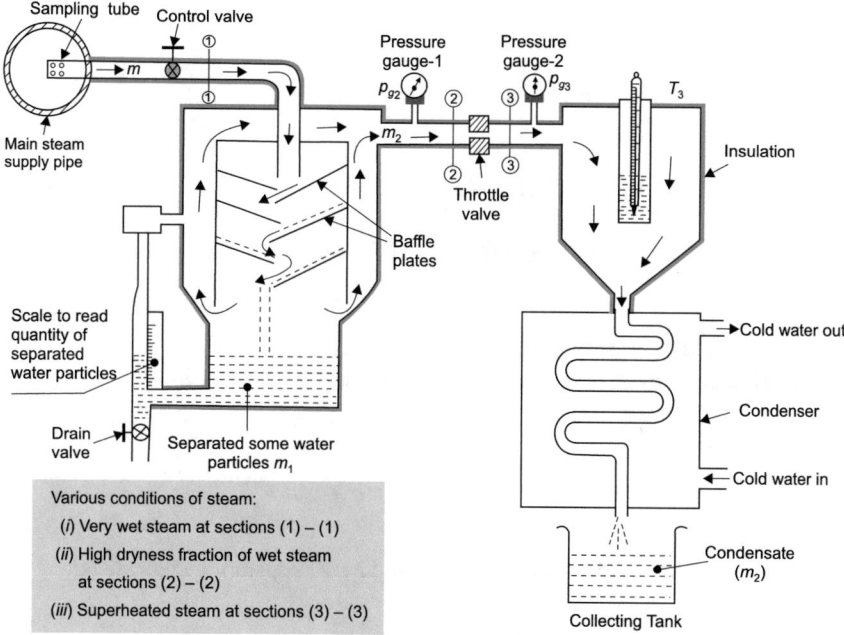

Fig. 1.36 Combined separating and throttling calorimeter

In this arrangement, both the separating and throttling calorimeters are connected in series. The steam from the main supply pipe is passed through the sampling tube to the separating calorimeter, where the maximum amount of water particles are separated and the dryness fraction of the steam increases. From the separating calorimeter, high dryness fraction of wet steam enters the throttle valve, where the steam after throttling becomes superheated. The superheated steam coming out from the throttling calorimeter is condensed in the condenser. The mass of the condensate is the mass of the high dryness fraction wet steam separated in the separated calorimeter. The processes are shown on the T-s, p-h, and h-s diagrams in Fig. 1.37. The dryness fraction of the wet steam is calculated as follows:

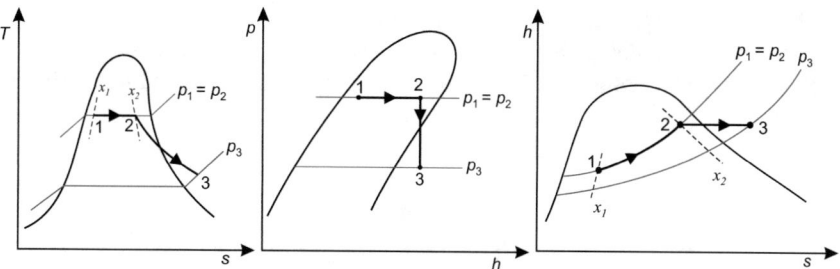

Fig. 1.37 Separating and throttling processes on T-s, p–h, and h–s diagrams

Let

$$m_1 = \text{mass of the sample steam}$$

or
mass of the wet steam at Sect. 1—1,

$$x_1 = \text{dryness fraction of the sample steam}$$

or
dryness fraction of the wet steam at Sect. 1—1.
Absolute pressure at Sect. 1—1 = absolute pressure at Sect. 2—2.

That is,

$$p_1 = p_2 = p_{g2} + p_{atm}$$

$$m_1 = \text{mass of water particles separated in the separating calorimeter,}$$

$$m_2 = \text{mass of the wet steam enters the throttling value from the separating calorimeter,}$$

$$x_2 = \text{dryness fraction of the wet steam before throttling valve,}$$

$p_2 = p_{g2} + p_{atm}$, absolute pressure of the superheated steam after the throttling,

$T_3 = $ temperature of the superheated steam measured by the mercury thermometer.

The specific enthalpy (h_3) of the superheated steam at state 3 is found out from superheated steam table, corresponding to the absolute pressure p_3 and temperature T_3.

From pressure-based saturated steam table, corresponding to the absolute pressure p_2, the properties h_f, h_g, and h_{fg} are found out.

∴ The specific enthalpy before throttling at state 2, $h_2 = h_f + x_2 h_{fg}$.

Since the specific enthalpy of the steam remains constant before and after throttling process,

i.e.,

$$h_2 = h_3$$
$$h_f + x_2 h_{fg} = h_3$$

or

$$x_2 = \frac{h_2 - h_f}{h_{fg}}$$

We know that the values of h_f, h_{fg}, and h_3, thus by using the above equation, we can find x_2.

Also $x_2 = \frac{m_g}{m_2}$.

or mass of the vapour: $m_g = x_2 m_2$.

The values of x_2 and m_2 are known. So by using x_2 and m_2, we can find m_g using above relation. Note that the mass of vapour m_g at Sects. 1—1, 2—2 and 3—3 remains constant.

Now, we are enable to calculate the dryness fraction of the wet steam at state 1 (i.e., sample steam).

Dryness fraction at state 1,

$$x_1 = \frac{\text{mass of the vapor}}{\text{mass of the wet steam}}$$
$$x_1 = \frac{m_g}{m}$$

where

$$m_g = x_2 m_2$$

and

$$m = m_1 + m_2$$

∴

$$x_1 = \frac{x_2 m_2}{m_1 + m_2}$$

4. **Electrical Calorimeter:**

The electrical calorimeter is used to measure the dryness fraction of any quality of the wet steam. This type of calorimeter has no limitation for dryness fraction measurement like separating calorimeter and throttling calorimeter. The sample of steam is passed in steady flow through an electric heater as shown in Fig. 1.38. The amount of heat energy Q is supplied to the wet steam by electric heater, the steam becomes superheated where the gauge pressure p_{g2} and temperature T_2 are measured. The dryness fraction of the wet steam can be calculated as follows:

Let p_{g1} = gauge pressure of the wet steam at Sect. 1—1

∴ Absolute pressure of the wet steam at Sect. 1—1,

$$p_1 = \text{gauge pressure} + \text{atmospheric pressure}$$

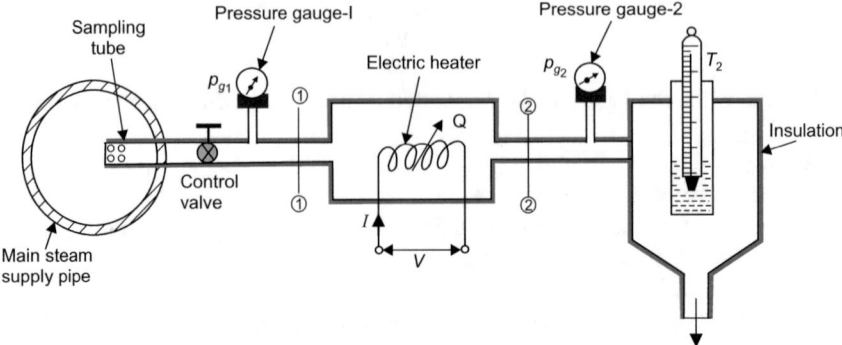

Fig. 1.38 Electrical calorimeter

$$= p_{g1} + p_{atm}$$

Heat supplied to the wet steam by electric heater:

$$Q = VI \ W$$

where

$$V = \text{voltage across the heating coil in volts,}$$

$$I = \text{current flow through the heating coil in amperes,}$$

or

$$Q = \frac{VI}{1000} \ \text{kW}$$

p_{g2} = gauge pressure of the superheated steam at Sect. 2—2.
∴ Absolute pressure of the superheated steam at Sect. 2—2,

$$p_2 = p_{g2} + p_{atm}$$
$$T_2 = \text{temperature of the superheatedsteam.}$$
$$m = \text{mass flow rate of the steam in kg/s}$$

Now applying the steady flow energy equation at Sects. 1—1 and 2—2, we get

$$m\left(h_1 + \frac{V_1^2}{2} + gz_1\right) + Q = m\left(h_2 + \frac{V_2^2}{2} + gz_2\right) + W$$

The change in potential and kinetic energies are neglected.
$W = 0$, no work interaction
$\therefore$

$$mh_1 + Q = mh_2 \qquad (1.3)$$

where h_2 is calculated by using superheated steam table, corresponding p_1 and T_1

$$Q = \frac{VI}{1000} \text{ kW}$$

and mass flow rate m is known
Hence, h_1 is calculated from the above Eq. (1.3)
also

$$h_1 = h_f + x_1 h_{fg}$$

where h_f and h_{fg} are found out from the saturated steam table, corresponding absolute pressure p_1
$\therefore$

$$x_1 = \frac{varvech_1 - h_f}{h_{fg}}$$

Hence, the dryness fraction x_1 of the wet steam can be determined by using electrical calorimeter (Fig. 1.39).

Problem 1.1: State the quality of steam in the following cases. Also mention the value of degree of superheated or dryness fraction.

(a) $p = 0.1$ bar, $T = 80\ °C$,
(b) $p = 0.1$ MPa, $h = 2500$ kJ/kg,
(c) $T = 100\ °C$, $s = 6.5$ kJ/kgK.

Solution:

(a) Given data

$$p = 0.1\,bar, T = 80\,°C$$

First, we will check whether the given data is of wet steam or of superheated steam.
From saturated steam table (pressure based)
at

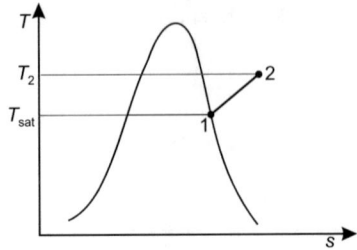

Fig. 1.39 T-s diagram for Problem 1.1

$$p = 0.1\,bar,\ T_{sat} = 45.8\,^\circ C$$

As $T > T_{sat}$ it means given data of superheated steam

$$\text{Degree of superheat} = T_2 - T_1$$
$$= T - T_{sat} \quad \because T_2 = T, T_1 = T_{sat}$$
$$= 80 - 45.8 = 34.2\,^\circ C$$

(b)   Given data:

$$p = 0.1\text{MPa},\ h = 2500\ \text{kJ/kg}$$
$$= 1\text{bar}$$

From saturated steam table (pressure based)
 at $p = 1$ bar
 we get $h_f = 417.5$ kJ/kg, $h_g = 2675.5$ kJ/kg
 So, $h < h_g$, given data of wet steam
 $\therefore$

$$h = h_f + x h_{fg} 2500$$
$$= 417.5 + x(2675.5 - 417.5)$$

or Dryness fraction: $x = \mathbf{0.9222}$

(c)   Given data:

$$T = 100\,^\circ C,\ s = 6.5\ \text{kJ/kgK}$$

From saturated steam table (temperature based),
 at $T = 100\ ^\circ C$
 we get $s_f = 1.3072$ kJ/kg K, $s_g = 7.3553$ kJ/kg K

So, $s < s_g$, given data of wet steam.

$s = s_f + x\, s_{fg}$

$6.5 = 1.3075 + x\,(7.3553 - 1.3072)$

or Dryness fraction: $x = \mathbf{0.8585}$

Problem 1.2: Determine the specific volume, specific enthalpy, and specific entropy of steam at 5 bar, 230 °C.

Solution: Given data:

$$p = 5\,bar, T = 230\,°C.$$

First of all, we will check whether steam is wet or superheated.
From saturated steam table (pressure based),
at $p = 5$ bar
we get $T_{sat} = 151.9\ °C$
As $T > T_{sat}$, it means the steam is superheated
From superheated steam table,
at $p = 5$ bar and $T = 230\ °C$

$p = 5$ bar	v, h, s	200 °C T_1	230 °C T	250 °C T_2
	v (m³/kg)	0.4249	?	0.4744
	h (kJ/kg)	2855.4	?	2960.7
	s (kJ/kgK)	7.059	?	7.271

The values of v, h, and s are not given directly in the superheated steam table, these values will be found with the help of interpolation.

Specific volume:

$$v = v_1 + \frac{(v_2 - v_1)(T - T_1)}{(T_2 - T_1)}$$

$$= 0.4249 + \frac{(0.4744 - 0.4249)(230 - 200)}{(250 - 200)}$$

$$= 0.4249 + 0.0297$$

$$= 0.4546\ \mathbf{m^3/kg}$$

Specific enthalpy:

$$h = h_1 + \frac{(h_2 - h_1)(T - T_1)}{(T_2 - T_1)}$$

$$= 2855.4 + \frac{(2960.7 - 2855.4)(230 - 200)}{(250 - 200)}$$

$$= 2855.4 + 63.18 = \mathbf{2918.58kJ/kg}$$

and specific entropy:

$$s = s_1 + \frac{(s_2 - s_1)(T - T_1)}{(T_2 - T_1)}$$

$$= 7.059 + \frac{(7.271 - 7.059)(230 - 200)}{(250 - 200)}$$

$$= 7.059 + 0.1272 = 7.1862\mathbf{kJ/kgK}$$

Problem 1.3: Determine the specific enthalpy and specific entropy of steam when the pressure 15 bar and the specific volume is 0.015 m³/kg.

Solution: Given data:

$p = 15$ bar.

and $v = 0.015$ m³/kg.

First, we will check whether the given data is of wet steam or of superheated steam.

From saturated steam table (pressure based),

at $p = 15$ bar,

we get $v_f = 0.001154$ m³/kg, $v_g = 0.132$ m³/kg.

So, $v_f < v < v_g$, it means given data lie between saturated vapour line and saturated liquid line, i.e., given data of wet steam.

Therefore,

$$v = v_f + xv_{fg}$$
$$v = v_f + x\left(v_g - v_f\right)$$
$$0.015 = 0.001154 + x(0.132 - 0.001154)$$

or $x = 0.1058$.

Also from saturated steam table at p $= 15$ bar.

we get

$$h_f = 844.9 \text{ kJ/kg, } h_{fg} = 1947.3 \text{ kJ/kg, } h_g = 2792.2 \text{ kJ/kg}$$
$$s_f = 2.315 \text{ kJ/kgK, } s_{fg} = 4.13 \text{ kJ/kgK, } s_g = 6.445 \text{ kJ/kgK}$$

Specific enthalpy:

$$h = h_f + xh_{fg}$$
$$= 8.44.19 + 0.1058 \times 1947.3 = \mathbf{1050.21kJ/kg}$$

Specific entropy:

$$s = s_f + x s_{fg}$$
$$= 2.315 + 0.1058 \times 4.13 = \textbf{2.752kJ/kgK}$$

Problem 8.4: Find the saturated temperature, change in specific volume, specific entropy during vapourization, and the latent heat of vapourization of steam at 10 bar.
  Solution: Given data:
  Pressure of steam: $p = 10$ bar.
  From saturated steam table (pressure based),
  at $p = 10$ bar.
  Saturated temperature: $T_{sat} = \textbf{179.9 °C.}$
  we get

$$v_f = 0.001127 \text{ m}^3/\text{kg}, \ v_g = 0.194 \text{ m}^3/\text{kg}$$
$$h_f = 762.8 \text{ kJ/kg}, \ h_g = 2778.1 \text{ kJ/kg}$$
$$s_f = 2.139 \text{ kJ/kgK}, \ s_g = 6.587 \text{ kJ/kgK}$$

Now, the change in specific volume: $v_{fg} = v_g - v_f = 0.194 - 0.001127 = $ **0.19287m³/kg**.

Change in specific entropy during vapourization:

$$s_{fg} = s_g - s_f$$
$$= 6.587 - 2.139 = \textbf{4.448kJ/kgK}$$

The latent heat of vapourization: $h_{fg} = h_g - h_f = 2778.1 - 762.8 = \textbf{2015.3kJ/kg.}$

Problem 1.5: A pressure cooker contains 2 kg of saturated steam at 5 bar. Find the quantity of heat which must be rejected so as to reduce the quality to 70% dry. Find also the pressure and temperature of the steam at new condition.
  Solution: Given data:
  Mass of saturated steam: $m = 2$ kg.
  Pressure of saturated steam at state 1: $p_1 = 5$ bar.
  Dryness fraction or quality at state 2: $x_2 = 70\% = 0.7$ (Fig. 1.40).
  From saturated steam table (pressure based),
  at $p_1 = 5$ bar.
  we get

$$T_{sat} = 151.9 \text{ °C}, \ v_{g1} = 0.375 \text{ m}^3/\text{kg}$$
$$h_{g1} = h_1 = 2748.7 \text{ kJ/kg}$$

Specific volume at state 1: $v_1 = v_g = 0.375 \text{ m}^3/\text{kg}$.

Specific volume at state 2: $v_2 = v_1 = 0.375 \text{ m}^3/\text{kg}$.

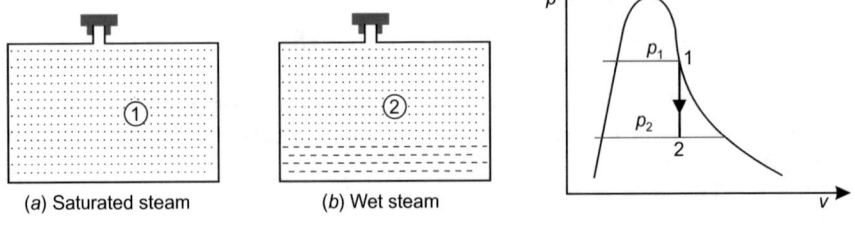

(a) Saturated steam   (b) Wet steam

Fig. 1.40 Schematic and p-v diagram for Problem 1.5

also $v_2 = v_{f2} + x_2(v_{g2} - v_{f2})$.

As $v_{g2} \gg v_{f2}$.

∴

$$v_2 = x_2 v_{g2}$$
$$0.375 = 0.7 \times v_{g2}$$

or

$$v_{g2} = 0.5357 \text{ m}^3/\text{kg}$$

From saturated steam table,
   at $v_{g2} = 0.5357 \text{ m}^3/\text{kg}$.

we get

$$p_2 = \mathbf{3.4 \text{ bar}}$$
$$T_{\text{sat}} = \mathbf{137.9^\circ C}$$
$$h_{f2} = 580 \text{ kJ/kg}, h_{fg2} = 2151.1 \text{ kJ/kg}$$
$$h_2 = h_{f2} + x_2 h_{fg_2}$$
$$= 580 + 0.7 \times 2151.2 = 580 + 1505.84$$
$$= 2085.84 \text{ kJ/kg}$$
$$u_2 = h_2 - p_2 v_2 \text{ where } h_2 \text{ in kJ/kg}, p_2 \text{ in kPa and } v_2 \text{ in m}^3/\text{kg}$$
$$= 2085.84 - 3.4 \times 10^2 \times 0.375$$
$$= 2085.84 - 127.5 = 1958.3 \text{ kJ/kg}$$
$$u_1 = h_1 - p_1 v_1 \text{ kJ/kg}$$

where

$$h_1 \text{ is in kJ/kg}$$
$$p_1 \text{ is in kPa}$$
$$v \text{ is in m}^3/\text{kg}$$

∴

$$u_1 = 2748.7 - 500 \times 0.375$$
$$= 2748.7 - 187.5 = 2561.2 \text{ kJ/kg}$$

Work interaction: $W_{1-2} = 0 . \therefore v_1 = v_2$.

According to the first law of thermodynamic for a process 1–2,

$$Q_{1-2} = (U_2 - U_1) + W_{1-2}$$
$$= m(u_2 - u_1) + 0$$
$$= 2(1958.34 - 2561.2) = -\textbf{1205.72kJ}$$

The $-ve$ sign shows that heat rejected from the system.

The pressure and temperature of the steam at new condition: p_2 and T_2

$$p_2 = \textbf{3.4} \text{ bar}$$
$$T_2 = T_{\text{sat 2}} = \textbf{137.9 °C}$$

Problem 1.6: A pressure cooker contains 2 kg of steam at 5 bar and 0.9 dry. Calculate the quantity of heat which must be rejected so as the quality of steam becomes 50%.

Solution: Given data:

Mass of steam: $m = 2$ kg.

At state 1,

Pressure: $p_1 = 5$ bar.

Dryness fraction: $x_1 = 0.9$

At state 2, dryness fraction: $x_2 = 50\% = 0.5$

From saturated steam table (pressure based) (Fig. 1.41),

at $p_1 = 5$ bar.

we get

$$v_{f1} = 0.00109 \text{ m}^3/\text{kg}$$
$$v_{g1} = 0.375 \text{ m}^3/\text{kg}$$
$$u_{f1} = 639.7 \text{ kJ/kg}$$
$$u_{g1} = 2561.2 \text{ kJ/kg}$$
$$h_{f1} = 640.2 \text{ kJ/kg}, h_{fg1} = 2108.5 \text{ kJ/kg}$$
$$v_1 = v_{f1} + x_1(v_{g1} - v_{f1})$$

Fig. 1.41 p-v dioagram for
Problem 1.6

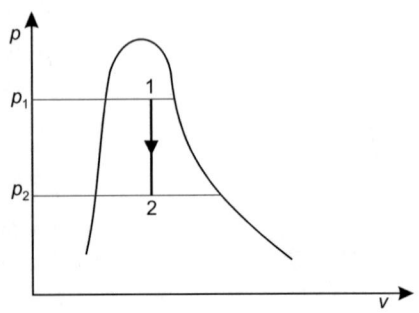

As

$$v_{g1} \gg v_{f1}$$
$$v_1 = x_1 v_{g1} = 0.9 \times 0.375$$
$$= 0.3375 \text{ m}^3/\text{kg}$$
$$= v_2$$
$$u_1 = u_{f1} + x_1 \left(u_{g1} - u_{f1} \right)$$
$$= 639.7 + .9(2561 - 639.7) = 2369.05 \text{ kJ/kg}$$

As

$$v_2 = v_1 = 0.3375 \text{ m}^3/\text{kg}$$
$$v_2 = x_2 v_{g2}$$
$$0.3375 = 0.5 \times v_{g2}$$

or

$$v_{g2} = 0.675 \text{ m}^3/\text{kg}$$

From saturated steam table (pressure based),

at

$$v_{g2} = 0.675 \text{ m}^3/\text{kg}$$

we get

$$u_{f2} = 546 \text{ kJ/kg}$$
$$u_{g2} = 2539.9 \text{ kJ/kg}$$

$$h_{f2} = 546.3 \text{ kJ/kg}, h_{fg2} = 2174.2.2 \text{ kJ/kg}$$
$$u_2 = u_{f2} + x_2(u_{g2} - u_{f2})$$
$$= 546 + 0.5(2539.9 - 546) = 1542.95 \text{ kJ/kg}$$

Work done: $W_{1-2} = 0 \therefore v_1 = v_2$.
According to the first law of thermodynamic for a process,

$$Q_{1-2} = (U_2 - U_1) + W_{1-2}$$
$$= m(u_2 - u_1) + 0$$
$$= 2(1542.95 - 2369.05) = -1652.2\text{kJ}$$

The $-ve$ sign shows that heat rejected from the system.
Note that the change in specific internal energy also determine as

$$u_2 - u_1 = (h_2 - p_2v_2) - (h_1 - p_1v_1)$$

where

$$h_2 = h_{f2} + x_2h_{fg2}$$
$$h_1 = h_{f1} + x_1h_{fg1}$$

Problem 1.7: A rigid tank with a volume of 2.5 m^3 contains 15 kg of saturated liquid–vapour mixture of water at 75 °C. Now the water is slowly heated. Determine the temperature at which the liquid in the tank is completely vapourized. Also, show the process on a T-v diagram with respect to saturation lines.
  Solution: Given data:

$$\text{Volume of tank} = \text{volume of liquid - vapor}$$
$$V = 2.5 \text{ m}^3$$

Mass of mixture: $m = 15$ kg.
  $\therefore$ Specific volume of mixture:

$$v_1 = \frac{V}{m} = \frac{2.5}{15} = 0.1666 \text{ m}^3/\text{kg}$$

Specific volume at state 1 = specific volume at state 2,
  i.e., $v_1 = v_2 = 0.1666$ m^3/kg.

By given condition, state 2 lies on saturated vapour line.
  at which the liquid in the tank is completely vapourized.

From saturated steam table (pressure based) (Fig. 1.42),

Fig. 1.42 T-s diagram for
Problem 1.7

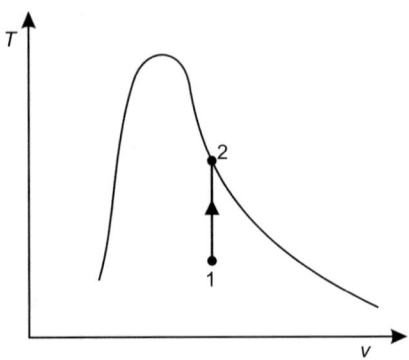

at $v_2 = v_g = 0.1666$ m^3/kg.

Temperature: °C	Specific volume at saturated vapour line v_g: m^3/kg
180	0.19405
T$_2$	0.1666
190	0.15654

By interpolation, we get

$$T_2 = 180 + \frac{(190 - 180) \times (0.1666 - 0.19405)}{(0.15654 - 0.19405)}$$

$$= 180 + 7.31 = \mathbf{187.31} \; °\mathbf{C}$$

Problem 1.8: A pressure cooker contains water at 100 °C with liquid volume being
one-tenth of vapour volume. At 100 °C, the specific volumes for saturated liquid
and saturated vapour are 0.001 m^3/kg and 1.67 m^3/kg, respectively. Determine the
dryness fraction of the mixture.

   Solution: Given data:
   Volume of liquid: $V_f = \frac{V_g}{10}$.
   or volume of vapour: $V_g = 10V_f$.
   Specific volume of saturated liquid: $v_f = 0.001$ m^3/kg.
   Specific volume of saturated vapour: $v_g = 1.6$ m^3/kg.
   Dryness fraction:

$$x = \frac{\text{mass of vapor: } m_g}{\text{mass of vapor } (m_g) + \text{mass of liquid } (m_f)}$$

$$x = \frac{m_g}{m_g + m_f}$$

where

$$m_g = \frac{V_g}{v_g} = \frac{10V_f}{1.6} = 6.25V_f$$

and

$$m_f = \frac{V_f}{v_f} = \frac{V_f}{0.001} = 1000V_f$$

$\therefore$

$$x = \frac{6.25V_f}{6.25V_f + 1000V_f} = \frac{6.25}{1006.25} = 6.21 \times 10^{-3}$$

Problem 1.9: A vessel of volume 0.04 m³ contains a mixture of saturated water and saturated steam at a temperature of 250 °C. The mass of the liquid present is 9 kg. Find the pressure, the mass, the specific volume, the enthalpy, entropy, and internal energy.

(*GGSIP Universiy, Delhi, Dec. 2001*).

Solution: Given data:

Volume of a vessel: $V = 0.04$ m³

Saturated temperature: $T_{sat} = 250°C$

Mass of liquid: $m_f = 9$ kg

From saturated steam table (temperature based),

at

$$T_{sat} = 250\,°C$$
$$p = 39.734\text{bar} = \mathbf{3973.4kPa}$$
$$v_f = 0.0125 \text{ m}^3/\text{kg},$$
$$v_g = 0.05013 \text{ m}^3/\text{kg}$$
$$h_f = 1085.36 \text{ kJ/kg}$$
$$h_{fg} = 1716.2 \text{ kJ/kg}$$
$$h_g = 2801.5 \text{ kJ/kg}$$
$$s_f = 2.7927 \text{ kJ/kgK}, s_{fg} = 3.1181 \text{ kJ/kgK}$$
$$s_g = 6.0730 \text{ kJ/kgK}$$

Now, volume of liquid:

$$V_f = m_f v_f = 9 \times 0.00125 = 0.01125 \text{ m}^3$$

Volume of vessel:

$$V = V_f + V_g$$

or volume of vapour:

$$V_g = V - V_f = 0.04 - 0.01125 = 0.02875 \text{ m}^3$$

also

$$V_g = m_g v_g = 0.02875$$
$$= m_g \times 0.05013$$

or

$$m_g = \frac{0.02875}{0.05013} = 0.5735 \text{ kg}$$

∴ Total mass of mixture:

$$m = m_f + m_g = 9 + 0.5735 = \mathbf{9.5735kg}$$

Dryness fraction of mixture:

$$x = \frac{m_g}{m_f + m_g} = \frac{m_g}{m} = \frac{0.5735}{9.5735} = 0.0599$$

Now, specific volume of mixture:

$$v = v_f + x$$
$$v_{fg} = v_f + x(v_g - v_f)$$
$$= 0.00125 + 0.0599 \times (0.05013 - 0.00125)$$
$$= \mathbf{0.004178 \text{ m}^3/\text{kg}}$$

and specific enthalpy of mixture:

$$h = h_f + x h_{fg}$$
$$= 1085.36 + 0.0599 \times 1716.2 = 1188.16 \text{ kJ/kg}$$

Therefore, enthalpy of mixture:

$$H = mh = 9.5735 \times 1188.16 = \mathbf{11374.85kJ}$$

Specific entropy of mixture:

$$s = s_f + x s_{fg}$$
$$= 2.7927 + 0.0599 \times 3.1181$$

$$= 2.9794 \text{ kJ/kgK}$$

Therefore, entropy of mixture: $S = ms = 9.5735 \times 2.9794 = \mathbf{28.523 kJ/K}$.
We know the specific enthalpy: $h = u + pv$
or specific internal energy:

$$u = h - pv$$
$$= 1188.16 - 3973.4 \times 0.004178$$
$$= 1171.559 \text{ kJ/kg}$$

Therefore, internal energy of mixture: $U = mu = 9.5735 \times 1173.559 = \mathbf{11215.92 kJ}$.

Note: Specific internal energy is also found out by the following relation, if u_f and u_{fg} are given in the steam table.

$$u = u_f + x u_{fg}$$

Then, internal energy of mixture: $U = mu$.

Problem 1.10: A certain heater operates under steady flow conditions receiving 4.2 kg/s of water at 75 °C temperature, specific enthalpy 331.93 kJ/kg. The water is heated by mixing with steam which is supplied to the heater at temperature 100.2 °C and specific enthalpy 2676 kJ/kg, the mixture leaves the heater as liquid water at temperature 100 °C and specific enthalpy 419 kJ/kg. How much steam must be supplied to the heater per hour?

Solution: Given data:
Water supplied to the heater at state 1,

$$m_1 = 4.2 \text{ kg/s}$$
$$T_1 = 75° \text{C}$$
$$h_1 = 313.93 \text{ kJ/kg}$$

Steam supplied to the heater at state 2,

$$m_2 = ?$$
$$T_2 = 100.2 °\text{C}$$
$$h_2 = 2676 \text{ kJ/kg}$$

Water leaving the heater at state 3 (Fig. 1.43),

$$m_3 = m_1 + m_2$$
$$T_2 = 100 °\text{C}$$
$$h_3 = 419 \text{ kJ/kg}$$

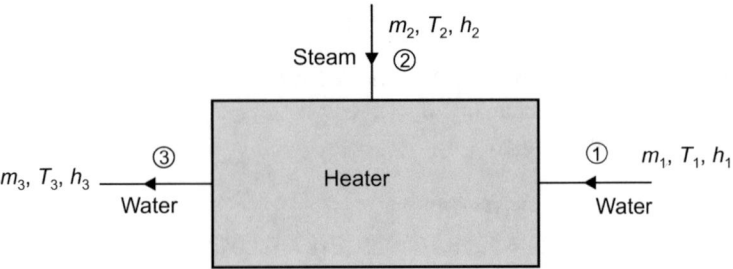

Fig. 1.43 Schematic for Problem 1.10

Applying the energy balance equation, we get

$$m_1 h_1 + m_2 h_2 = m_3 h_3$$
$$m_1 h_1 + m_2 h_2 = (m_1 + m_2) h_3$$
$$4.2 \times 313.93 + m_2 \times 2676 = (4.2 + m_2) \times 419$$
$$1318.50 + 2676 m_2 = 1759.8 + 419 m_2$$
$$\text{or} \quad 2676 m_2 - 419 m_2 = 1759.8 - 1318.50$$
$$2257 m_2 = 441.3$$
$$\text{or} \quad m_2 = \frac{441.3}{2287} = 0.19552 \text{ kg/s}$$
$$= 0.19552 \times 3600 \text{ kg/hr} = \textbf{703.87kg/hr}$$

703.87 kg of steam must be supplied to the heater per hour

Problem 1.11: A rigid tank contains 3 kg of saturated liquid–vapour mixture at 80 °C. If 2 kg of water is in the liquid form and the rest in the vapour form, determine the pressure of the mixture and the volume of the tank.

Solution: Given data:

Mass of the saturated liquid–vapour mixture,

$m = 3$ kg

Temperature of the mixture: $T_{sat} = 80°C$

Mass of the liquid: $m_f = 2$ kg

∴ Mass of vapour: $m_g = m - m_f = 3 - 2 = 1$ kg

Dryness fraction: $x = \frac{m_g}{m} = \frac{1}{3} = 0.3333$.

From saturated steam table (temperature based),

at

$$T_{sat} = 80 °C$$
$$p = \textbf{47.39kPa}$$
$$v_f = 0.001029 \text{ m}^3/\text{kg}$$

$$v_g = 3.403 \text{ m}^3/\text{kg}$$

Specific volume of the mixture: v

$$
\begin{aligned}
v &= v_f + x\left(v_g - v_f\right)\\
&= 0.001029 + 0.3333(3.403 - 0.001029)\\
&= 0.001029 + 1.133876 = 1.1349 \text{ m}^3/\text{kg}
\end{aligned}
$$

also

$$
\begin{aligned}
v &= \frac{V}{m}\\
1.1349 &= \frac{V}{3}
\end{aligned}
$$

or $V = \mathbf{3.4047 \text{ m}^3}$

The volume of the tank is **3.4047 m³**.

Problem 1.12: One kg of steam with a quality of 20% is heated at a constant pressure of 200 kPa until the temperature reaches 400 °C. Determine the work done by the steam.

Solution: Given data:

Mass of steam: $m = 1$ kg.

Quality of steam: $x = 20\% = 0.20$.

Pressure of steam: $p = 200$ Pa $= 2$ bar.

Final temperature: $T_2 = 400$ °C.

From saturated steam table
(Pressure based),
at $p = 2$ bar (Fig. 1.44).
we have

$$
\begin{aligned}
T_{\text{sat}} &= 120.2 \text{ °C}\\
v_f &= 0.00161 \text{ m}^3/\text{kg}\\
v_g &= 0.886 \text{ m}^3/\text{kg}
\end{aligned}
$$

Fig. 1.44 T-s diagram for Problem 1.12

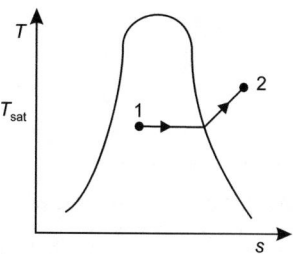

$$v_1 = v_f + x(v_g - v_f)$$
$$= 0.001061 + 0.20(0.86 - 0.001061)$$
$$= 0.7180 \text{ m}^3/\text{kg}$$

From superheated steam table,
  at $p = 2$ bar and $T_2 = 400 \, °C$
  we have $v_2 = 1.5493 \text{ m}^3/\text{kg}$
  Work done:

$$W = p(V_2 - V_1)$$
$$= mp(v_2 - v_1)$$
$$= 1 \times 200(1.5493 - 0.1780)$$
$$= 2.74. \ 26 \text{ kJ}$$

Problem 1.13: Steam initially at 15 bar, 300 °C expands reversibly and adiabatically in a steam turbine to 40 °C. Determine the ideal work output of the turbine per kg of steam.

  Solution: Given data:
  At inlet of a steam turbine,

$$p_1 = 15 \text{ bar}$$
$$T_1 = 300 \, °C$$

First, we will check whether the steam at inlet of the turbine is wet or superheated.
  From saturated steam table (pressure based),
    at

$$p_1 = 15 \text{bar}$$
$$T_{sat} = 198.3 \, °C$$

As $T_{sat} < T_1$, it means that the steam at inlet of the turbine is superheated.
  From superheated steam table,
    at

$$p_1 = 15 \text{bar}, \quad T_1 = 300 \, °C$$
$$h_1 = 3037.6 \text{ kJ/kg}$$
$$s_1 = 6.918 \text{ kJ/kgK}$$

At exit of a steam turbine,

Fig. 1.45 T-s diagram for
Problem 1.13

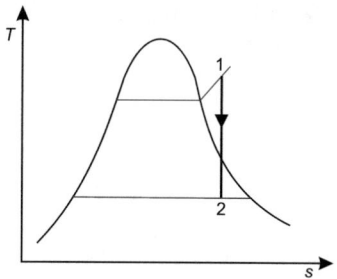

$$T_2 = 40 \,°C$$
$$s_2 = s_1 = 6.918 \text{ kJ/kgK}$$

From saturated steam table (temperature based),
   at

$$T_2 = 40 \,°C$$
$$s_g = 8.257 \text{ kJ/kgK}$$

As $s_g > s_2$, it means that the steam at exit is wet (Fig. 1.45).
   Other required properties of the saturated steam at $T_2 = 40 \,°C$

$$s_f = 0.573 \text{ kJ/kgK}$$
$$h_f = 167.6 \text{ kJ/kg}, h_{f\,g} = 2406.7 \text{ kJ/kg}$$
$$s_2 = s_f + x_2(s_g - s_f)$$
$$6.918 = 0.573 + x_2(8.257 - 0.573)$$

or $6.918 - 0.573 = 7.684x_2$

   or $7.684x_2 = 6.345$

   or $x_2 = 0.8257$

$$h_2 = h_f + x_2h_{fg} = 167.6 + 0.8257 \times 2406.7$$
$$= 167.6 + 1987.21 = 2154.81 \text{ kJ/kg}$$

Work output: $w_{1-2} = h_1 - h_2 = 0307.6 - 2154.81 = \textbf{882.79 kJ/kg.}$

Problem 1.14: Saturated water vapour at 60 °C has its pressure decreased to increase
the volume by 10% while keeping the temperature constant. To what pressure should
it be expanded?
   Solution: Given data:
   Temperature of saturated water vapour:
   $T_1 = 60 \,°C$.
   From saturated steam table (temperature based),

Fig. 1.46 T-s diagram for
Problem 1.14

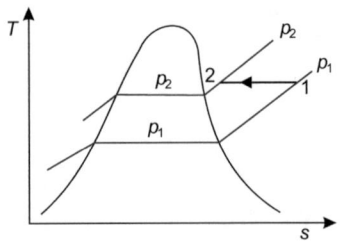

at

$$T_1 = 60\ ^\circ\text{C}, \quad p_1 = 19.94\text{kPa}$$
$$v_1 = v_g$$
$$\quad = 7.671\ \text{m}^3/\text{kg}$$
$$v_2 = 1.1v_1$$
$$\quad = 1.1 \times 7.671$$
$$\quad = 8.4381\ \text{m}^3/\text{kg}$$
$$T_2 = T_1 = 60\ ^\circ\text{C}$$

From superheated steam table (Fig. 1.46),
at

$$T_2 = 60\ ^\circ\text{C}$$
$$v_2 = 8.4381\ \text{m}^3/\text{kg}$$

For process 1–2,

$$p_1 v_1 = p_2 v_2$$
$$19.94 \times 7.671 = p_2 \times 8.4381$$

or $p_2 = \mathbf{18.127\ kPa.}$

Problem 1.15: 1 kg of steam undergoes a reversible isothermal process from 20 bar
and 250 °C to pressure of 30 bar. Determine the heat flow; whether it is supplied or
rejected and sketch the process on a T-s diagram.

   Solution: Given data for a reversible isothermal process:

$$m = 1\ \text{kg}$$
$$p_1 = 20\text{bar} = 2000\text{kPa}$$
$$T_1 = 250\ ^\circ\text{C} = T_2$$
$$p_2 = 30\text{bar}.$$

Fig. 1.47 T-s diagram for
Problem 1.15

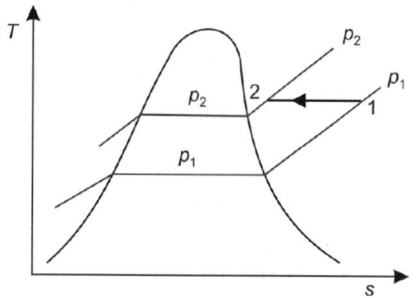

From saturated steam table (pressure based),
at

$$p_1 = 20 \text{ bar}$$
$$T_{\text{sat}} = 212.4 \,°\text{C}$$

As $T_1 > T_{sat}$, the steam is superheated.

From superheated steam table (Fig. 1.47),
at $p_1 = 20$ bar, $T_1 = 250\,°$C.
we get

$$v_1 = 0.1114 \text{ m}^3/\text{kg}$$
$$u_1 = 2679.6 \text{ kJ/kg}$$
$$h_1 = 2902.5 \text{ kJ/kg}$$

From saturated steam table on pressure based,
at

$$p_2 = 30 \text{ bar}$$
$$T_{\text{sat}} = 233.9 \,°\text{C}$$

As $T_2 > T_{sat}$, the steam is superheated.
From superheated steam table,
at $p_2 = 30$ bar, $T_2 = 250\,°$C.
we get

$$v_2 = 0.07058 \text{ m}^3/\text{kg}$$
$$u_2 = 2644 \text{ kJ/kg}$$
$$h_2 = 2855.8 \text{ kJ/kg}$$

Heat flow:

$$Q_{1-2} = m(u_2 - u_1) + m p_1 v_1 \log_e \frac{v_2}{v_1} \text{ kJ}$$

where m is in kg,
$\quad u_1$ and u_2 are in kJ/kg,
$\quad p_1$ is in kPa,
$\quad v_1$ and v_2 are in m³/kg.
$\therefore$

$$Q_{1-2} = 1(2644 \times 2679.6) + 1 \times 200 \times .1114 \log_e \frac{0.07058}{0.1114}$$

$$= -35.6 - 101.68 = -\mathbf{137.28 \, kJ}$$

The $-ve$ sign indicates the heat rejected by the process.

Problem 1.16: A 110 mm diameter cylinder contains 100 cm³ of water at 60 °C. A 50 kg piston sits on top of the water. If heat is added until the temperature is 200 °C, determine the work done by the system.

Solution: Given data:

Diameter of cylinder: $d = 110$ mm.
$$A = \frac{\pi}{4} d^2$$
$\therefore$ Cross-sectional area:
$$= \frac{3.14}{4} \times (0.11)^2 \quad \cdot$$
$$= 9.498 \times 10^{-3} \text{ m}^2$$

Initial volume of water: $V_1 = 100 \text{ cm}^3 = 10^{-4} \text{ m}^3$.
$\quad$ Temperature of water: $T_1 = 60$ °C.
$\quad$ Mass of piston: $M = 50$ kg.
$\quad$ Final temperature of the system: $T_2 = 200$ °C (Fig. 1.48).

Fig. 1.48 Schematic for Problem 1.16

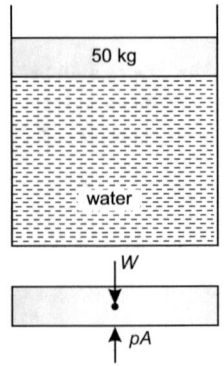

First of all, we will check whether water is saturated or subcooled.

Applying the force balance equation,

$$W = pA$$
$$mg = pA$$
$$50 \times 9.81 = p \times 9.49 \times 10^{-3}$$
$$= 51.64 \text{kPa (Gauge pressure)}$$

Absolute pressure of water:

$$p_1 = p + p_{\text{atm}}$$
$$= 51.64 + 101.325 = 152.96 \text{kPa} = 1.529 \text{bar}$$

From saturated steam table (Pressure based),
at

$$p_1 = 1.529 \text{bar}$$
$$T_{\text{sat}} = 111.4\,^\circ\text{C}$$

As $T_{\text{sat}} > T_1$, condition of water is subcooled.
From saturated steam table (Temperature based),
at $T_1 = 60\,^\circ\text{C}$.
we have $v_f = v_1 = 0.001017 \text{ m}^3/\text{kg}$.
Mass: $m = \frac{V_1}{v_1} = \frac{10^{-4}}{0.001017} = 0.0983 \text{ kg}$ (Fig. 1.49).
From superheated stream table,
at

$$T_2 = 200\,^\circ\text{C},\ p_2 = p_1 = 1.529 \text{bar}$$
$$p = 1 \text{bar},\ T = 200\,^\circ\text{C},\ v = 2.172 \text{ m}^3/\text{kg}$$
$$p = 2 \text{bar},\ T = 200\,^\circ\text{C},\ v = 1.0803 \text{ m}^3/\text{kg}$$

Fig. 1.49 T-s diagram for
Problem 1.16

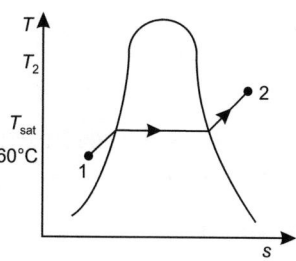

By interpolation,

$$v_2 = 2.172 - \frac{(2.72 - 1.0803) \times (2 - 1.529)}{2 - 1}$$
$$= 2.172 - 0.514 = 1.658 \ \text{m}^3/\text{kg}$$

Work done:

$$W = p(V_2 - V_1) = mp(v_2 - v_1)$$
$$= 0.0983 \times 152.96(1.658 - 0.001017)$$
$$= \textbf{24.91kJ}.$$

Problem 1.17: An insulated vessel is divided into two compartments, one containing 2 kg of dry saturated steam at 2 bar and the other 3 kg of steam, 0.8 quality at 5 bar. If the partition between the compartments is removed and the steam is mixed thoroughly, determine the final pressure, steam quality, and entropy in the process.

Solution: Given data:

For compartment 1

$$m_1 = 2 \ \text{kg}$$
$$x_1 = 1$$
$$p_1 = 2\text{bar}$$

From saturated steam table (pressure based),
at $p_1 = 2$ bar
we get

$$v_{g1} = v_1 = 0.886 \ \text{m}^3/\text{kg}$$
$$s_{g1} = s_1 = 7.127 \ \text{kJ/kgK}$$
$$u_{g1} = u_1 = 2529.5 \ \text{kJ/kg}$$

For compartment 2 (Fig. 1.50),

$$m_2 = 3 \ \text{kg}$$
$$x_2 = 0.8$$
$$p_2 = 5\text{bar}$$

From saturated steam table (pressure based),
at $p_2 = 5$ bar

we get

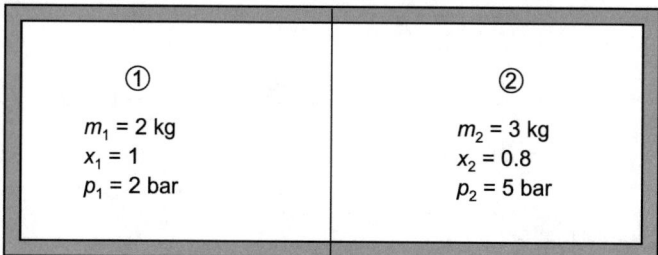

Fig. 1.50   Schematic for Problem 1.17

$$v_{f2} = 0.001093 \text{ m}^3/\text{kg}, \; v_{g2} = 0.375 \text{ m}^3/\text{kg}$$
$$u_{f2} = 639.7 \text{ kJ/kg}, \; u_{g2} = 2561.2 \text{ kJ/kg}$$
$$s_{f2} = 1.861 \text{ kJ/kgK}, \; s_{fg2} = 4.961 \text{ kJ/kgK}$$

$$v_2 = v_{f2} + x_2(v_{g2} - v_{ff})$$
$$= 0.001093 + 0.8(0.375 - 0.001093)$$
$$= 0.3002 \text{ m}^2/\text{kg}$$

$$u_2 = u_{f2} + x_2(u_{g2} - u_{f2})$$
$$= 639.7 + 0.8(2561.2 - 639.7) = 2176.9 \text{ kJ/kg}$$
$$s_2 = s_{f2} + x_2 s_{f\,g2}$$
$$= 0.001093 + 0.8 \times 4.961 = 3.9698 \text{ kJ/kgK}$$

After removed the partition,

   Mass: $m_3 = m_1 + m_2 = 2 + 3 = 5$ kg.

Volume:

$$V_3 = V_1 + V_2$$
$$= m_1 v_1 + m_2 v_2 = 2 \times 0.886 + 3 \times 0.3002$$
$$V_3 = 2.6726 \text{ m}^3$$
$$\text{also} \quad V_3 = m_3 v_3$$

$\therefore 2.6726 = 5 \times v_3$

or $v_3 = 0.5345 \text{ m}^3/\text{kg}$.

   By energy balance, we get

$$m_3 u_3 = m_1 u_1 + m_2 u_2$$

$$u_3 = \frac{m_1 u_1 + m_2 u_2}{m_3}$$

$$= \frac{2 \times 2529.5 + 3 \times 2176.9}{5} = 2317.94 \text{ kJ/kg}$$

$$= h_3$$

From the Mollier diagram,

at $h_3 = 2317.94$ kJ/kg, $v_3 = 0.5345$ m^3/kg.

we get $p_3 = \mathbf{2.7}$ bar, $x_3 = \mathbf{0.813}$, $s_3 = 6.03$ kJ/kgK.

The entropy change during the process:

$$= m_3 s_3 - (m_1 s_1 + m_2 s_2)$$
$$= 5 \times 6.03 - (2 \times 7.127 + 3 \times 3.9698)$$
$$= 3015 - 26.16 = \mathbf{3.99 kJ/K}$$

Problem 1.18: 0.5 kg of steam expands reversibly at a constant pressure of 6 bar until the volume is doubled. Find the final temperature, work done and heat transferred if the initial dryness fraction of steam is 0.7.

Solution: Given data:

Mass of the steam:

$$m = 0.5 \text{ } kg$$

Initial pressure of steam:

$$p_1 = 6 \text{ } bar$$

Dryness fraction:

$$x_1 = 0.7..$$

From saturated steam table (pressure based),

at

$$p_1 = 6 \text{ } bar$$

we get

$$T_{\text{sat } 1} = 158.9^\circ C$$
$$v_{g1} = 0.316 \text{ m}^3/\text{kg}$$
$$h_{f1} = 670.6 \text{ kJ/kg}$$
$$h_{fg1} = 2086.3 \text{ kJ/kg}$$

Specific volume at state 1,$v_1 = x_1 v_{g1} = 0.7 \times 0.316 = 0.2212$ m³/kg

Specific volume at state 2,$v_2 = 2v_1 = 2 \times 0.2212 = 0.4424$ m³/kg

Pressure at state 2,$p_2 = p_2 = 6$ bar

  As $v_2 > v_{g1}$, it means that steam at state 2 is superheated.
  From superheated steam table,
  at

$$p_2 = 6 \text{ bar,}$$
$$v_2 = 0.4424 \text{ m}^3/\text{kg}$$

we get

$$T_a = 300 \text{ °C}$$
$$T_b = 350 \text{ °C}$$
$$v_a = 0.4344 \text{ m}^3/\text{kg}$$
$$v_b = 0.4742 \text{ m}^3/\text{kg}$$
$$h_a = 3061.6 \text{ kJ/kg}$$
$$h_b = 3165.7 \text{ kJ/kg}$$

By interpolation, we get (Fig. 1.51)

$$T_2 = T_a + (T_b - T_a)\frac{(v_2 - v_a)}{(v_b - v_a)}$$
$$= 300 + (350 - 300)\frac{(0.4424 - 0.4344)}{(0.4742 - 0.4344)}$$
$$= 300 + 10.05 = \mathbf{310.05} \text{ °C}$$

$$h_2 = h_a + (h_b - h_a)\frac{(v_2 - v_a)}{(v_b - v_a)}$$

Fig. 1.51 T-s diagram for Problem 1.18

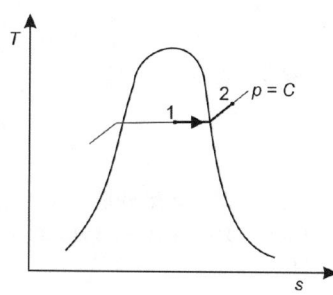

$$= 3061.6 + (3165.7 - 3061.6)\frac{(0.4424 - 0.4344)}{(0.4742 - 0.4344)}$$

$$= 3061.6 + 20.92 = 3082.52 \text{ kJ/kg}$$

$$h_1 = h_{f1} + x_1 h_{fg1}$$
$$= 670.6 + 0.7 \times 2086.3 = 2131.01 \text{ kJ/kg}$$

Work done:

$$W_{1-2} = mp(v_2 - v_1)$$
$$= 0.5 \times 600(0.4424 - 0.2212) = 66.36 \text{kJ}$$

Heat transferred:

$$Q_{1-2} = m(h_2 - h_1)$$
$$= 0.5(3082.52 - 2131.01) = \mathbf{475.75kJ}$$

Problem 1.19: Steam at 18 bar and 300 °C passes through a pipe at a velocity of 100 m/s. If steam flows at the rate of 500 kg/hr, find the diameter of the pipe.
   Solution: Given data:
Pressure of steam:

$$p = 18 \ bar.$$

Temperature of steam:

$$T = 300 \degree C.$$

First of all, we will check whether steam is wet or superheated.
From saturated steam table (pressure based),
at

$$p = 18 \ bar.$$

we get

$$T_{sat} = 207.2 \degree C.$$

As $T > T_{sat}$ it means that the given steam is superheated.
From superheated steam table,
at

$$p = 18 \, \text{bar}, \quad T = 300 \, °C.$$

we get

$$v = 0.1402 \, m^3 / kg.$$

Density of steam: $\rho = \frac{1}{v} = \frac{1}{0.1402} = 7.132 \, \text{kg/m}^3$.
Velocity of steam: $V = 100$ m/s.
Mass flow rate: $m = 500$ kg/hr

$$m = \frac{500}{3600} \, \text{kg/s}$$

also $m = \rho A V$ from continuity equation.

$\therefore \frac{500}{3600} = 7.132 \times A \times 100.$

or $A = 1.947 \times 10^{-4} \, \text{m}^2$

also $A = \frac{\pi}{d} d^2$.

$\therefore 1.947 \times 10^{-4} = \frac{\pi}{4} d^2 = \frac{3.14}{4} d^2.$

or

$$d^2 = 2.480 \times 10^{-4}$$
$$d = 0.01574 \, \text{m} = \textbf{15.74mm}$$

Problem 1.20: 600 kg of steam at a pressure of 16 bar and 0.92 dry is generated by a boiler in one hour. This steam then passes through a superheater where its temperature raises to 400 °C. If the temperature of feedwater is 40 °C, determine.

(i)   total heat supplied to water for producing wet steam.
(ii)   total heat absorbed in superheater.

Take specific heat for superheated steam is 2.1 kJ/kgK and for water is 4.18 kJ/kgK
Solution: Given data:
Mass flow rate of the steam: $m_s = 600$ kg/hr
Pressure: $p = 16$ bar
Dryness fraction of wet steam at state 2:
$x_2 = 0.92$
Temperature of superheated steam: $T_3 = 400$ °C
Temperature of feedwater: $T_1 = 40$ °C
Specific heat for superheated steam:
$c_{ps} = 2.1$ kJ/kgK
Specific heat for water: $c_{pw} = 4.18$ kJ/kgK
From saturated steam table (pressure based) (Fig. 1.52),

Fig. 1.52 T-s diagram for
Problem 1.20

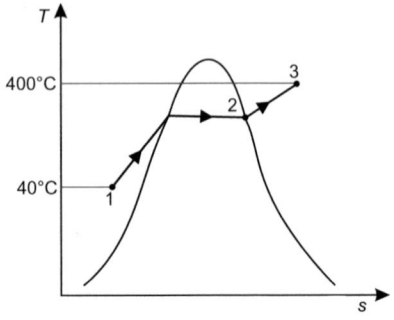

at

$$p = 16 \text{ bar}$$
$$T_{\text{sat}} = 201.4 \,^{\circ}\text{C}$$
$$h_f = 858.8 \text{ kJ/kg}, h_{fg} = 1935.2 \text{ kJ/kg}$$
$$h_g = 2794 \text{ kJ/kg}$$

$\therefore$

$$h_2 = h_f + x_2$$
$$h_{fg} = 858.8 + 0.92 \times 1935.2 = 2639.18 \text{ kJ/kg}$$

(i)   Total heat required to water for producing wet steam:

$$Q_{1-2} = m_s \left[c_{pw}(T_{\text{sat}} - T_1) + (h_2 - h_f) \right]$$
$$= 600[4.18(201.4 - 40) + (2639.18 - 858.8]$$
$$= 600[674.65 + 1780.38]$$
$$= 600[2455.03] = 1473018 \text{ kJ/hr} = \mathbf{409.17kJ/s}$$

(ii)   Total heat absorbed in preheater:

$$Q_{2-3} = m_s \left[(h_g - h_2) + c_{ps}(T_3 - T_{sat}) \right]$$
$$= 600[(2794 - 2639.18) + 2.1(400 - 201.4)]$$
$$= 600[154.82417.06] = 343128 \text{ kJ/hr} = \mathbf{95.31kJ/s}$$

Problem 1.21: A closed system containing dry saturated steam undergoes expansion according to the law $pv^n = C$ from an initial pressure of 10 bar to a final pressure of 2 bar. If the steam is finally wet with dryness fraction 0.85, determine the index of expansion and the work done.

   Solution: Given data:

Fig. 1.53 p-v diagram for
Problem 1.21

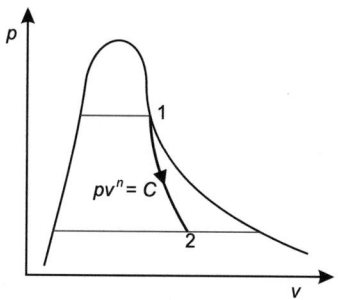

Initial condition of the steam is dry saturated:
$p_1 = 10$ bar (Fig. 1.53)
From saturated steam table (pressure based),
at

$$p_1 = 10 \text{ bar}$$
$$v_{g1} = v_1 = 0.194 \text{ m}^3/\text{kg}$$

Final condition of the steam is wet

$$p_2 = 2 \text{ bar}, x_2 = 0.85$$

From saturated steam table (pressure based),
at $p_2 = 2$ bar
we get

$$u_{f2} = 0.00106 \text{ m}^3/\text{kg}$$
$$v_{g2} = 0.886 \text{ m}^3/\text{kg}$$

Specific volume at state 2,

$$v_2 = v_{g2} + x_2\left(v_{g2} - v_{f2}\right)$$
$$= 0.00106 + 0.85(0.886 - 0.00106)$$
$$= 0.7532 \text{ m}^3/\text{kg}$$

Process 1–2 follows the law of $pv^n = C$
$$\therefore p_1 v_1^n = p_2 v_2^n$$

Taking $\log_e$ both sides, we get

$$\log_e p_1 + n \log_e v_1 = \log_e p_2 + n \log_e v_2$$

$$n\left[\log_e v_1 - \log_e v_2\right] = \log_e p_2 - \log_e p_1$$

$$n \log_e \frac{v_1}{v_2} = \log_e \frac{p_2}{p_1}$$

or

$$n = \frac{\log_e \frac{p_2}{p_1}}{\log_e \frac{v_1}{v_2}} = \frac{\log_e \frac{2}{10}}{\log_e \frac{0.194}{0.7532}}$$

$$= \frac{-1.609}{-1.356} = 1.18$$

The index of expansion: $n = 1.18$

Work done: $w_{1-2} = \frac{p_1 v_1 - p_2 v_2}{n-1}$ kJ/kg.

where

$$p_1 = 1000 \text{kPa}$$
$$p_2 = 200 \text{kPa}$$
$$v_1 = 0.194 \text{ m}^3/\text{kg}$$
$$v_2 = 0.7532 \text{ m}^3/\text{kg}$$

$$\therefore w_{1-2} = \frac{1000 \times 0.94 - 200 \times 0.7532}{1.18 - 1} = \mathbf{240.88 kJ/kg}.$$

Problem 1.22: Steam at 250 kPa, 0.8 dry expands to 100 kPa in the following two processes in a piston cylinder arrangement. The processes are (a) $pv = $ constant and (b) $pv^{1.2} = $ constant. Determine the work done, heat transferred, and change in entropy in each case. Use the properties given below

p	T	v_f	v_g	h_f	h_{fg}	h_g	s_f	s_{fg}	s_g
kPa	°C	m³/kg	m³/kg	kJ/kg	kJ/kg	kJ/kg	kJ/kgK	kJ/kgK	kJ/kgK
100	99.62	0.001043	1.694	417.44	2258.06	2675.5	1.3025	6.0568	7.3593
250	127.43	0.001067	0.7187	535.34	2181.56	2716.9	1.6072	5.4454	7.0526

Solution: Given data:

At initial state,

$$p_1 = 250 \text{ kpa}$$
$$x_1 = 0.8$$

At final state,

$$p_2 = 100 \text{ kpa}$$

Fig. 1.54 Process pv = C

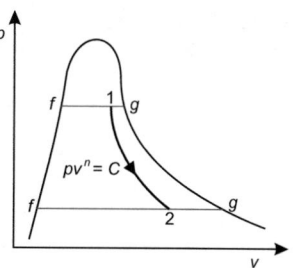

$$p_2 = 100 \text{ kPa}$$

(a) $pv = $ constant, hyperbolic process (Fig. 1.54)

At

$$p_1 = 250 \text{kPa}$$
$$v_f = 0.001067 \text{ m}^3/\text{kg}, \ v_g = 0.7187 \text{ m}^3/\text{kg}$$
$$h_f = 535.34 \text{ kJ/kg}, \ s_{f\,g} = 5.4454 \text{ kJ/kgK}$$
$$v_1 = v_f + x_1(v_g - v_f)$$
$$= 0.001067 + 0.8(0.7187 - 0.001067)$$
$$= 0.5751 \text{ m}^3/\text{kg}$$
$$p_1 v_1 = p_2 v_2$$
$$250 \times 0.5751 = 100 \times v_2$$

or

$$v_2 = 1.4377 \text{ m}^3/\text{kg}$$

At

$$p_2 = 100 \text{kPa}$$
$$T_{\text{sat}} = 99.62 ^\circ\text{C} = (273 + 99.62)\text{K} = 372.62 \text{ K}$$
$$v_f = 0.001043 \text{ m}^3/\text{kg}, \ v_g = 1.694 \text{ m}^3/\text{kg}$$
$$h_f = 417.44 \text{ kJ/kg}, \ h_{fg} = 2258.06 \text{ kJ/kg}, \ h_g = 2675.50 \text{ kJ/kg}$$
$$s_f = 1.3025 \text{ kJ/kgK}, \ s_{fg} = 6.0568 \text{ kJ/kgK}, \ s_g = 7.3593 \text{ kJ/kgK}$$

As $v_2 < v_g$ at $p_2 = 100$ kPa, it means that the state 2 of the steam is wet (Fig. 1.55).
also

Fig. 1.55 p-v diagram for
Problem 1.22

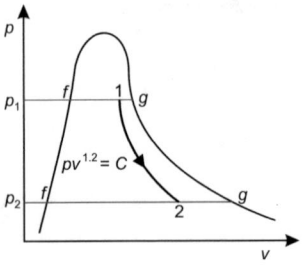

$$v_2 = v_f + x_2(v_g + v_f) \text{ at } p_2 = 100\text{kPa}$$
$$1.4277 = 0.001043 + x_2(1.694 - 0.001043)$$

or

$$x_2 = 0.8486$$
$$h_2 = h_f + x_2 h_{fg} \quad \text{at } p_2 = 100\text{kPa}$$
$$= 417.44 + x_2 h_{fg}$$
$$= 417.44 + 0.8486 \times 2258.06$$
$$= 2333.62 \text{ kJ/kg}$$

and

$$h_1 = h_f + x_1 h_{fg} \quad \text{at } p_1 = 250\text{kPa}$$
$$= 535.34 + 0.8 \times 2181.56$$
$$= 2280.58 \text{ kJ/kg}$$

Work done:

$$w_{1-2} = p_1 v_1 \log_e \frac{v_2}{v_1}$$
$$= 250 \times 0.5751 \log_e \frac{1.4377}{0.5751}$$
$$= \mathbf{131.73kJ/kg}$$

Heat transfer:

$$q_{1-2} = w_{1-2} + (u_2 - u_1)$$
$$= 131.73 + [(h_2 - p_2 v_2) - (h_1 - p_1 v_1)]$$
$$= 131.73 + h_2 - h_1$$

$$= 131.73 + 2333.62 - 2280.58$$
$$= \mathbf{184.77 kJ/kg}$$

Change in entropy:

$$\Delta s = s_2 - s_1$$

where

$$s_1 = s_f + x_1 s_{fg} \text{ at } p_1 = 250 \text{kPa}$$
$$= 1.6072 + 0.8 \times 5.4454 = 5.963 \text{ kJ/kgK}$$

and

$$s_2 = s_f + x_2 s_{fg} \text{ at } p_2 = 100 \text{kPa}$$
$$= 1.3025 + 0.8486 \times 6.0568$$
$$= 6.442 \text{ kJ/kgK}$$

$$\therefore \Delta s = 6.442 - 5.963 = \mathbf{0.479 kJ/kgK}.$$

(b)
$$pv^{1.2} = \text{constant}$$
$$p_1 v_1^{1.2} = p_2 v_2^{1.2}$$

or $\left(\dfrac{v_2}{v_1}\right)^{1.2} = \dfrac{p_1}{p_2}.$

or
$$\dfrac{v_2}{v_1} = \left(\dfrac{p_1}{p_2}\right)^{1/1.2}$$
$$\dfrac{v_2}{0.5751} = \left(\dfrac{250}{100}\right)^{0.8333} = 2.145$$

or $v_2 = 2.145 \times 0.571 = 1.2247 \text{ m}^3/\text{kg}.$

As $v_2 < v_g$ at $p_2 = 100$ kPa, it means that the state 2 of the steam is wet.
also

$$v_2 = v_f + x_2 v_{fg} \text{ at } p_2 = 100 \text{kPa}$$
$$1.2247 = 0.00143 + x_2(1.694 - 0.001043)$$

or

$$x_2 = 0.7227$$
$$h_2 = h_f + x_2 h_{fg} \quad \text{at } p_2 = 100 \text{kPa}$$
$$= 417.44 + 0.7227 \times 2258.06$$

$$= 2049.33 \text{ kJ/kg}$$
$$s_2 = s_f + x_2 s_{fg}$$
$$= 1.3025 + 0.7227 \times 6.0568$$

Work done:

$$
\begin{aligned}
w_{1-2} &= \frac{p_1 v_1 - p_2 v_2}{n - 1} \\
&= \frac{250 \times 0.5751 - 100 \times 1.2247}{1.2 - 1} \\
&= \mathbf{106.525 kJ/kg}
\end{aligned}
$$

Heat transfer:

$$
\begin{aligned}
q_{1-2} &= w_{1-2} + (u_2 - u_1) \\
&= 106.525 + [(h_2 - p_2 v_2) - (h_1 - p_1 v_1)] \\
&= 106.25 + [(2049.33 - 100 \times 1.2247) \\
&\quad - (2280.58 - 250 \times 0.5751)] \\
&= 106.52 + [1926.86 - 2136.80] \\
&= 106.525 - 209.94 = \mathbf{-103.41 kJ/kg}
\end{aligned}
$$

The $-ve$ sign shows that heat transfer from system to surroundings.

(i)   Change in entropy:

$$
\begin{aligned}
\Delta s &= s_2 - s_1 \\
&= 5.5425 - 61.479 = \mathbf{-0.6054 kJ/kgK}
\end{aligned}
$$

Problem 1.23: Steam at 10 bar, 200 °C is expanded to 5 bar in a piston-cylinder arrangement as per the following relation. Find the work done and heat transfer in each case:

(i) $pv^{1.25} = \text{constant}$,
(ii) $pv = 1$ constant.

   Solution: Given data:
   At initial state,

$$
\begin{aligned}
p_1 &= 10 \text{ bar} \\
T_1 &= 200 \,°C
\end{aligned}
$$

At final state,

$$p_2 = 5 \text{ bar.}$$

First, we will check whether the initial state is wet steam or superheated steam. From saturated steam table (pressure based), at

$$p_1 = 10 \text{ bar}$$
$$T_{sat} = 179.9\,°C$$

As

$$T_2 > T_{sat},$$

it means that initial state of the steam is superheated. From superheated steam table (FIg. 1.56),

at

$$p_1 = 10 \text{ bar}, T_1 = 200\,°C$$
$$v_1 = 0.2060 \text{ m}^3/\text{kg}$$
$$h_1 = 2827.9 \text{ kJ/kg}$$
$$s_1 = 6.694 \text{ kJ/kgK}$$
$$u_1 = 2621.9 \text{ kJ/kg}$$

Case-I: Steam is expanded to follows the law of $pv^{1.25} = C$.

$$p_1 v_1^{1.25} = p_2 v_2^{1.25}$$
$$10 \times (0.2060)^{1.25} = 5 v_2^{1.25}$$

or $v_2^{1.25} = 0.2775$.
   or $v_2 = (0.2775)^{1/1.25} = 0.3586 \text{ m}^3/\text{kg}$

From saturated steam table (pressure based),
   at

$$p_2 = 5 \text{ bar}$$
$$v_g = 0.3749 \text{ m}^3/\text{kg}$$

As

$$v_g > v_2, \quad \text{it means that final state of the steam is wet.}$$

Other required properties at $p_2 = 5$ bar.

Fig. 1.56 T-s diagram for
Problem 1.23

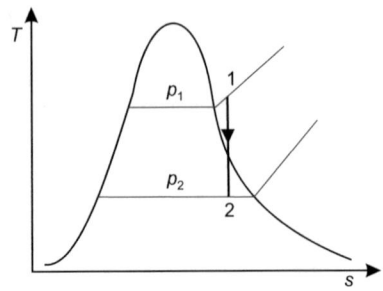

$$v_f = 0.00109 \text{ m}^3/\text{kg}$$
$$u_f = 6.39.7 \text{ kJ/kg}$$
$$u_g = 2561.2 \text{ kJ/kg}$$
$$h_f = 640.2 \text{ kJ/kg}$$
$$h_{fg} = 2108.5 \text{ kJ/kg}$$
$$v_2 = v_f + x_2(v_g - v_f)$$
$$0.3586 = 0.00109 + x_2(0.375 - 0.00109)$$
$$0.3586 = 0.00109 + 0.3739 x_3$$

or

$$0.3739 \, x_2 = 0.3575$$
$$x_2 = 0.9561$$

Specific internal energy:

$$u_2 = u_f + x_2(u_g - u_f)$$
$$u_2 = 639.7 + 0.9561(2561.2 - 639.7)$$
$$= 639.7 + 1837.14 = 2476.84 \text{ kJ/kg}$$

Work done:

$$w_{1-2} = \frac{p_1 v_1 - p_2 v_2}{n - 1} \text{ kJ/kg}$$

where p_1 and p_2 are in kPa; v_1 and v_2 are in m^3/kg.

$$\therefore = \frac{1000 \times 0.2060 - 500 \times 0.3586}{1.25 - 1} = \mathbf{106.8 kJ/kg.}$$

Heat transfer:

$$q_{1-2} = (u_2 - u_1) + w_{1-2}$$
$$= (2476.84 - 2621.9) + 106.8$$
$$= -145.06 + 106.8 = -38.26 \text{ kJ/kg}$$

The $-ve$ sign shows that heat transfer from the system.

Case-II: Steam is expanded to follow the law of $pv = C$

$$p_1 v_1 = p_2 v_2$$
$$10 \times 0.2060 = 5 \times v_2$$

or

$$v_2 = 0.412 \text{ m}^3/\text{kg}.$$

From saturated steam table (pressure based),
at

$$p_2 = 5 \text{ bar}$$
$$v_g = 0.3749 \text{ m}^3/\text{kg}$$

As $v_2 > v_g$, it means that final state of the steam in this process is superheated (Fig. 1.57).

From superheated steam table,
at

$$p_2 = 5 \text{ bar}, \quad v_2 = 0.412 \text{ m}^3/\text{kg}.$$

$p_2 = 5$ bar	$T_{sat} = 151.9$ °C	$T = 200$ °C
	$v_g = 0.3749$ m³/kg	$v = 0.4249$ m³/kg
	$h_g = 2748.7$ kJ/kg	$h = 2855.4$ kJ/kg

Fig. 1.57 T-s diagram for Problem 1.23

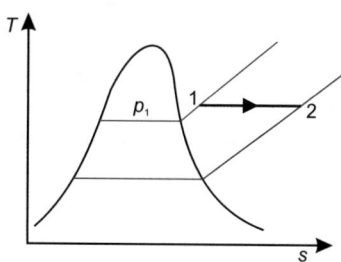

The value of v_2 at pressure p_2 lies between v_g and v.
Hence, the specific enthalpy corresponding v_2 is determined by interpolation

$$h_2 = h_g + (h - h_g) \times \frac{(v_2 - v_g)}{(v - v_g)}$$

$$= 2748.7 + (2855.4 - 2748.7) \times \frac{(0.412 - 0.3749)}{(0.4249 - 0.3749)}$$

$$= 2748.7 + 79.17 = 2827.87 \text{ kJ/kg}$$

Work done:

$$w_{1-2} = p_1 v_1 \log_e \frac{v_2}{v_1}$$

$$= 1000 \times 0.2060 \times \log_e \frac{0.412}{0.2060}$$

$$= 142.78 \text{ kJ/kg}$$

Heat transfer:

$$q_{1-2} = (u_2 - u_1) = w_{1-2}$$

where

$$u_1 = h_1 - p_1 v_1$$

and

$$u_2 = h_2 - p_2 v_2$$

$\therefore$

$$q_{1-2} = (h_2 - p_2 v_2) - (h_1 - p_1 v_1) + w_{1-2}$$
$$= (h_2 - h_1) + w_{1-2} \quad \therefore p_1 v_1 = p_2 v_2$$
$$= (2827.87 - 2827.9) + 142.78 = \mathbf{142.75 kJ/kg}$$

The $+ve$ sign shows that heat transfer to the system.
   OR
For superheated steam, the hyperbolic process, $pv = C$ is identical as isothermal
process.
   $\therefore w_{1-2} = q_{1-2} = p_1 v_1 \log_e \frac{v_2}{v_1}$.

Problem 1.24: Steam flows in a pipeline at 2 MPa. After expanding to 0.5 bar in a throttling calorimeter, the temperature is found to be 100 °C. Find the quality of steam in the pipe. Show the process on a *T-s* diagram.

Solution: Given data:

Pressure of steam in pipeline:$p_1 = 2$ MPa $= 20$ bar.

From saturated steam table (pressure based),

at 20 bar,

$$T_{sat} = 212.4 \,°C$$
$$h_f = 908.8 \text{ kJ.kg}, \quad h_{fg} = 1890.7 \text{ kJ/kg}$$

at 0.5 bar,

$$T_{sat} = 81.3 \,°C$$

Temperature of steam after throttling:

$$T_2 = 100 \,°C \text{ (given)}$$

So, $T_2 > T_{sat}$, the condition of steam is superheated after throttling process. From superheated steam table,

at

$$p_2 = 0.5 \text{ bar}, \quad T_2 = 100 \,°C$$
$$h = 2682.5 \text{ kJ/kg}$$

We know, the condition in throttling process (Fig. 1.58),

$$h_1 = h_2$$

also

$$h_1 = (h_f + x_1 \, h_{fg})_{20 \text{ bar}}$$

or

$$h_2 = 908.8 + x_1 \times 1890.7$$
$$2682.5 = 908.8 + 1890.7 \, x_1$$

or Dryness fraction:

$$x_1 = 0.9381$$

or

Fig. 1.58   T-s diagram for
Problem 1.24

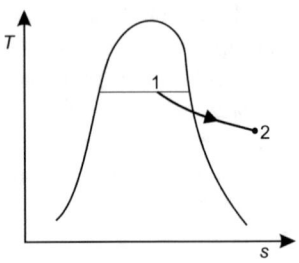

$$\text{Quality of steam} = x_1 \times 100\%$$
$$= 0.9381 \times 100\% = \mathbf{93.81\%}$$

Problem 1.25: A sample of steam from a boiler at 30 bar is put through a throttling calorimeter in which the pressure and temperature are found to be 1 bar and 120 °C. Find the quality of steam taken from a boiler.

Solution: Given data:

At initial state, before the throttling

$$p_1 = 20 \text{ bar}$$

From saturated steam table (pressure based) (Fig. 1.59), at

$$p_1 = 30 \text{ bar}$$
$$h_{f1} = 1008.4 \text{ kJ/kg}$$
$$h_{fg1} = 1795.7 \text{ kJ/kg}$$

Specific enthalpy at state 1,

$$h_1 = h_{f1} + x_1 h_{fg1}$$

At final state, after the throttling

Fig. 1.59   h-s diagram for
Problem 1.25

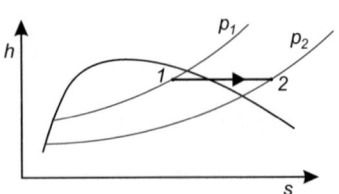

$$p_2 = 1 \text{ bar}, \quad T_2 = 120\,°\text{C}.$$

From superheated steam table,
at

$$p_2 = 1 \text{ bar}, \quad T_2 = 120\,°\text{C}$$

at

$$p_2 = 1 \text{ bar}$$

	$T = 100°\text{C}$	$T = 150°\text{C}$
h kJ/kg	2676.2	2776.4

By interpolation

$$
\begin{aligned}
h_2 &= h_{100} + (h_{150} - h_{100}) \times \frac{(T_2 - T_{100})}{(T_{50} - T_{100})} \\
&= 2676.2 + (2776.4 - 2676.2) \times \frac{(120 - 100)}{(150 - 100)} \\
&= 2676.2 + 40.08 = 2716.28 \text{ kJ/kg}
\end{aligned}
$$

For throttling process,

$$
\begin{aligned}
h_1 &= h_2 \\
h_{f1} + x_1 h_{fg1} &= 2716.28 \\
1008.4 + x_1 \times 1795.7 &= 2716.28 \\
1795.7 x_1 &= 2716.28 - 1008.4 = 1707.88
\end{aligned}
$$

or

$$
\begin{aligned}
x_1 &= \frac{1707.88}{1795.7} \\
&= 0.9510 = \mathbf{95.10}
\end{aligned}
$$

Problem 1.26: The pressure in a main steam pipe is 12 bar. A sample of steam is drawn off and passed through a throttling calorimeter, the pressure and temperature at exit from the calorimeter being 1 bar and 140 °C, respectively. Determine the dryness fraction of the steam in the main pipe, starting any assumptions made in the throttling process.

Solution: Given data:

Pressure of the wet steam: $p_1 = 12$ bar (before throttling)

Fig. 1.60 h-s diagram for
Problem 1.26

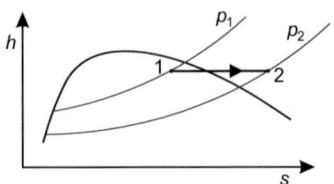

Pressure of the superheated steam: $p_2 = 1$ bar (after throttling)
Temperature of the superheated steam:
$T_2 = 140\,°C$
From saturated steam table (pressure based) (Fig. 1.60),
at

$$p_1 = 12 \text{ bar}$$
$$h_f = 798.6 \text{ kJ/kg}$$
$$h_{fg} = 1986.2 \text{ kJ/kg}$$
$$h_g = 2784.8 \text{ kJ/kg}$$
$$\therefore \quad h_1 = h_f + x_1 h_{fg}$$
$$h_1 = 798.6 + x \times 1986.2 \qquad\qquad (1.4)$$

From superheated steam table,
at $p_2 = 1$ bar, $T_2 = 140\,°C$.

		100 °C	150 °C
$p = 1$ bar	h in kJ/kg	2676.2	2776.2

By interpolation,

$$h_2 = h_{140\,°C} = h_{100\,°C} + (h_{150\,°C} - h_{100\,°C}) \times \frac{(140 - 100)}{(150 - 100)}$$

$$= 2676.2 + (2776.4 - 2676.2) \times \frac{40}{50}$$

$$= 2676.2 + 80.16 = 2756.36 \text{ kJ/kg}$$

For throttling process,

$$h_1 = h_2$$
$$h_1 = 2753.36 \text{ kJ/kg}$$

Substituting the value of h_1 in the above Eq. (1.4), we get

$$2756.36 = 798.6 + x \times 1986.21986.2$$
$$\text{or} \quad x = 2756.36 - 798.6 = 1957.76$$

or

$$x = \frac{1957.76}{1986.2} = 0.9856 = \mathbf{98.56\%}$$

Problem 1.27: In a separating and throttling calorimeter, the steam main pressure is 7 bar absolute and the temperature after throttling is 130 °C. The pressure after throttling is 0.013 bar gauge, the barometer reading being 75 cm of Hg. At the separate 0.20 kg of water is trapped and 1.8 kg of steam passed through the throttling calorimeter. Determine the dryness fraction of the steam in the steam main.

Solution: Given data:

Pressure main steam: $p_1 = 7$ bar

Temperature after throttling: $T_3 = 130$ °C

Gauge pressure after throttling: $p_{g3} = 0.013$ bar

Barometer reading: $h = 75$ cm of Hg $= 0.75$ m of Hg

$\therefore$

$$\text{Atmospheric pressure:} \quad p_{atm} = (\rho g h)_{Hg}$$
$$= 13600 \times 9.81 \times 0.75n$$
$$= 100062 \text{ Pa} = 1 \text{ bar}$$

$\therefore$

Absolute pressure after throttling,

$$p_3 = p_{g3} + p_{atm}$$
$$= 0.013 + 1 = 1.013 \text{bar}$$

Mass of water separate in the separator,

$$m_1 = 0.2 \text{ kg}$$

Mass of wet steam passing the throttling calorimeter,

$$m_2 = 1.8 \text{ kg}$$

From superheated steam table (Fig. 1.61),
at

$$p_3 = 1.013 \text{ bar}, \quad T_3 = 130 \text{ °C}$$

we get

$$T_{sat} = 100\,°C\,,\,T = 130\,°C$$

$$h_s = 2676\text{ kJ/kg}\,,\,h = 2776.3\text{ kJ/kg}$$

By interpolation

$$h_3 = h_s + (h - h_s)\frac{(T_3 - T_{sat})}{(T - T_{sat})}$$
$$= 2676 + (2776.3 - 2676)\frac{(1300 - 100)}{(300 - 100)}$$
$$= 2676 + 100.3 \times 0.15$$
$$= 2676 + 15.04 = 2691.04\text{ kJ/kg}$$

From saturated steam table (pressure based),
    at

$$p_2 = p_1 = 7\text{ bar}$$

we get

$$h_f = 697.2\ kJ/\text{kg},\ h_{fg} = 2066.3\ kJ/\text{kg}$$

Specific enthalpy at state 2,

$$h_2 = h_f + x_2\,h_{fg}$$

where

$$x_2 = \text{ dryness fraction at state 2}$$
$$h_2 = 697.2 + x_2 \times 2066.3$$

Fig. 1.61 h-s diagram for Problem 1.27

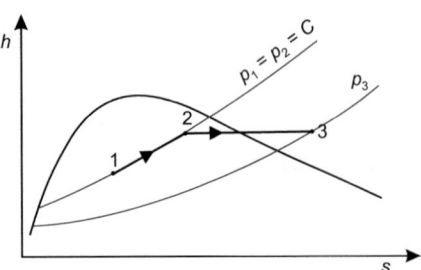

also

$$h_2 = h_3$$

$h_2 = h_3$

$$h_3 = 697.2 + x_2 \times 2066.3$$
$$\therefore \quad 2691.04 = 697.2 + x_2 \times 2066.3$$

or

$$x_2 = 0.9649$$

also

$$x_2 = \frac{m_{v2}}{m_{v2}} 0.9649$$
$$= \frac{m_{v2}}{1.8}$$

or $m_{v2} = 1.736$ kg

Dryness fraction at state 1, *i.e.*, in the steam main:

$$x_1 = \frac{m_{v1}}{m}$$

where

$$m_{v1} = m_{v2} = 1.736\text{kg}$$

and

$$m = m_1 + m_2 = 0.2 + 1.8 = 2\,\text{kg}$$

$$\therefore$$

$$x_1 = \frac{1.736}{2} = \mathbf{0.868}$$

1. A substance that has a fixed chemical composition throughout in each phase during a thermodynamics process is called a **pure substance**, for example, Water, carbon dioxide, nitrogen, etc. Water is one of the pure substances which exists in there different phases, *i.e.*, in the solid phase as ice, in the liquid phase as water and in the gases phase as steam, and in all the three phases it retains the same chemical composition, *i.e.*, H_2O.

2. **Vapourization** is a process of phase change from the liquid phases to vapour phase. It is a more general term and applies to both evaporation and boiling.

Evaporation is a process of phase change from the liquid to vapour that occurs below the boiling point where particles of the liquid moves from the free surface of liquid to the vapour state.
   Condition for evaporation,

$$p_v < p$$

where

$$p_v = \text{vapor pressure of the liquid}$$
$$p = \text{pressure of the liquid or surrounding pressure}$$

Boiling is a process of phase change from the liquid to vapour at fixed temperature and at given pressure. Boiling takes place when the vapour pressure p_v becomes equal to or greater than the pressure on liquid free surface.
   Condition for boiling,

$$p_v \geq p \text{ at } T = C$$

3. **Saturated Liquid State.** The state of liquid at which a vapour just begins to form is called saturated liquid state. For water, saturated liquid state lies at 1 atm (101.325 kPa) pressure and 100 °C temperature.
4. **Sensible Heat.** The amount of heat required to raise the temperature of the substance at constant pressure when phase change does not occur is called sensible heat.
5. **Saturated Temperature.** It is the temperature at which the liquid starts boiling at the given pressure. It is denoted by T_{sat}. For water, at pressure of 101.325 kPa, $T_{sat} = 100$ °C.
6. **Saturated Pressure.** It is the pressure at which the liquid starts boiling at a given temperature. It is denoted by p_{sat}. For water, at temperature of 100 °C, $p_{sat} = 101.325$ kPa.
7. **Compressed Liquid or Subcooled Liquid.** The state of liquid at which it is not about to vapourize is called compressed liquid or subcooled liquid. The difference between the temperature of subcooled liquid and saturated temperature at given pressure is called degree of subcooling.
8. **Saturated VAPOUR State.** The state of vapour at which a condensation just begins on loss of heat from the vapour is called saturated vapour state.
9. **Latent Heat.** The amount of heat required to change the phase from liquid to vapour (or vapour to liquid) at constant temperature and pressure is called the latent heat.

10. **Wet Steam.** The mixture of water-vapour is called wet steam. This steam lies between saturated liquid state and saturated vapour state.
11. **Dry Saturated Steam.** When all the liquid water in wet steam is just to vapourization is called dry saturated steam. This steam lies at saturated vapour state.
12. **Superheated Steam.** The steam obtained by heating the dry saturated steam above the saturated temperature is called superheated steam. The difference between the temperature of superheated steam and saturated temperature of steam at given pressure is called degree of superheating.
13. The **Saturated Liquid Line** is obtained by joining the saturated liquid states. The saturated vapour line is obtained by joining the saturated vapour states.
14. **Critical Point:**
(a) The critical point is the state at which there is no distinction between the saturated liquid and saturated vapour.
(b) The point at which the saturated liquid line and the saturated vapour line meet with increase in temperature or pressure is called critical point.
(c) The point at which the latent heat of vapourization is zero is called critical point.

The temperature, pressure, and volume at the critical point are called critical temperature T_c, critical pressure p_c, and critical volume v_c, respectively.

For water

$$T_c = 374.15 \,°C$$
$$p_c = 22.12 MPa = 221.2 bar$$
$$v_c = 0.003155 \, m^3/kg$$

15. **p-v-T Surface.** All the information on both the p–v and the p–T diagram can be shown on a single diagram if the three coordinates p, v, and T are plotted in three-dimensional space. The three-dimensional space among p, v and T is called p–v-T surface.
16. **Triple Point.** The triple point is state where all the three phases like solid, liquid, and vapour coexist in equilibrium. For water, the values of temperature and pressure at triple point are as follows:

$$T_{TP} = 0.01°C = (273.15 + 0.01)K = 273.16 \, K$$
$$\text{and} \quad p_{TP} = 611.2 \, Pa = 0.6112 kPa = 0.006112 bar$$

17. **Dryness Fraction.** The dryness fraction of the wet steam is defined as the ratio of mass of vapour to the total mass of the wet steam. It is denoted by x.

Dryness fraction:

$$x = \frac{\text{mass of vapour}}{\text{total mass of the wet steam}}$$
$$x = \frac{m_{vapour}}{m}$$

where

$$m = m_{\text{liquid}} + m_{\text{vapor}}$$
$$= m_f + m_g$$

$\therefore$

$$x = \frac{m_g}{m_f + m_g}$$

.

Specific volume of the wet steam:

$$v = v_f + x v_{fg}$$
$$= v_f + x\left(v_g - v_f\right)$$

As

$$v_g \gg v_f$$

$\therefore$

$$v = x v_g$$

Specific enthalpy of the wet steam: $h = h_f + x h_{fg}$
Specific entropy of the wet steam: $s = s_f + x s_{fg}$

18.   Thermodynamic Process for the Steam.

(a)   Isothermal process:

$T = C$

(i)   For wet-wet steam process:

Work done: $\underset{1-2}{w} = p v_g (x_2 - x_1)$.

Heat transfer $\underset{1-2}{q} = (x_2 - x_1) h_{fg}$.

Change in specific internal energy:

$$u_2 - u_1 = (h_2 - h_1) - p v_g (x_2 - x_1)$$

(ii)   For wet-superheated steam process:

$$q_{1-2} = h_2 - h_1$$
$$u_2 - u_1 = (h_2 - h_1) - \left(p_2 v_2 - p_1 x_1 v_{g1}\right)$$
$$w_{1-2} = p_2 v_2 - p_1 x_1 v_{g1}$$

(iii)   For superheated-superheated steam process:

$$\underset{1-2}{q} = h_2 - h_1$$
$$u_2 - u_1 = (h_2 - h_1) - (p_2 v_2 - p_1 v_1)$$
$$\underset{1-2}{w} = p_2 v_2 - p_1 v_1$$

(b)   Isobaric process:

$p = c$

(i)   For wet-wet steam process:

$$\underset{1-2}{q} = h_2 - h_1 = (x_2 - x_1) h_{fg}$$
$$u_2 - u_1 = (h_2 - h_1) - p v_g (x_2 - x_1)$$
$$w_{1-2} = p v_g (x_2 - x_1)$$

(ii)   For wet-superheated steam process:

$$q_{1-2} = h_2 - h_1$$
$$u_2 - u_1 = (h_2 - h_1) - \left(v_2 - x v_g\right)$$
$$q_{1-2} = p\left(v_2 - x_1 v_g\right)$$

(iii)   For superheated-superheated steam process:

$$\underset{1-2}{q}\, h_2 - h_1$$
$$u_2 - u_1 = (h_2 - h_1) - p(v_2 - v_1)$$
$$\underset{1-2}{w} = p(v_2 - v_1)$$

(c)   Isochoric process:$v = C$

(i)   For wet-wet steam process:

$$\underset{1-2}{w} = 0$$
$$u_2 - u_1 = (h_2 - h_1) - v(p_2 - p_1)$$
$$\underset{1-2}{q} = (u_2 - u_1)$$

(ii)   For wet-superheated steam process:

$$w_{1-2} = 0$$
$$u_2 - u_1 = (h_2 - h_1) - v(p_2 - p_1)$$
$$q_{1-2} = u_2 - u_1$$

(iii)   For superheated-superheated steam process:

$$w_{1-2} = 0$$
$$u_2 - u_1 = (h_2 - h_1) - v(p_2 - p_1)$$
$$q_{1-2} = u_2 - u_1$$

(d)   **Adiabatic process:** $q_{1-2} = 0$

(i)   For wet-wet process

$$q = 0$$
$$u_2 - u_1 = (h_2 - h_1) - \left(p_2 x_2 v_{g2} - p_1 x_1 v_{g1}\right)$$
$$w_{1-2} = -(u_2 - u_1) = u_1 - u_2$$

(ii)   For wet-superheated steam process:

$$q_{1-2} = 0$$
$$u_2 - u_1 = (h_2 - h_1) - \left(p_2 v_2 - p_1 x_1 v_{g1}\right)$$

Work done: $w_{1-2} = u_1 - u_2$.

(iii)   For superheated-superheated steam process:

$$q_{1-2} = 0$$
$$u_2 - u_1 = (h_2 - h_1) - (p_2 v_2 - p_1 v_1)$$
$$w_{1-2} = u_1 - u_2$$

(e)   Polytropic process:$pv^n = c$

(i)   For wet-wet steam process:

$$w_{1-2} = \frac{p_1 x_1 v_{g1} - p_2 x_2 v_{g2}}{n-1}$$
$$u_2 - u_1 = (h_2 - h_1) - \left(p_2 x_2 v_{g2} - p_1 x_1 v_{g1}\right)$$
$$q_{1-2} = (u_2 - u_1) + \frac{p_1 x_1 v_{g1} - p_2 x_2 v_{g2}}{n-1}$$

(ii) For wet-superheated steam process:

$$\underset{1-2}{w} = \frac{p_1 x_1 v_{g1} - p_2 v_2}{n - 1}$$

$$u_2 - u_1 = (h_2 - h_1) - (p_2 v_2 - p_1 x_1 v_{g1})$$

$$\underset{1-2}{q} = (u_2 - u_1) + \frac{p_1 x_1 v_{g1} - p_2 v_2}{n - 1}$$

(iii) For superheated-superheated steam process:

$$\underset{1-2}{w} = \frac{p_1 v_1 - p_2 v_2}{n - 1}$$

$$u_2 - u_1 = (h_2 - h_1) - (p_2 v_2 - p_1 v_1)$$

$$\underset{1-2}{q} = (u_2 - u_1) + \frac{p_1 v_1 - p_2 v_2}{n - 1}$$

(f) Hyperbolic process:$pv = c$

(i) For wet-wet steam process:

$u_2 - u_1 = h_2 - h_1$

(ii) For wet-superheated steam process:

$$u_2 - u_1 = h_2 - h_1$$
$$= h_2 - h_f - x_1 h_{fg1}$$

(iii) For superheated-superheated steam process:

$$\underset{1-2}{q} = p_1 v_1 \log_e \frac{v_2}{v_1}$$

$$u_1 - u_2 = h_2 - h_1$$

$$\underset{1-2}{q} = p_1 v_1 \log_e \left(\frac{v_2}{v_1}\right) + (u_2 - u_1)$$

19. **Throttling process:**$h_1 = h_2, \underset{1-2}{q} = \underset{1-2}{w} = 0.$

(i) For wet-wet steam process:

$$h_1 = h_2 h_{f1} + x_1$$
$$h_{fg1} = h_{f2} + x_2 h_{fg2}$$

(ii) For wet-superheated steam process:

$$h_1 = h_2 h_{f1} + x_1$$

$$h_{fg1} = h_{g2} + c_{ps}(T_2 - T_{\text{sat }2})$$

(iii)   For superheated-superheated steam process:

$$h_1 = h_2$$
$$h_{g1} + c_{ps}(T_2 - T_{sat1}) = h_{g2} + c_{ps}(T_2 - T_{sat2})$$

20.   Experimental methods for determination of dryness fraction.
(i)   Separating calorimeter,
(ii)   Throttling calorimeter,
(iii)   Combined separating and throttling calorimeter,
(iv)   Electrical calorimeter.
1.   What is a pure substance? Can air be treated as a pure substance?
2.   Define the following terms:

(i)   Vapourizing,
(ii)   Evaporation, and
(iii)   Boiling.

3.   Discuss the formation of steam from water at 25 °C with the help of T-v and T-h diagrams at atmospheric pressure.
4.   Define the following terms :

(i)   Saturated liquid state,
(ii)   Saturated vapour state,
(iii)   Saturated temperature,
(iv)   Subcooled liquid,
(v)   Degree of subcooling,
(vi)   Superheated vapour,
(vii)   Degree of superheated,
(viii)   Sensible heat, and
(ix)   Latent heat.

5.   What is the meaning of wet steam, dry saturated steam, and superheated steam?
6.   Discuss the graphical representation of T-v diagram for generation of superheated steam from −20 °C of ice.
7.   What do you understand by triple point? Write down the values of pressure and temperature of water on the triple point.
8.   What is the critical point? Write down the values of the following terms on the critical point of water

(i)   latent heat,
(ii)   temperature, and
(iii)   pressure.

9.   What is the difference between saturated liquid and saturated vapour.
10.   Can liquid water at 100 °C be called saturated liquid? Why or why not?

11.   Which process is possible below the critical point? Which process is impossible above the critical point?
12.   It is true that water boils at higher temperature at higher pressure? Explain.
13.   What is quality? Does it have any meaning in the superheated vapour region?
14.   Which of the following state requires more energy to completely vapourizing 1 kg of saturated liquid (*i*) at 100 kPa and (*ii*) at 500 kPa. Why?
15.   Why do the isobars on Mollier diagram diverge from one another?
16.   What is dryness fraction of steam? What are the different methods of measurement of dryness fraction?
17.   Sketch a Mollier diagram and indicate the throttling and isochoric processes for the following end states of steam :

(i)   wet-wet steam process,
(ii)   wet-superheated steam process, and
(iii)   superheated-superheated steam process.

18.   State the difference between isothermal and hyperbolic process as applied to steam.
19.   How throttling process helps in estimating the dryness fraction of wet steam?
20.   With a neat sketch of a combined separating and throttling calorimeter, explain how dryness fraction of the steam is determined?
21.   Why does a throttling calorimeter not measure the quality if the steam is very wet? How is the quality measured then?
22.   What do you understand by the degree of subcooling and the degree of superheat?

1.   Determine the dryness fraction of steam which has 0.25 kg of liquid water in 2 kg of steam.[**Ans.** 0.875]
2.   Determine the state of steam, *i.e.*, whether it is wet, dry, or superheated in the following cases:

(i) $p = 5$ bar, $v = 0.210$ m^3/kg,
(ii) $p = 10$ bar, $T = 200$ °C,
(iii) $p = 20$ bar, $v = 0.0996$ m^3/kg.

[**Ans.** (*i*) wet steam (*ii*) superheated steam (*iii*) dry saturated steam]

3.   Determine the specific volume, specific enthalpy, and specific entropy of steam at 18 bar and 85% quality.[**Ans.** 0.09367 m^3/kg, 2510.34 kJ/kg, 5.7818 kJ/kgK]
4.   Determine the specific volume and specific entropy for superheated steam at 250 kPa and 500 °C.[**Ans.** 1.484 m^3/kg, 8.419 kJ/kgK]
5.   10 kg of steam with a quality of 40% is contained in a tank at 1 MPa. What is the volume of the tank.[**Ans.** 0.7827 m^3]
6.   Find the specific enthalpy and specific entropy of the steam at 3 MPa and the specific volume is 0.055 m^3/kg.[**Ans.** 2483.21 kJ/kg, 5.554 kJ/kgK]
7.   5 kg of steam at 8 bar and 0.5 dryness fraction is heated, so that it becomes

(i)   0.90 dryness fraction at 8 bar,

    (ii)   dry saturated at constant pressure,
    (iii)   superheated to 350 °C at constant pressure. Take $c_{ps} = 2.1$ kJ/kgK,
    (iv)   superheated to 250 °C at 10 bar.

Determine the amount of heat supplied to each case.
  [**Ans.** (*i*) 4096 kJ, (*ii*) 5120 kJ, (*iii*) 7005.8 kJ, (*iv*) 5987.5 kJ]

8.   A piston-cylinder device contains steam at 500 kPa and 50% quality. The initial volume is 0.04 m^3. Determine the amount of heat additions required to increase the volume to 0.1554 m^3 at constant pressure.[**Ans.** 387.27 kJ]

9.   A rigid container is filled with steam at 600 kPa and 200 °C. At what temperature will the steam begin to condense when it is cooled? Determine the corresponding pressure.
    [**Ans.** 154.42 °C, 535.5 kPa]

10.   Steam at 10 bar pressure and 0.9 dryness fraction is cooled at constant volume at 160 °C. What will be its final condition?[**Ans.** $p_2 = 6.178$ bar, $x_2 = 0.5636$]
    Hints: Considering specific volume of the liquid: v_f

11.   A pressure cooker contains 2 kg of steam at 5 bar pressure and 0.9 dryness fraction. Find the quantity of steam which must be transferred so as the quality of steam becomes 0.6 dry.[**Ans.** –1222.88 kJ]
    Hints: $v_g \gg v_f$, neglecting the v_f

12.   A piston-cylinder arrangement contains 1 kg of wet steam of quality 0.8 at 10 bar. As heat supplied at constant pressure, the temperature of steam rises to 350 °C. Determine the work done and the heat supplied.[**Ans.** 151.58 kJ, 952.43 kJ]

13.   3 kg of steam initially saturated at 60 bar expands reversibly and isothermally to 10 bar in a cylinder. Find the heat transferred and the work done.[**Ans.** 1565 kJ, 1075 kJ]

14.   0.12 kg of steam initially saturated at 10 bar expands reversibly in a cylinder until the pressure is 1 bar. The volume is then found to be 0.17 m^3. Assuming that the process is polytropic, find the index of expansion and the heat transferred during the process. Also find the change of entropy.[**Ans.** $n = 1.16$, –10.4 kJ, – 0.0264 kJ/K]

15.   Steam at 10 bar and 0.9 dryness is throttled to a pressure of 2 bar. Using steam tables evaluate the final dryness fraction or degree of superheated. Estimate the change of entropy during this process.[**Ans.** $x = 0.94$, 0.65 kJ/kgK]

16.   A pressure cooker contains 1.5 kg of saturated steam at 5 bar. Find the quantity of heat which must be rejected so as to reduce the quality to 0.60 dryness fraction. Determine the pressure and temperature of the steam at the new state.
    [**Ans.** –770 kJ, 2.9 bar, 132.35 °C]

17.   One kg of steam at a pressure of 1 bar absolute is compressed in a cylinder to a pressure of 2 bar absolute, the law of compression being $pv^{1.25} = $ constant. Find the final condition of steam and heat which passes through the cylinder walls in kJ. The steam is 0.85 dry at the commencement of the compression.[**Ans.** $x_2 = 0.934$, 135.58 kJ]

18. The steam in the cylinder of an engine expands according to the law $pv^{1.1} =$ constant. The initial and final steam pressure are respectively 12 bar absolute and 4 bar absolute while the initial steam dryness fraction is 0.95. Determine per kg of steam:

    (i)   the final volume,
    (ii)   the work done, and
    (iii)   the heated interchange between the steam and the cylinder walls.

[**Ans.** (*i*) 0.42 m³/kg (*ii*) 169.18 kJ/kg (*iii*) 49.38 kJ/kg]

19. One kg of water at 95 °C is heated under a constant pressure of 17.2 bar absolute until is it converted into steam 0.98 dry. Determine the work done, heated added, and change in internal energy and entropy.[**Ans.** 2360 kJ, 195 kJ, 2160 kJ, 5.05 kJ/kgK]

20. Steam at a pressure of 0.5 bar absolute and 0.9 dry is expanded adiabatically to an absolute pressure of 0.06 bar. Calculate the final condition of steam and check the result from the *h-s* chart. [**Ans.** $x = 0.824$]

21. Steam at a pressure of 5 bar and a temperature of 200 °C is expanded adiabatically to a pressure of 0.7 bar absolute. Find the final condition of steam from the *h-s* chart and check the result by calculation. Assume c_{ps} for superheated steam as 2.2 kJ/kgK.

    [**Ans.** 0.93]

22. Steam enters an engine at a pressure of 12 bar absolute with 67 °C of superheated and exhausted at 0.15 bar absolute and 0.94 dry. Estimate the drop in total heat and volume per kg of steam at admission and exhaust.[**Ans.** 490 kJ, 0.1905 m³/kg]

23. Steam at a pressure of 14 bar absolute and 100 °C superheated expands adiabatically until it becomes just saturated. What will be the lower pressure? Find the value of *n*, assuming the law of expansion $pv^n = $ constant.[**Ans.** 1.24]

24. Steam at a pressure of 20 bar absolute and dryness fraction 0.8 is throttled to a pressure of 0.5 bar. Represent this operation by a horizontal line on the Mollier diagram and read off the final condition of the steam. Check this result by calculation.

    [**Ans.** $x = 0.905$]

25. Using Mollier diagram find the final condition of steam if it is throttled from pressure of 17 bar and dryness fraction 0.9 to a pressure of 0.35 bar. Check this result by calculation with the help of steam tables.[**Ans.** 0.989]

26. One kg of steam is at pressure of 10 bar absolute and temperature 230 °C Find the density of steam. If the steam expands in an engine cylinder hyperbolically to a pressure of 3 bar absolute, determine the condition of steam after expansion.

    [**Ans.** 4.53 kg/m³, 71.35 °C of superheat]

27. Two kg of wet steam at 10 bar absolute and 90% dry are expanded according to the law $pv = $ constant to a pressure of 1 bar absolute. What is the final condition of steam? Calculate the change in internal energy of steam.

[**Ans.** (*a*) Steam is superheated to a temperature of 112 °C, (*b*) 257 kJ]

28.   One kg of steam having a pressure of 9 bar absolute and dryness 0.9 is expanded in a cylinder to a pressure of 0.4 bar absolute. If the expansion is hyperbolic, determine the quality of heat which passes through the cylinder walls to the steam.[**Ans.** 668 kJ]

29.   Steam at 17.5 bar absolute pressure and 0.98 dry is reduced to 4 bar absolute by throttling. It is then taken to an engine from which it is exhausted at 1.12 bar absolute and 0.9 dry. Assuming no loss of heat by radiation, etc.

Determine

(a)   The quality of steam after throttling.
(b)   The work done in the engine per kg of steam.

[**Ans.** (*a*) Steam is superheated to a temperature of 152 °C, (*b*) 176 kJ/kg]

30.   In carrying out the test for finding the dryness fraction with a throttling calorimeter, it was found that the steam passing to the engine was at a pressure of 12 bar absolute and that a sample after being reduced to 1 bar in the calorimeter was at a temperature of 120 °C. Estimate the dryness fraction of the steam assuming $c_{ps} = 2.1$ kJ/kgK.
      [**Ans.** 0.965]

31.   Calculate the minimum dryness fraction of steam, which may be determined by a throttling calorimeter, if the steam to be tested has an absolute pressure of 14 bar and the absolute pressure of steam after throttling is 1 bar absolute.
      [**Ans.** 0.945]

32.   Following observations were recorded in an experiment on a separating and throttling calorimeter:
      Initial absolute pressure= 13.8 bar,
      Absolute pressure after throttling= 1.02 bar,
      Temperature after throttling= 120 °C,
      Water collected in separating calorimeter= 0.45 kg,
      Steam condensed after throttling= 6.75 kg.
      Determine the dryness fraction of steam sample. Take $c_{ps} = 2.1$ kJ/kgK.
      [**Ans.** 0.904]

Chapter 2
Vapor Power Cycles

Nomenclature

The following is a list of the nomenclature in this chapter:

m_s	kg/s	Mass flow rate of the steam
Q_{4-1}	kW	Heat supplied to the boiler
q_{4-1}	kJ/kg	Heat supplied to the boiler per unit mass
W_{1-2}	kW	Power output of the turbine
w_{1-2}	kJ/kg	Work output of the turbine per unit mass
h	kJ/kg	Specific enthalpy
w_{3-4} or w_p	kJ/kg	Work input to the pump per unit mass
η_{th}	—	Thermal efficiency of the cycle
FWH	—	Feedwater heater
LPT	—	Low-pressure turbine
HPL	—	High-pressure turbine

2.1 Introduction

The cycles used to convert heat into work are called the power cycle. If the working fluid in the cycle is gas, it is called gas power cycle or air power cycle, for example, Otto cycle, diesel cycle, Bryton cycle, etc. When the working fluid in the cycle is vapour, it is called vapour power cycle, for example, Rankine cycle. In the vapour power cycle, the working fluid is alternatively vapourized and condensed.

In vapour power cycles, steam is the most commonly used working fluid because steam has many desirable properties such as low cost, easily available, and high latent heat of vapourization. Steam power plants are referred to as coal plants, natural gas

© The Author(s) 2022
S. Kumar, *Thermal Engineering Volume 2*,
https://doi.org/10.1007/978-3-030-89216-6_2

plants, nuclear plant or geothermal plants, depending on the type of fuel used to supply heat to the steam.

2.2   The Rankine Cycle

In vapour power cycle, the working fluid changes phase from liquid to vapour and vapour to liquid. The main components of simple vapour power cycle are as follows (Fig. 2.1):

1. Turbine
2. Condenser
3. Feed pump
4. Boiler

The simplest vapour power cycle is called the Rankine cycle. The Rankine cycle is an idealized cycle in which friction losses in each of the four components are neglected. The losses usually are quite small and will be neglected completely in our initial analysis. The Rankine cycle is composed of the four ideal processes shown on the $T-s$ diagram in Fig. 2.2.

The turbine, condenser, feed pump, and boiler associated with the Rankine cycle are steady-flow devices, and thus all the processes of this cycle can be analysed as steady-flow process. The change in kinetic and potential energy of the steam is usually small compared with the work and heat transfer terms and is neglected.

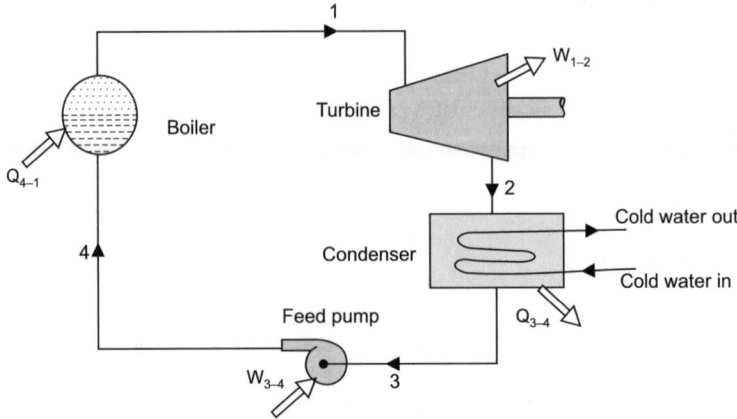

Fig. 2.1 Main components of simple vapor power cycle

Fig. 2.2 Simple ideal Rankine cycle on the $T-s$ diagram

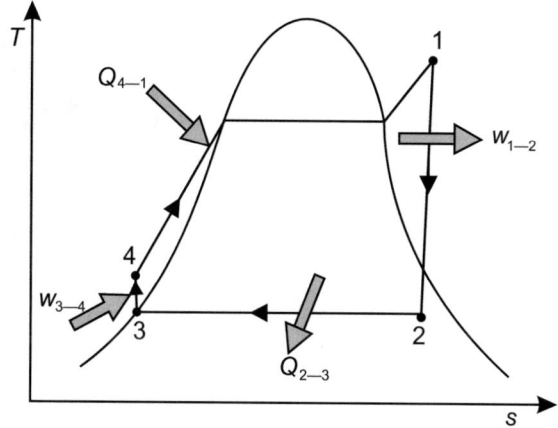

2.2.1 Turbine

Process 1–2, isentropic expansion in a turbine.

The superheated steam coming from the boiler at state 1 enters the turbine, where it expands isentropically and produces work. The pressure and temperature of steam drop during this process and exit at state 2. This process is shown on $T-s$ diagram in Fig. 2.2.

Applying the steady flow energy equation,

$$m_s \left[h_1 + \frac{V_1^2}{2} + g z_1 \right] + Q_{1-2} = m_s \left[h_2 + \frac{V_2^2}{2} + g z_2 \right] + W_{1-2}$$

Turbine can be analysed with the following assumptions:

(a) Heat transfer: $Q_{1-2} = 0$ for adiabatic isentropic expansion,
(b) Change in kinetic energy is neglected, and
(c) Change in potential energy is neglected
Above equation becomes (Fig. 2.3)

$$m_s h_1 = m_s h_2 + W_{1-2}$$

or

$$W_{1-2} = m_s (h_1 - h_2) \, \text{kW}$$

where m_s is mass flow rate of the steam in kg/s,
h_1 and h_2 are in kJ/kg.
For unit mass flow rate

Fig. 2.3 Turbine

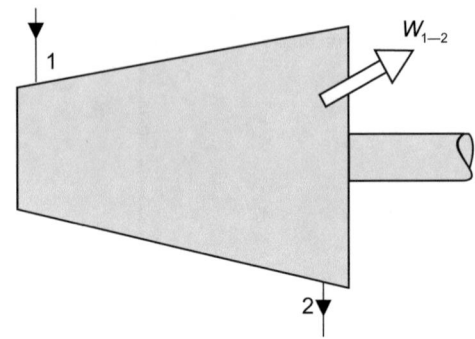

$$w_{1-2} = h_1 - h_2 \, \text{kJ/kg}$$

2.2.2 Condenser

Process 2–3, constant pressure heat rejection in a condenser.

Inlet of the condenser at state 2, steam is usually at saturated liquid–vapour mixture with a high quality. Steam is condensed at constant pressure in the condenser by rejecting latent heat to a cooling medium such as cold water. Steam leaves the condenser as saturated liquid at state 3 (Fig. 2.4).

Applying the steady flow energy equation

$$m_s \left(h_2 + \frac{V_2^2}{2} + g z_2 \right) + Q_{2-3} = m_s \left(h_3 + \frac{V_3^2}{2} + g z_3 \right) + W_{2-3}$$

Condenser can be analysed with the following assumptions.

(a) Work done: $W_{1-2} = 0$
(b) Change in KE is neglected
(c) Change in PE is neglected
     Above equation becomes

Fig. 2.4 Condenser

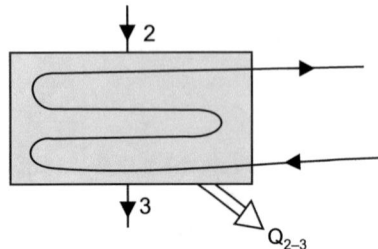

$$m_s h_2 + Q_{2-3} = m_s h_3$$

or

$$Q_{2-3} = m_s(h_3 - h_2)\,\text{kW}$$

For unit mass flow rate

$$q_{2-3} = (h_3 - h_2)\,\text{kJ/kg}$$

As $h_2 > h3$, q_{2-3} is always $-ve$ which indicates that heat transfer from the condenser.

Above equation is written as.

Heat loss:

$$q_{2-3} = h_2 - h_3$$

2.2.3 Feed Pump

Process 3–4, isentropic compression in a feed pump.

The saturated water coming from the condenser at state 3 enters the feed pump, where it is compressed isentropically to the operating pressure of the boiler (Fig. 2.5).

The water pressure increases during isentropic compression process which a slight decrease in the specific volume of water. For the simplification, water is considered as incompressible, i.e., density or specific volume of water remains constant. This process is shown on $T-s$ diagram in Fig. 2.2.

Applying the steady flow energy equation

$$m_s\left[h_3 + \frac{V_3^2}{2} + gz_3\right] + Q_{3-4} = \left[h_4 + \frac{V_4^2}{2} + gz_4\right] + W_{3-4}$$

Feed pump can be analysed with following assumptions:

(a) Heat transfer: $Q_{3-4} = 0$
(b) Change in kinetic and potential energy are neglected

Above equation becomes

Fig. 2.5 Feed pump

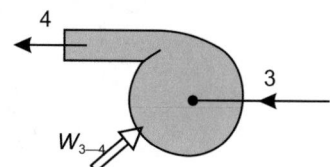

$$m_s h_3 = m_s h_4 + W_{3-4}$$

or

$$W_{3-4} = m_s (h_3 - h_4) \, \textbf{kW}$$

where m_s is in kg/s,
$\quad$ h_3 and h_4 are in kJ/kg.
$\quad$ For unit mass flow rate

$$w_{3-4} = h_3 - h_4 \, \textbf{kJ/kg}$$

By definition of specific enthalpy,

$$h = u + pv$$

On differentiating, we get

$$dh = du + vdp \qquad \because v = c \qquad (5.1)$$

From first law of thermodynamic,

$$\delta q = du + pdv$$

where

$$\delta q = 0 \qquad \text{adiabatic flow}$$

$$dv = 0 \qquad \because v = c$$

$\therefore$

$$du = 0$$

Equation (5.1) becomes

$$dh = vdp$$

Integrating between states 3 and 4 gives

$$\int_3^4 dh = \int_3^4 v_3 dp = v_3 \int_3^4 dp$$
$$h_4 - h_3 = v_3 (p_4 - p_3)$$

or

$$h_3 - h_4 = v_3(p_3 - p_4)$$

$\therefore$

$$w_{3-4} = h_3 - h_4 = v_3(p_3 - p_4)$$
$$= h_3 - h_4 = v_3(p_2 - p_1) \quad \because p_2 = p_3,\, p_1 = p_4$$

As $h_4 > h_3$ or $p1 > p_2$, w_{3-4} is always $-ve$ which shows the work required to drive the pump.

Above equation is written as.

Work input:

$$w_{3-4} = h_4 - h_3 = v_3(p_1 - p_2)$$
$$w_{3-4} = \boldsymbol{h_4 - h_3}\, \mathbf{kJ/kg} = \boldsymbol{v_3(p_1 - p_2)}\mathbf{kJ/kg}$$

where

h_3 and h_4 are in kJ/kg

v_3 is in m^3/kg

p_1 and p_2 are in kPa.

2.2.4 *Boiler*

Process 4–1, constant pressure heat addition in a boiler.

Water enters the boiler as a compressed liquid at state 4 and leaves as a superheated vapour at state 1 (Fig. 2.6). The boiler is used to convert liquid water into superheated steam at constant pressure by application of heat. This process is shown on $T-s$ diagram in Fig. 2.2.

Applying the steady flow energy equation

Fig. 2.6 Boiler

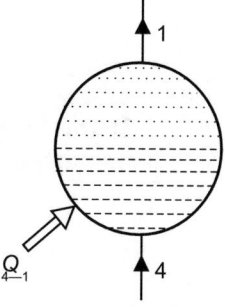

$$m_s\left[h_4 + \frac{V_4^2}{2} + gz_4\right] + Q_{4-1} = m_s\left[h_1 + \frac{V_1^2}{2} + gz_1\right] + W_{4-1}$$

Boiler can be analysed with following assumptions:

(a)   Work transfer:

$$W_{4-1} = 0$$

(b)   Change in kinetic and potential energy are neglected.
      Above equation becomes

$$m_s h_4 + Q_{4-1} = m_s h_1$$

or

$$Q_{4-1} = m_s (h_1 - h_4)\, \mathbf{kW}$$

where
   m_s is in kg/s,
   h_1 and h_2 are in kJ/kg.
   For unit mass flow rate.
   Heat supplied:

$$q_{4-1} = h_1 - h_4\, \mathbf{kJ/kg}$$

Net work output of the Rankine cycle,

$$w_{net} = \text{turbine output work} - \text{pump input work}$$
$$= w_{1-2} - w_{3-4}$$

The thermal efficiency of the Rankine cycle,

$$\eta_{th} = \frac{\text{net work output}}{\text{heat supplied}} = \frac{w_{net}}{q_{4-1}} = \frac{w_{1-2} - w_{3-4}}{q_{4-1}}$$

where
   w_{1-2}, w_{3-4} and q_{4-1} are expressed as positive quantities.
   The pump work is usually quite small, compared to the turbine work and can most often be neglected.
   $\therefore$

$$\eta_{th} = \frac{w_{1-2}}{q_{4-1}}$$

Fig. 2.7 Thermal efficiency represented on T-s diagram

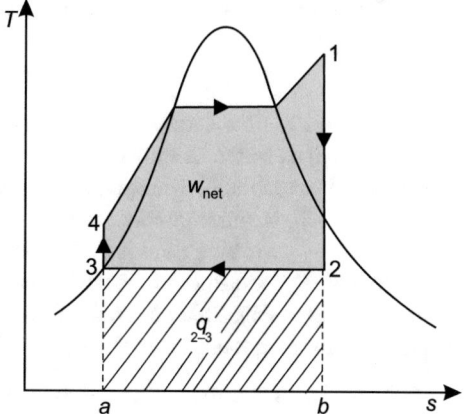

This is the relation most often used for the thermal efficiency of Rankine cycle.

Work ratio: It is defined as the ratio of the net work output to the turbine work output.

Mathematically,

$$\text{Work ratio} = \frac{\text{net work output}}{\text{turbine work output}} = \frac{w_{net}}{w_T}$$

The thermal efficiency of the Rankine cycle is represented graphically on the $T-s$ diagram as shown in Fig. 2.7. The net work output is represented by area 1–2–3–4–1. The heat supplied to the working substance is represented by area 1–b–a–4–1. Thus, the thermal efficiency of the Rankine cycle is

$$\eta_{th} = \frac{\text{area } 1 - 2 - 3 - 4 - 1}{\text{area } 1 - b - a - 4 - 1}$$

That is, desired output is divided by the energy input. Obviously, the thermal efficiency can be improved by increasing the numerator or by decreasing the denominator. This can be done by increasing the pump output pressure p_1 (i.e., boiler pressure), increasing the boiler outlet temperature T_1, or decreasing the turbine output pressure p_2 (i.e., condenser pressure).

2.3 Why Is Carnot Cycle Not Practical for a Vapour Power Plant?

Consider a steady flow Carnot cycle executed within the saturation dome of a pure substance such as water, as shown in Fig. 2.8. In the cycles in Fig. 2.8a, b, c, process $1'-2'$, steam is expanded isentropically in a turbine.

Process $2'-3'$, steam is condensed isothermally in a condenser,

Process $3'-4'$, working fluid is compressed isentropically by a pump,

Process $4'-1'$, working fluid is heated isothermally in a boiler.

In the cycle in Fig. 2.8a due to the following problems, the Carnot cycle cannot be practical for a vapour power plant.

(i)   In the isentropic expansion of the steam (process $1'-2'$) in a turbine, the quality of the steam decreases as shown in Fig. 2.8a. Thus, the turbine has to handle steam with low quality, i.e., steam with a large liquid droplets.

The liquid droplets when strike the blades causes erosion, as a result of damage to the turbine blades. Thus, for safe operation (i.e., without erosion of blades), the quality of the steam exiting the turbine cannot be recommended less than 90%.

(ii)   The isentropic compression of liquid (process $3'-4'$) in a pump requires a very large work.

(iii) The sensible heat addition in a process $4'-f$ at constant temperature in the whole process $4'-1'$ is impossible.

In the cycle in Fig. 2.8b, due to the following problem, the Carnot cycle cannot practical for a vapour power plant.

(i)   It is impossible to maintain the quality of the steam at point $3'$ during the condensation process in the condenser.

(ii)   It is very difficult to manufacture and maintain a pump that can handle both liquid and vapour at point $3'$ and convert it to saturated liquid at the outlet point $4'$.

(iii) The pump work is very large.

(iv)   The superheating steam (process $g-1'$) in a boiler at constant temperature is impossible.

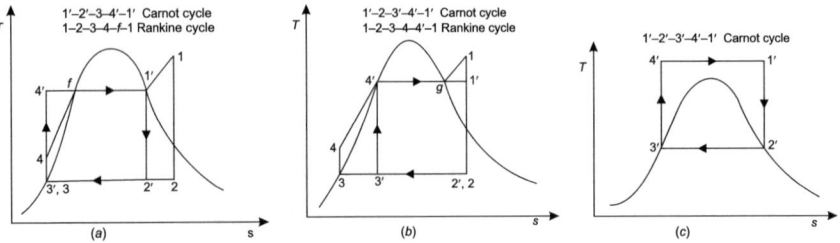

Fig. 2.8 Carnot vapor cycle

In the cycle in Fig. 2.8c, some of the problem associated with it can be avoided by executing the cycle. However, the following problems are associated with it.

(i) Isentropic compression of the liquid to very high pressure which require the large amount of pump work.
(ii) Isothermal heat transfer at variable pressure in the boiler is not possible.

2.4 Mean Temperature of Heat Supplied

In the Rankine cycle, heat is supplied at constant pressure and variable temperatures as shown by process 4–1 in Fig. 2.9. There is a large difference between temperature of feedwater at state 4 and the superheated at state 1. If T_m is the mean temperature of heat supplied, so that the area under 4–1 is equal to the area under 5–6, then heat supplied

$$q_{4-1} = q_{5-6}$$
$$(h_1 - h_4) = T_m(s_1 - s_4)$$

or

$$T_m = \frac{h_1 - h_4}{s_1 - s_4}$$

Heat rejected:

$$q_{2-3} = T_2(s_2 - s_3)$$
$$= T_2(s_1 - s_4) \quad \therefore s_2 = s_1, s_3 = s_4$$

Net work output:

Fig. 2.9 Mean temperature of heat supplieds

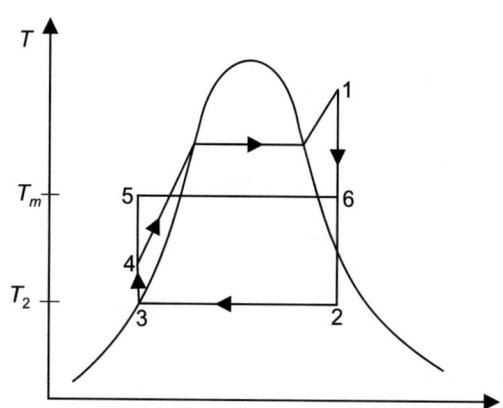

$$w_{net} = q_{4-1} - q_{2-3}$$

Thermal efficiency of the Rankine cycle,

$$\eta_{Rankine} = \frac{w_{net}}{q_{4-1}} = \frac{q_{4-1} - q_{2-3}}{q_{4-1}} = 1 - \frac{q_{2-3}}{q_{4-1}} = 1 - \frac{T_2(s_1 - s_4)}{T_m(s_1 - s_4)}$$

$$\eta_{Rankine} = 1 - \frac{T_2}{T_m}$$

Therefore, the two methods for improving the thermal efficiency of the Rankine cycle are as follows:

(i)   increase the mean temperature T_m at which heat is supplied to the working fluid in the boiler.
(ii) decrease the temperature T_2 at which heat is rejected from the working fluid in the condenser.

2.5   Improvements of the Rankine Cycle Efficiency

The efficiency of the Rankine cycle can be improved by varying the following parameters:

1   **Increasing the boiler pressure**: The efficiency of the Rankine cycle can be improved by increasing the boiler pressure while maintaining the constant maximum temperature and the minimum pressure in the cycle. The net increase in work output is the cross-hatched area minus the shaded area of Fig. 2.10a, a relatively small. The heat supplied decreases by the shaded area minus the cross-hatched area of Fig. 2.10b. This is a significant decrease, and it leads to a significant increase in efficiency. The disadvantage of raising the boiler

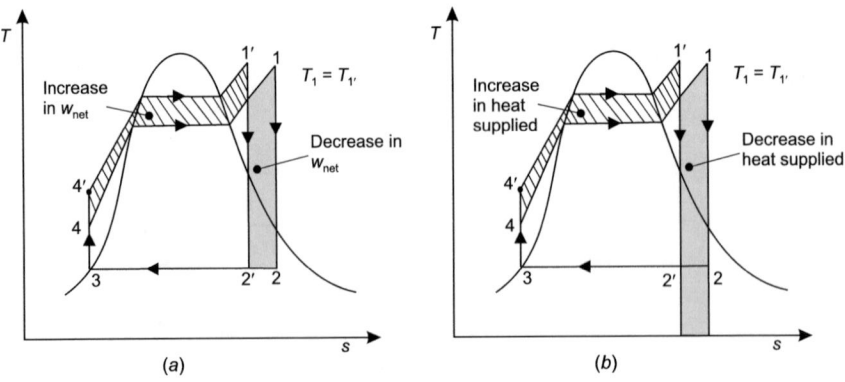

Fig. 2.10 Effect of increased pressure on the Rankine cycle

Fig. 2.11 Effect of increased temperature of the steam at inlet of the turbine

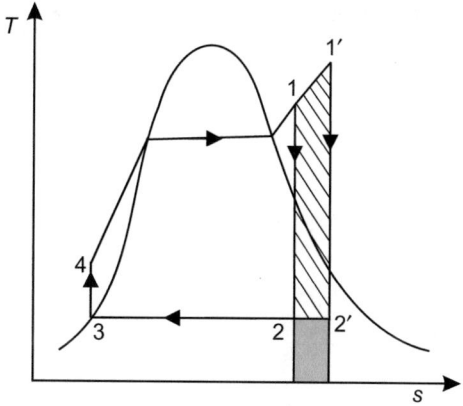

pressure is that the quality of the steam exiting the turbine to fall below 90%. The liquid droplets that strike the blades causes erosion, as a result of damage to the turbine blades and impaired turbine efficiency.

2   **Increasing temperature of the steam at inlet of the turbine/Superheating the steam to high temperature**: It also results in an improvement in thermal efficiency of the Rankine cycle. In Fig. 2.11, the net work is increased by the cross-hatched area, and the heat input is increased by the area under the process 1–1′ (sum of the cross-hatched area and the shaded area). Thus, both the net work and heat input increases as a result of superheating the steam to the higher temperature. The overall effect is an increase in the thermal efficiency.

Due to metallurgical considerations, there is a limit in the maximum temperature that can be attained in the boiler. Temperature is limited to about 600 °C by the turbine blades near the turbine inlet; however, ceramic blades with imbedded metal may allow much higher temperature.

Superheating the steam to higher temperature has another advantage that the quality of state 2′ is obviously increased, this reduces water droplet formation in the turbine.

3   **Decreasing the condenser pressure**: Decreasing the condenser pressure also results in increased Rankine cycle efficiency. In Fig. 2.12, the net work will increase a significant amount, represented by the cross-hatched area, and the heat input will increase a slight amount because state 4′ will move to a slightly lower enthalpy than that of state 4, the net result will be an increase in the Rankine cycle efficiency. The heat transfer process that occurs in the condenser limits the lower pressure. The heat is rejected by transferring heat to cooling water at about 20 °C, the heat transfer process requires a temperature difference of at least 10 °C between the cooling water and the steam. Hence, a temperature at least 30 °C is required in the condenser, this corresponds to a minimum condenser pressure [see the saturated steam table (temperature based)] of approximately 4.25 kPa. That is, decrease in condenser pressure depends on the temperature of the cooling

Fig. 2.12 Effect of decreased condenser pressure

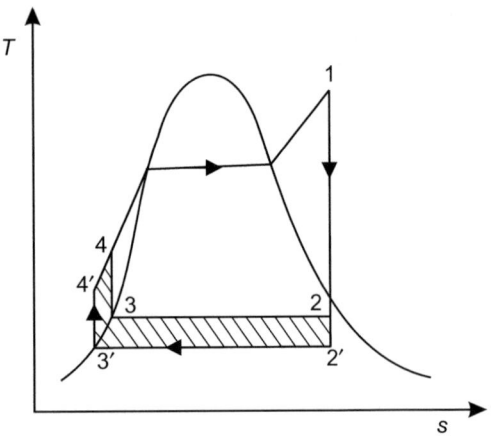

water and the temperature difference required between the steam and the cooling water.

The disadvantage of decreasing the condenser pressure is that the quality of the steam exiting the turbine decreases which causes the blades erosion and the turbine efficiency to fall. Another disadvantage of this process, it creates the possibility of air leakage in the condenser.

2.6 The Reheat Cycle

In the previous section, we discussed when Rankine cycles are operated with a high boiler pressure or a low condenser pressure, it is difficult to prevent liquid-droplet formation in the turbine exit (i.e., low-pressure portion of the turbine). This problem is overcome by reheat cycle. The reheat cycle is used to prevent liquid-droplet formation in the turbine exit. In the reheat cycle, steam is reheated after expanding partially in the high-pressure turbine. This is done by the steam partial expansion in the high-pressure turbine. The steam is then sent back to the boiler where it is reheated at constant pressure, usually to the inlet temperature of the high-pressure turbine. Steam is then supplied to the low-pressure turbine in which it expands isentropically to the condenser pressure.

The mean temperature can be increased by increasing the number of stages. The use of more than two reheat stages is not practical. The theoretical improvement in efficiency from the second reheat (Fig. 2.13) is about half of that which results from a single reheat. If the turbine inlet pressure is not high enough, double reheat would result in superheated exhaust. This is undesirable as it would cause the average temperature of heat rejection to increase and thus the cycle efficiency to decrease. Therefore, double reheat is used only on supercritical pressure ($p_{Bioler} > 221.2$ bar) power plant. A third reheat stage would increase the cycle efficiency by about half of the improvement attained by the second reheat. Thus, third stage reheat increases a

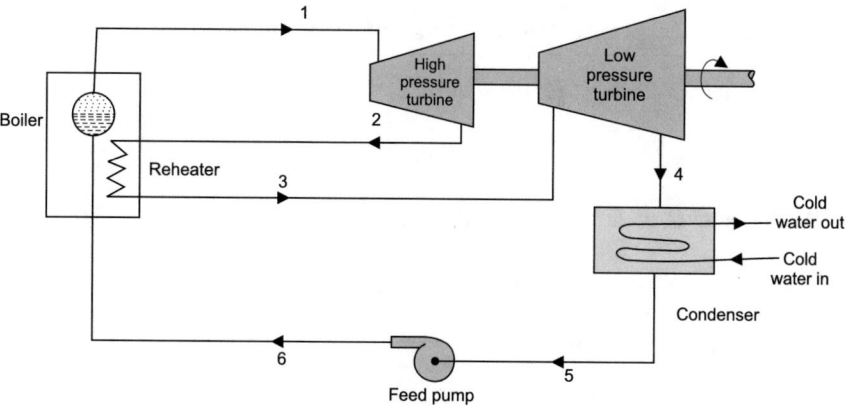

Fig. 2.13 Reheat cycle

significant investment in additional equipment, and the use of such equipment does not economically be justified by too small gain in cycle efficiency (Fig. 2.14).

Heat supplied in a boiler:

$$Q_{6-1} = m_s(h_1 - h_6)$$

Heat supplied in a reheater:

$$Q_{2-3} = m_s(h_3 - h_2)$$

Net heat supplied:

$$Q_1 = Q_{6-1} + Q_{2-3}$$

Fig. 2.14 T-s diagram of Reheat cycle

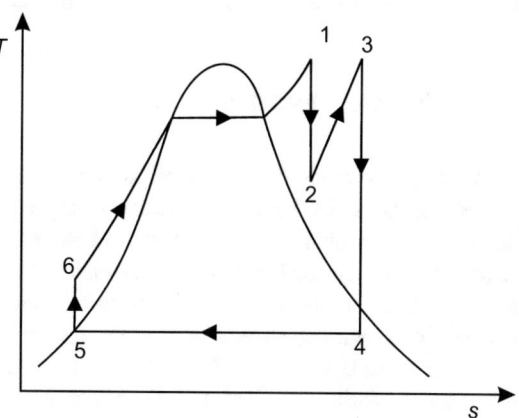

$$= m_s(h_1 - h_6) + m_s(h_3 - h_2)$$

Work output of the high-pressure turbine:

$$W_{1-2} = m_s(h_1 - h_2)$$

Work output of the low-pressure turbine:

$$W_{3-4} = m_s(h_3 - h_4)$$

Work input to a feed pump

$$W_{5-6} = m_s(h_6 - h_5)$$

Net work output:

$$W_{net} = W_{1-2} + W_{3-4} - W_{5-6}$$

Thermal efficiency:

$$\eta_{th} = \frac{\text{net work output}}{\text{net heat supplied}} = \frac{W_{net}}{Q_1}$$

Advantages of a reheat cycle:

(i)   It increases the thermal efficiency of the cycle when the cycle works at optimum values of the boiler pressure and the reheat pressure.
(ii)   It increases the net work output.
(iii) It decreases the moisture content of the steam at the low-pressure turbine exit, reducing the turbine blade erosion.
(iv)   It increases the average temperature of heat addition.

2.7   The Regenerative Cycle

In the simple Rankine cycle, as well as in the reheat cycle, a considerable amount of heat energy is used to heat the feedwater from temperature T_4 to its saturated temperature. The cross-hatched area in Fig. 2.15 represents this necessary heat energy. To reduce this energy, the fed water is preheated before it enters the boiler by extracting some of the steam as it expands in the turbine (for example, at state 2 in Fig. 2.16a and mixing it with the water as it exits the condensate pump, thereby preheating the water from temperature T_5 to T_6 as shown in Fig. 2.16b. A cycle that utilizes this type of heating is called a regenerative cycle, and the process is referred to as regeneration.

Fig. 2.15 Simple Rankine cycle in T-s diagram

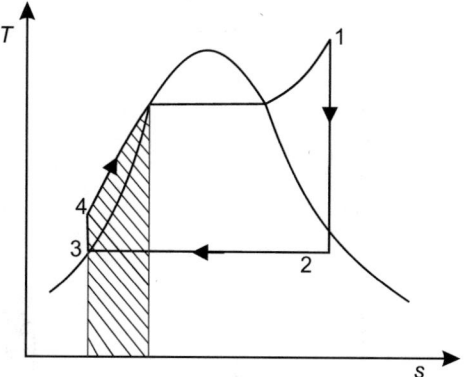

A schematic representation of the main components of such a cycle is shown in Fig. 2.16a. The water entering the boiler is called feedwater, and the device used to mix the extracted steam and the condenser water is called a regenerator, or a feedwater heater. When the condenser water is mixed directly with the extracting steam, it is done in an open feedwater heater, as shown in Fig. 2.16a.

Let

1 kg of steam supplied to the turbine or leaving from the boiler.
m_1 kg is extracted at state 2 for feedwater heater. $(1 - m_1)$ kg is expanded completely in the remaining part of the turbine up to state 3.

The heat and work interactions of a regenerative cycle with one feedwater heater can be expressed per unit mass of steam flowing through the boiler as follows:

Heat supplied in the boiler:

$$q_{7-1} = h_1 - h_7$$

Pump-I work input:

$$w_{4-5} = (1 - m_1)(h_5 - h_4)$$

also

$$= v_4(p_2 - p_3)$$

Pump-II work input:

$$w_{6-7} = h_7 - h_6$$

also

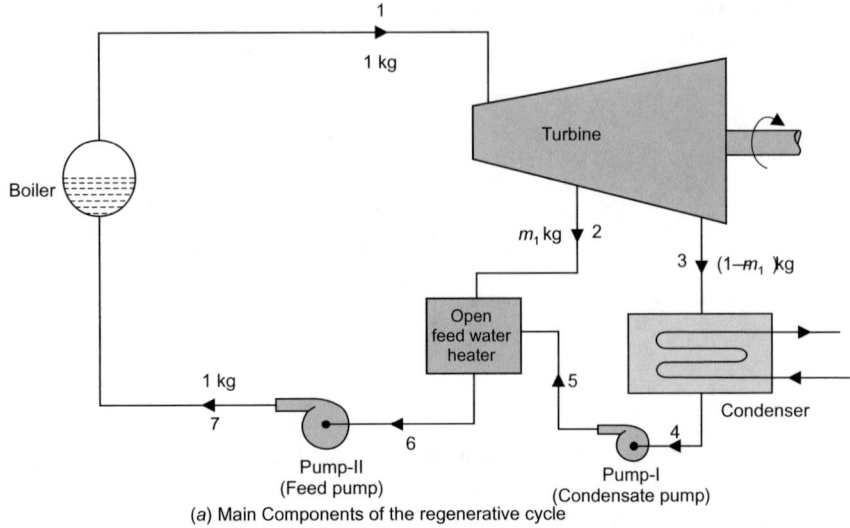

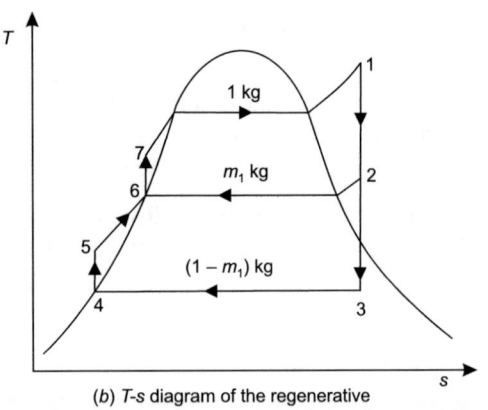

(b) T-s diagram of the regenerative

Fig. 2.16 Regenerative cycle

$$w_{6-7} = v_6(p_1 - p_2)$$

Net pumps input work:

$$w_p = w_{4-5} + w_{6-7}$$
$$= (1 - m_1)(h_5 - h_4) + (h_7 - h_6)$$

Turbine output work:

$$w_{1-3} = 1(h_1 - h_2) + (1 - m_1)(h_2 - h_3)$$

Fig. 2.17 An open
feedwater heater

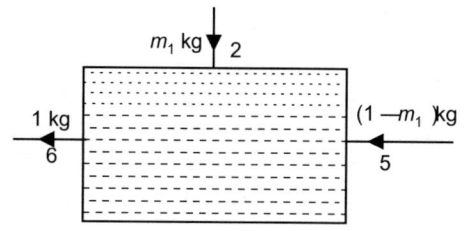

Net work output:

$$w_{net} = w_{1-3} - w_p$$

For an open feedwater heater:

Applying the energy balance equation (Fig. 2.17),
We get

$$m_1 h_2 + (1 - m_1)h_5 = h_6$$
$$m_1 h_2 + h_5 - m_1 h_5 = h_6$$
$$m_1(h_2 - h_5) = h_6 - h_5$$

or

$$m_1 = \frac{h_6 - h_5}{h_2 - h_5}$$

The thermal efficiency of the regenerative cycle,

$$\eta_{th} = \frac{\text{net work output}}{\text{heat supplied}} = \frac{w_{net}}{q_{7-1}}$$

Advantages of a regenerative cycle:

(i) It increases the thermal efficiency of the cycle.
(ii) Water particles escape with the bled steam through extraction belts. This reduces the moisture content in the steam, thus decreasing the turbine blade erosion.
(iii) Feed heating by bled steam decreases the quantity of steam reaching the condenser. Thus in turn reduces the size of the condenser and the cooling water requirements.
(iv) Steam extraction for feed heating reduces the flow rate significantly in the low-pressure stages, thus allowing the blades to be shorter. This is a great advantage in large units where designing long turbine blades is very critical.

Closed Feedwater Heater

The closed feedwater heater is a heat exchanger in which the water passes through the tubes and the steam surrounds the tubes, condensing on the outer surfaces. The condensate formed at temperature T_7 is pumped with a pump-I into the main feed-water line, as shown in Fig. 2.18a. The schematic of a steam power plant with one closed feedwater heater and the $T-s$ diagram of the cycle are shown in Fig. 2.18. The disadvantage of such system is that it is more expensive and its heat transfer characteristic is not as desirable as heat transfer in which the steam and water are simply mixed, as in the open feedwater heater. Mass and energy balances are also required when analysing a closed feedwater heater.

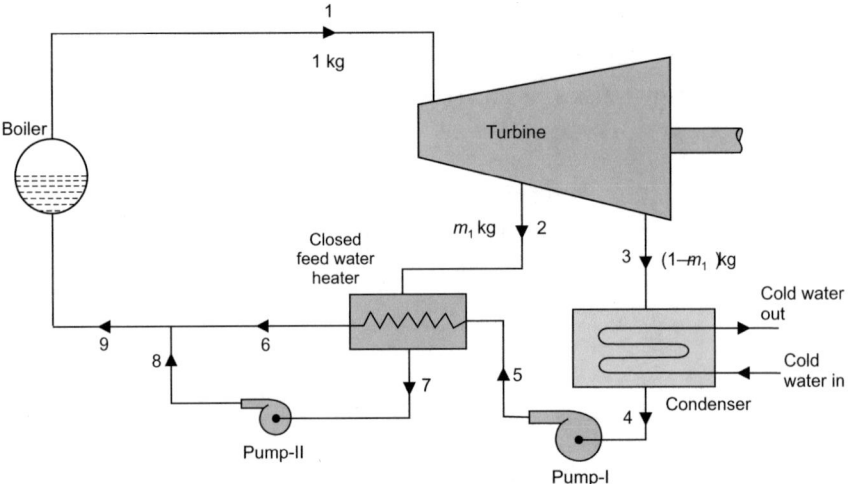

(a) Main Components of the regenerative Rankine cycle with a closed feed water heater.

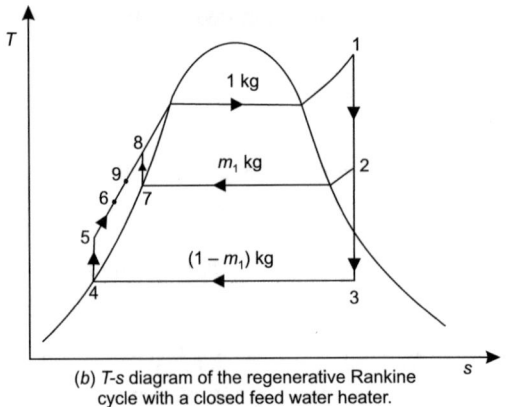

(b) T-s diagram of the regenerative Rankine cycle with a closed feed water heater.

Fig. 2.18 Regenerative Rankine cycle with a closed FWH

2.8 Combined Gas-Vapour Power Cycle

The combined cycle means that both gas turbine cycle and vapour power cycle are combined together. The exhaust of the gas turbine at high temperature is supplied to the boiler to raise steam which can be utilized for power generation in a steam turbine. A schematic representation of the main components of the combined cycle is shown in Fig. 2.19 and $T-s$ diagram is shown in Fig. 2.20. The combined cycle increases the efficiency without increasing the initial cost greatly. Consequently, many new power plants operate on combined cycle, and many more existing vapour or gas turbine plants are being converted to combined cycle power plants.

Advantages of the combined cycle plants over the conventional steam plants are as follows:

(i)   High thermal efficiency. Some recent combined cycle power plants have achieved efficiencies above 60%.
(ii)   High total power.
(iii)   Lower cooling water requirements. Gas turbine needs no cooling. Only cooling water required is for the condenser.
(iv)   Lower specific investment cost.

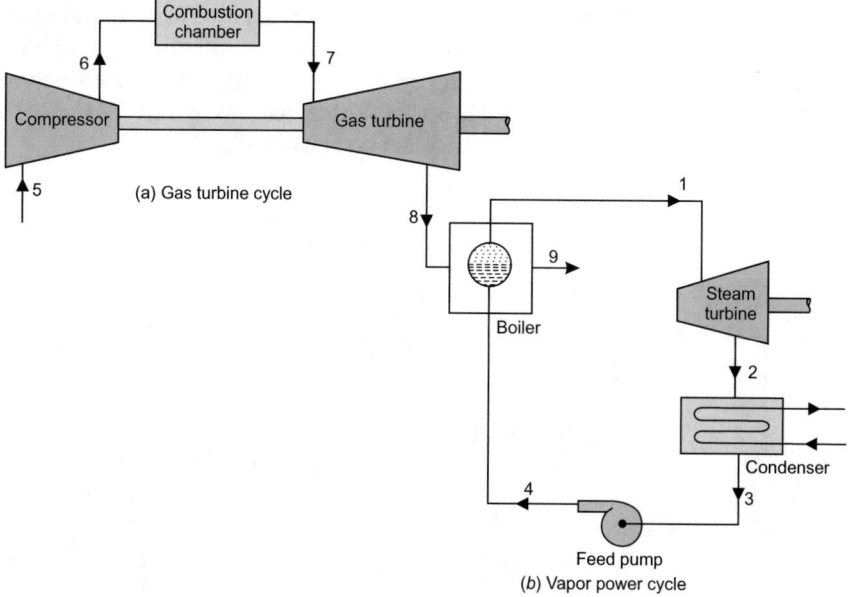

Fig. 2.19 Combined cycle

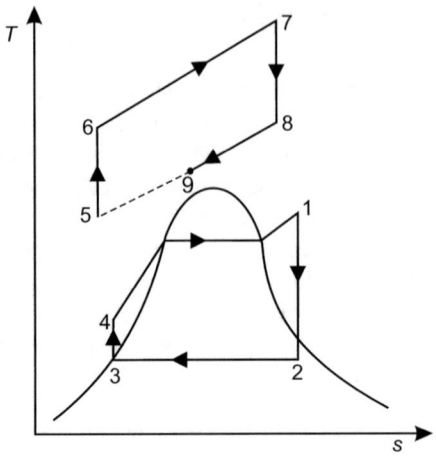

Fig. 2.20 $T-s$ diagram of the gas–vapor combined cycle

2.9 Binary Vapour Cycle

The binary vapour cycle means that two Rankine cycles are used in series. Mercury is used in one cycle, and steam is used in other cycles. Mercury which has high saturation temperature is used in topping cycle, and heat released in the condenser of topping cycle is used to generate steam in the bottoming cycle (Fig. 2.21).

The binary vapour cycle on a $T-s$ diagram is shown in Fig. 2.22.

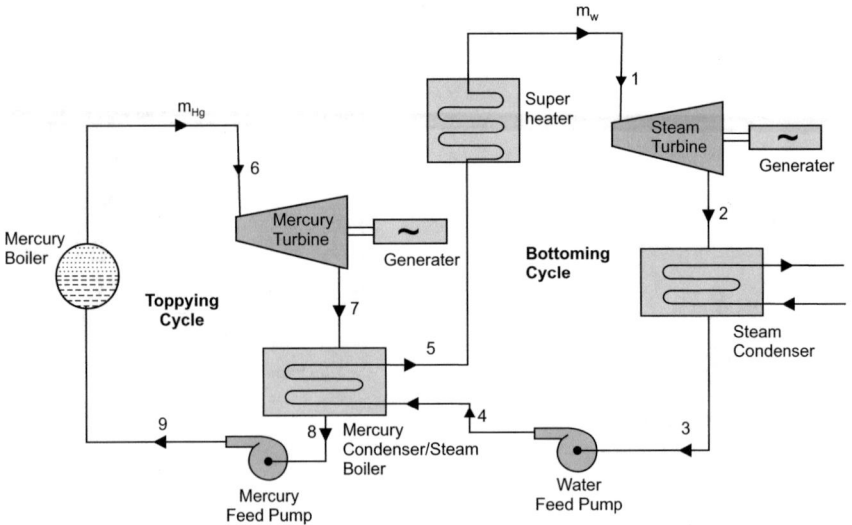

Fig. 2.21 Binary vapor cycle

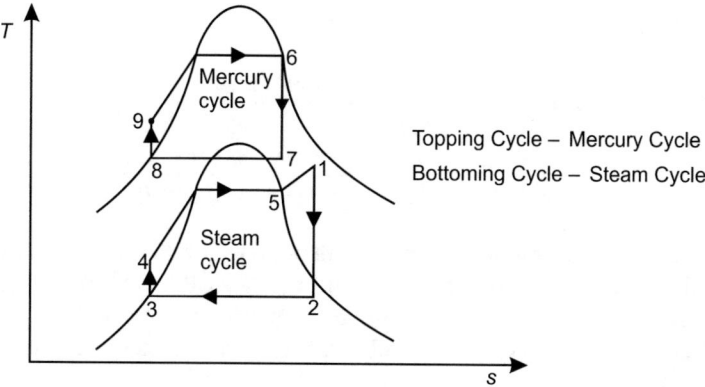

Fig. 2.22 Binary vapor cycle in T-s diagram

Topping cycle–Mercury cycle

Process 6–7: Isentropic expansion of mercury in a mercury turbine.
Process 7–8: Heat rejection at constant pressure in a mercury condenser is utilized saturated steam for bottoming cycle.
Process 8–9: Isentiopic compression of liquid mercury in a mercury feed pump.
Process 9–6: Heat addition at constant pressure in a mercury boiler.

Bottoming cycle–steam cycle

Process 1–2: Isentropic expansion of steam in a steam turbine.
Process 2–3: Heat rejection at constant pressure in a steam condenser.
Process 3–4: Isentropic compression of water in a water feed pump.
Process 4–5: Heat gain at constant processes in a mercury condenser in raising saturated steam by the heat rejected by the condensation of mercury.
Process 5–1: Heat addition at constant pressure to formation of superheated steam in a superheater.

Let

$$m_{\text{Hg}} = \text{mass flow rate of the mercury in the topping cycle}$$
$$m_w = \text{mass flow rate of the steam in the bottoming cycle}$$

Net work output:

$$W_{\text{Net}} = \text{work output in the mercury cycle} + \text{work output}$$
$$W_{\text{Net}} = m_{\text{Hg}}(h_6 - h_7) + m_w(h_1 - h_2)$$

Net heat supplied in the binary cycle:

$$Q_s = \text{Heat supplied in the mercury boiler} + \text{Heat supplied in the superheater}$$
$$= m_{\text{Hg}}(h_6 - h_9) + m_w(h_1 - h_5)$$

Cycle efficiency:

$$\eta = \frac{\text{Net work output: } W_{\text{Net}}}{\text{Net heat supplied in the binary cycle: } Q_s}$$
$$= \frac{W_{Net}}{Q} = \frac{m_{Hg}(h_6 - h_7) + m_w(h_1 - h_2)}{m_{Hg}(h_6 - h_9) + m_w(h_1 - h_5)}$$

Problem 2.1: Consider a steady flow Carnot cycle with water as the working fluid. The maximum and minimum temperatures in cycle are 350 °C and 60 °C. The quality of water is 0.891 at the beginning of the heat rejection process and 0.1 at the end. Show the cycle on a $T-s$ diagram relative to the saturation lines, and determine.

(i)   the thermal efficiency,
(ii)   the pressure at the turbine inlet, and
(iii)   the net work output.

Solution: Given data:
    Maximum temperature at turbine inlet:

$$T_1 = 350\,^\circ\text{C}$$

Minimum temperature at condenser or turbine exit:

$$T_2 = 60^\circ C$$

Dryness fraction at turbine exit:

$$x_2 = 0.891$$

Dryness fraction at condenser exit:

$$x_3 = 0.1$$

From saturated steam table (temperature based),

at

$$T_2 = 60^\circ$$

we get

$$v_f = 0.001017\,\text{m}^3/\text{kg}, \qquad v_g = 7.671\,\text{m}^3/\text{kg}$$
$$h_f = 251.13\,\text{kJ/kg}, \qquad h_{fg} = 2358.5\,\text{kJ/kg}$$
$$s_f = 0.8312\,\text{kJ/kgK}, \qquad s_{fg} = 7.0782\,\text{kJ/kgK}$$
$$s_2 = s_f + x_2 s_{fg} = 0.8312 + 0.891 \times 7.0782$$

$$= 7.1378 \text{ kJ/kgK}$$
$$h_2 = h_f + x_2 h_{fg} = 251.13 + 0.891 \times 2358.5$$
$$= 2352.55 \text{ kJ/kg}$$

As

$$s_2 = s_1$$

From saturated steam table (temperature based),

at

$$T_1 = 350\,°\text{C}$$

$$s_g = 5.2112 \text{ kJ/kgK}$$

As $s_g < s_1$, it means steam is superheated at turbine inlet.
From superheated steam table,
at 350 °C, $s_1 = s_2 = 7.1378$ kJ/kgK

$$p_1 = 14 \text{ bar}$$

Specific enthalpy at the end of the condenser:

$$h_3 = h_f + x_3 h_{fg} \qquad \text{at} \quad T_2 = 60\,°\text{C}$$
$$= 251.13 + 0.1 \times 2358.5$$
$$= 486.98 \text{ kJ/kg}$$

(a) Thermal efficiency: η_{th}

$$\eta_{th} = 1 - \frac{T_2}{T_1} = 1 - \frac{(60 + 273)}{(350 + 273)}$$
$$= 1 - 0.5345 = 0.4655 = \mathbf{46.55\%}$$

(b) Pressure at the turbine inlet: p_1 (Fig. 2.23)

$$p_1 = \mathbf{14\,bar}$$

(c) Net work output: w_{net}

$$\eta_{th} = 1 - \frac{T_2}{T_1} = 1 - \frac{q_R}{q_A}$$

Fig. 2.23 T-s diagram for
Problem 2.1

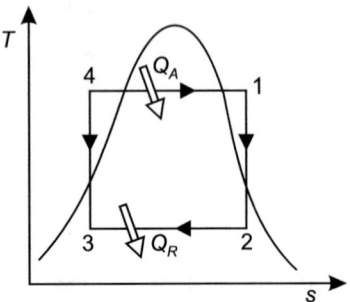

$$\frac{T_2}{T_1} = \frac{q_R}{q_A}$$

or

$$q_A = \frac{T_1}{T_2} q_R = \frac{T_1}{T_2} \times (h_2 - h_3)$$
$$= \frac{623}{333} \times (2352.55 - 486.98) = 3490.24 \text{ kJ/kg}$$

Net work output:

$$w_{net} = q_A - q_R = 3490.24 - 1865.57 = \mathbf{1624.67 \text{ kJ/kg}}.$$

Problem 2.2: A steam power plant operates on a simple ideal Rankine cycle between
the pressure limits of 3 MPa and 50 kPa. The temperature of the steam at the turbine
inlet is 300 °C, and the mass flow rate of steam through the cycle is 35 kg/s. Show
the cycle on a $T-s$ diagram with respect to saturation lines, and determine (*a*) the
thermal efficiency of the cycle (*b*) the net power output of the power plant, and (*c*)
the work ratio.

Solution: Given data:
  Boiler pressure:

$$p_1 = 3 \text{ MPa} = 30 \text{ bar} = 3 \times 10^3 \text{ kPa}$$

Condenser pressure:

$$p_2 = 50 \text{ kPa}$$

Temperature of the steam at the turbine inlet (Fig. 2.24):

$$T_1 = 300 \,^\circ\text{C}$$

Fig. 2.24 T-s diagram for
Problem 2.2

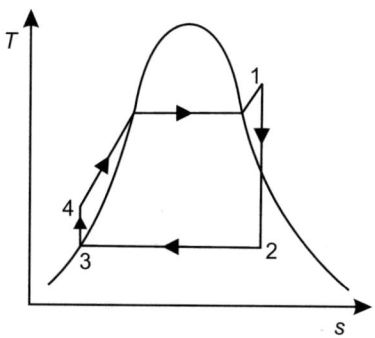

Mass flow rate:

$$m = 35 \, \text{kg/s}$$

From saturated steam table (pressure based),
At

$$p_1 = 30 \, \text{bar}$$
$$T_{\text{sat}} = 233.9 \, ^\circ\text{C}$$

So, $T_1 > T_{sat}$, it means the steam is superheated at the turbine inlet.
From superheated steam table,
At

$$p_1 = 30 \, \text{bar}, \, T_1 = 300 \, ^\circ\text{C}$$
$$h_1 = 2993.5 \, \text{kJ/kg}$$
$$s_1 = 6.539 \, \text{kJ/kgK}$$
$$\text{Entropy at state 1} = \text{entropy at state 2}$$

i.e.,

$$s_1 = s_2 = 6.539 \, \text{kJ/kgK}$$

From saturated steam table (pressure based),
At

$$p_2 = 50 \, \text{kPa} = 0.5 \, \text{bar}$$
$$s_g = 7.594 \, \text{kJ/kgK}$$

As $s_g > s_2$, it means the condition of steam is wet at turbine exit.
Other properties:

$$h_f = h_3 = 340.6\,\text{kJ/kg}, \qquad h_{fg} = 2305.4\,\text{kJ/kg}$$
$$s_f = 1.091\,\text{kJ/kgK}$$
$$v_f = 0.001\,\text{m}^3/\text{kg} = v_3$$
$$s_2 = s_f + x_2 s_{fg} \qquad \text{at} \quad p_2 = p_3 = 0.5\,\text{bar}$$
$$6.539 = 1.091 + x_2(7.594 - 1.091)$$

or

$$x_2 = 0.8377$$
$$h_2 = h_f + x_2 h_{fg} \qquad \text{at} \quad p_2 = 0.5\text{bar}$$
$$= 340.6 + 0.8377 \times 2305.4 = 2271.83\,\text{kJ/kg}$$

Pump work:

$$w_P = v_3(p_1 - p_2)$$
$$= 0.001(3 \times 10^3 - 50) = 2.95\,\text{kJ/kg}$$

also

$$w_P = h_4 - h_3$$

$\therefore$

$$2.95 = h_4 - 340.6$$

or

$$h_4 = 343.55\,\text{kJ/kg}$$

Turbine work:

$$w_T = h_1 - h_2$$
$$= 2993.5 - 2271.83 = 721.67\,\text{kJ/kg}$$

Net work output:

$$w_{net} = w_T - w_P$$
$$= 721.67 - 2.95 = 718.72\,\text{kJ/kg}$$

Heat supplied:

$$q_{4-1} = h_1 - h_4$$
$$= 2993.5 - 343.55 = 2649.95 \text{ kJ/kg}$$

(a)   Thermal efficiency:

$$\eta_{th} = \frac{w_{net}}{q_{4-1}} = \frac{718.72}{2649.50} = 0.2712 = \mathbf{27.12\%}$$

(b)   Net power output of the plant:

$$P = mw_{net}$$
$$= 35 \times 718.72 = 25155.2 \text{ kW} = \mathbf{25.155\,MW}$$

(c) Work ratio $= \dfrac{\text{net work output}}{\text{turbine work}}$

$$= \frac{w_{net}}{w_T} = \frac{718.72}{721.67} = \mathbf{0.9959}$$

Problem 2.3: A coal-fired power plant, operating on a simple ideal Rankine cycle, has a boiler pressure of 32 bar and a condenser pressure of 0.75 bar. The exit state from the steam turbine is dry saturated vapour. Given saturated water properties:

Pressure bar	Specific volume of liquid (V_f) m³/kg	Specific enthaply		Specific Entropy	
		Liquid (h_f) kJ/kg	Vapour (h_g) kJ/kg	Liquid (s_f) kJ/kgK	Vapour (s_g) kJ/kgK
0.75	0.001037	384.36	2662.96	1.2129	7.4563
32	0.001224	1026	2803	2.678	6.160

Superheat steam properties: at 32 bar and 592 °C, $h = 3663$ kJ/kg, $s = 7.4563$ kJ/kgK.

Determine: (*i*) the specific heat addition in the boiler, and (*ii*) the specific pump work input.

Solution: Given data:

Boiler pressure:

$$p_1 = 32 \text{ bar}$$
$$= 3200 \text{ kPa}$$

Condenser pressure:

$$p_2 = 0.75 \text{ bar}$$

Fig. 2.25 T-s diagram for
Problem 2.3

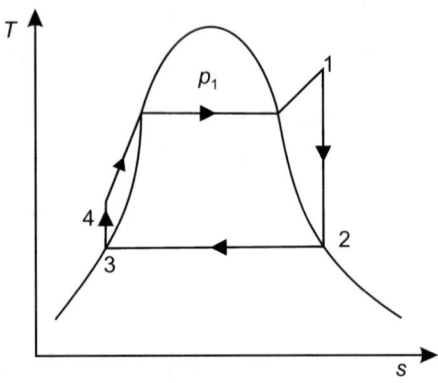

$$= 75 \, \text{kPa}$$

The $T-s$ diagram of given problem is shown in Fig. 2.25.
The condition of steam of inlet of the steam turbine is superheated.
From superheated steam properties at 32 bar and 592 °C,
we get

$$h_1 = 3663 \, \text{kJ/kg}$$

The exit state from the steam turbine is dry saturated vapour.
From saturated water properties at

$$p_2 = 0.75 \, \text{bar}$$
$$h_2 = h_g$$
$$= 2662.96 \, \text{kJ/kg}$$
$$h_3 = h_f = 384.36 \, \text{kJ/kg}$$
$$v_3 = v_f = 0.001037 \, \text{m}^3/\text{kg}$$

Specific pump work input:

$$w_{3-4} = v_3(p_2 - p_1) \, \text{kJ/kg}$$

where v_3 is in m³/kg,
   p_1 and p_2 are in kPa.
   $\therefore$

$$w_{3-4} = 0.001037(3200 - 75) = 3.24 \, \text{kJ/kg}$$

also

$$q_{3-4} = 0, \quad \text{for adiabatic flow}$$

$$\therefore$$

$$3.24 = h_4 - 384.36$$

or

$$h_4 = 3.24 + 384.36 = 387.6\,\text{kJ/kg}$$

(i) Specific heat addition:

$$q_{4-1} = h_1 - h_4 = 3663 - 387.6 = \mathbf{3275.4\,kJ/kg}$$

(ii) Specific pump work input:

$$w_{3-4} = 3.24\,\text{kJ/kg}$$

Problem 2.4: The following pertaining of a steam power plant are given for each state corresponding to figure shown below. Determine heat transfer in each process and turbine work (Fig. 2.26).

State	1	2	3	4
Pressure (MPa)	2	1.9	1.5	1.4
Enthalpy (kJ/kg)	3020	3000	2300	200

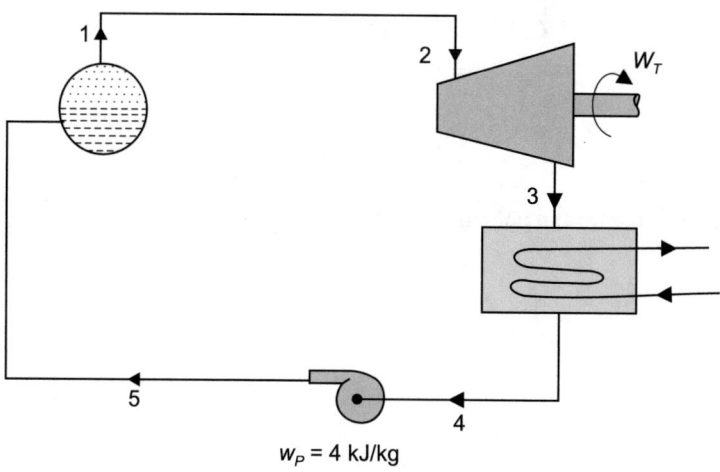

Fig. 2.26 Schematic for Problem 2.4

Solution: Given data:

At state 1,

$$p_1 = 2\,\text{MPa} = 200\,\text{kPa}$$
$$h_1 = 3020\,\text{kJ/kg}$$

At state 2,

$$p_2 = 1.9\,\text{MPa} = 1900\,\text{kPa}$$
$$h_2 = 3000\,\text{kJ/kg}$$

At state 3,

$$p_3 = 1.5\,\text{MPa} = 1500\,\text{kPa}$$
$$h_3 = 2300\,\text{kJ/kg}$$

At state 4,

$$p_4 = 1.4\,\text{MPa} = 1400\,\text{kPa}$$
$$h_4 = 200\,\text{kJ/kg}$$

Pump work:

$$w_p = 4\,\text{kJ/kg}$$

Applying the steady flow energy equation between states 1 and 2, we get

$$h_1 + \frac{V_1^2}{2} + g z_1 + q_{1-2} = h_2 + \frac{V_2^2}{2} + g z_2 + w_{1-2}$$

where

$$w_{1-2} = 0$$

Change in kinetic and potential energy are neglected

$$\therefore$$

$$h_1 + q_{1-2} = h_2$$

or

$$q_{1-2} = h_2 - h_1 = 3000 - 3020 = \mathbf{-20\,kJ/kg}$$

Process 2–3 (for turbine):

Applying the steady flow energy equation, we get

$$h_2 + \frac{V_2^2}{2} + gz_2 + q_{2-3} = h_3 + \frac{V_3^2}{2} + gz_2 + w_{2-3}$$

where

$$q_{2-3} = 0$$

Change in kinetic and potential energy are neglected

$$\therefore$$

$$h_2 = h_3 + w_{2-3}$$

or

$$w_{2-3} = h_2 - h_3$$
$$= 3000 - 2300$$
$$= \mathbf{700\ kJ/kg}$$
$$= w_T, \text{ turbine work}$$

Process 3–4 (for condenser):

Applying the steady flow energy equation, we get

$$h_3 + \frac{V_3^2}{2} + gz_3 + q_{3-4} = h_4 + \frac{V_4^2}{2} + gz_4 + w_{3-4}$$

where $w_{3-4} = 0$, no work done.

Change in kinetic and potential energies are neglected

$$\therefore$$

$$h_3 + q_{3-4} = h_4$$

or

$$q_{3-4} = h_4 - h_3 = 200 - 2300$$
$$= \mathbf{-2100\ kJ/kg}$$

Process 4–5 (for feed pump):

Pump work:

$$w_P = 4 \, \text{kJ/kg}$$

also

$$w_P = h_5 - h_4$$

$\therefore$

$$4 = h_5 - 200$$

or

$$h_5 = 204 \, \text{kJ/kg}$$

Applying the steady flow energy equation, we get

$$h_4 + \frac{V_4^2}{2} + gz_4 + q_{4-5} = h_5 + \frac{V_5^2}{2} + gz_5 + w_{4-5}$$

Change is kinetic and potential energy are neglected

$\therefore$

$$h_4 + q_{4-5} = h_5 + w_{4-5}$$
$$200 + q_{4-5} = 204 - 4$$

where $w_{4-5} = w_p = 4 \, \text{kJ/kg}$, $-ve$ sign indicates work done on the system.

or

$$q_{4-5} = 204 - 4 - 200 = 0$$

Net heat transfer:

$$\begin{aligned} q_\text{net} &= q_{1-2} + q_{2+3} + q_{3-4} + q_{4-5} \\ &= -20 + 0 - 2100 + 0 \\ &= \mathbf{-2120 \, kJ/kg} \end{aligned}$$

and Turbine work:

$$w_T = w_{2-3} = \mathbf{700 \, kJ/kg}$$

Problem 2.5: Consider a steam power plant that operates on a simple ideal Rankine cycle and has a net power output of 45 MW. Steam enters the turbine at 7 MPa and

500 °C and is cooled in the condensers at a pressure of 10 kPa by running cooling water from a lake through the tubes of the condenser at a rate of 2000 kg/s. Show the cycle on a T–s diagram with respect to saturation lines, and determine;

(a) the thermal efficiency of the cycle,
(b) the mass flow rate of the steam, and
(c) the temperature rise of the cooling water.

Solution: Given data for simple ideal Rankine cycle.
  Net power output:

$$P = 45\,\text{MW}$$
$$= 45 \times 10^3 \text{kW}$$

Turbine inlet:

$$p_1 = 7\,\text{MPa} = 70\,\text{bar}$$
$$= 7000\,\text{kPa}$$
$$T_1 = 500\,^\circ\text{C}$$

Condenser pressure:

$$p_2 = 10\,\text{kPa}$$

Mass flow rate of cooling water:

$$m_w = 2000\,\text{kg/s}$$

From saturated steam table (pressure based),
  At

$$p_1 = 70\,\text{bar},$$

Fig. 2.27 T-s diagram for Problem 2.5

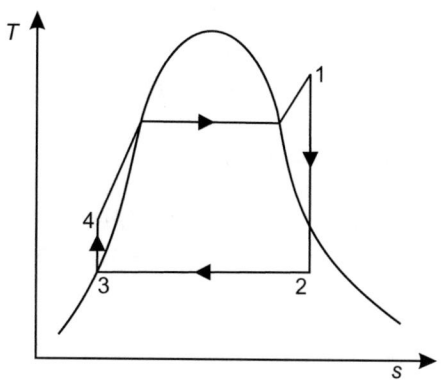

$$T_{sat} = 285.9\,°C$$

As $T_1 > T_{sat}$, it means the steam is superheated.
From superheated steam table,
At

$$p_1 = 70\,\text{bar},\ T_1 = 500\,°C$$
$$h_1 = 3410.3\,\text{kJ/kg},\ s_1 = 6.798\,\text{kJ/kgK}$$
$$\text{Entropy at state } 1 = \text{entropy at state } 2$$

i.e.,

$$s_1 = s_2 = 6.798\,\text{kJ/kgK}$$

From saturated steam table (pressure based),
At

$$p_2 = 10\,\text{kPa} = 0.1\,\text{bar}$$
$$s_g = 8.1502\,\text{kJ/kgK}$$

As $s_g > s_2$, it means the condition of steam is wet at turbine exit.
Other properties:

$$s_f = 0.6493\,\text{kJ/kgK},\ s_{fg} = 7.5009\,\text{kJ/kgK}$$
$$h_f = h_3 = 191.83\,\text{kJ/kg},\ h_{fg} = 2392.8\,\text{kJ/kg}$$
$$v_f = v_3 = 0.00101\,\text{m}^3/\text{kg}$$
$$s_2 = s_1 = s_f + x_2 s_{fg}$$
$$6.793 = 0.6493 + x_2 + 7.5009$$

or

$$x_2 = 0.8190$$
$$h_2 = h_f + x_2 h_{fg}$$
$$= 191.83 + 0.8190 \times 23.92.8 = 2153.53\,\text{kJ/kg}$$

Pump work:

$$w_P = v_3(p_1 - p_2)$$
$$= 0.00101(7000 - 10) = 7.06\,\text{kJ/kg}.$$

also

$$w_P = h_4 - h_3$$

$\therefore$

$$7.06 = h_4 - 191.83$$

or

$$h_4 = 198.89 \, \text{kJ/kg}$$

Turbine work:

$$w_T = h_1 - h_2$$
$$= 3410.3 - 2153.53 = 1256.77 \, \text{kJ/kg}$$

Net work output:

$$w_{net} = w_T - w_P = 1256.77 - 7.06 = 1249.71 \, \text{kJ/kg}$$

Heat addition:

$$q_{4-1} = h_1 - h_4 = 3410.3 - 198.89 = 3211.41 \, \text{kJ/kg}$$

(a) Thermal efficiency:

$$\eta_{th} = \frac{w_{net}}{q_{4-1}} = \mathbf{0.3891} = \mathbf{38.91\%}$$

(b) Mass flow rate of the steam: m_s
Net power output:

$$P = m_s w_{net}$$
$$45 \times 10^3 = m_s \times 1249.71$$

or

$$m_s = \mathbf{36 \, kg/s}$$

(c) Temperature rise of the cooling water: $(\Delta T)_w$
By energy balance equation

Heat lost by the condenser = heat gained by the cooling water
$$m_s(h_2 - h_3) = m_w c (\Delta T)_w$$
$$36(2153.53 - 191.83) = 2000 \times 4.2 \times (\Delta T)_w$$

or

$$(\Delta T)_w = \mathbf{8.40\,°C\,or\,K}$$

Problem 2.6: Repeat Problem 2.5 assuming an isentropic efficiency 87% for both the turbine and the pump.

Solution: Given data:
   Isentropic efficiency for the turbine:

$$\eta_T = 0.87$$

Isentropic efficiency for the pump:

$$\eta_P = 0.87$$

From the previous problem,

$$h_1 = 3410.3\ \text{kJ/kg}$$
$$h_{2s} = 2153.53\ \text{kJ/kg}$$
$$h_3 = 191.83\ \text{kJ/kg}$$
$$h_{4s} = 198.89\ \text{kJ/kg}$$

For turbine:

$$\eta_T = \frac{(\Delta h)_{act}}{(\Delta h)_{Isen}} = \frac{h_1 - h_2}{h_1 - h_{2s}}$$
$$0.87 = \frac{1685.43}{4103.83}$$

or

Fig. 2.28 T-s diagram for Problem 2.6

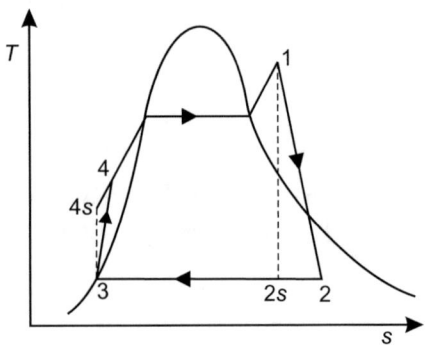

$$h_2 = 2316.91 \text{ kJ/kg}$$

For pump:

$$\eta_P = \frac{(\Delta h)_{Isen}}{(\Delta h)_{act}}$$

$$\eta_P = \frac{h_{4s} - h_3}{h_4 - h_3}$$

$$0.87 = \frac{198.89 - 191.83}{h_4 - 191.83}$$

or

$$h_4 = 199.94 \text{ kJ/kg}$$

Pump work:

$$w_P = h_4 - h_3$$

Remember
$w_P = v_3(p_1 - p_2)$ only for isentropic process

$$w_P = 199.94 - 191.83 = 8.11 \text{ kJ/kg}$$

Turbine work:

$$w_T = h_1 - h_2 = 3410.3 - 2316.91 = 1093.39 \text{ kJ/kg}$$

Net work output:

$$w_{npt} = w_T - w_P = 1093.39 - 8.11 = 1085.28 \text{ kJ/kg}$$

Heat addition:

$$q_{4-1} = h_1 - h_4$$
$$= 3410.3 - 199.94 = 3210.36 \text{ kJ/kg}$$

(a)   Thermal efficiency:

$$\eta_{th} = \frac{w_{net}}{q_{4-1}} = \frac{1085.28}{3210.36} = 0.3380 = \mathbf{33.8\%}$$

(b)   Mass flow rate of the steam: m_s
        Net power output:

$$P = m_s w_{net}$$
$$45 \times 10^3 = m_s \times 1085.28$$

or

$$m_s = \mathbf{41.46\,kg/s}$$

(c)   Temperature rise of the cooling water: $(\Delta T)_w$
        By energy balance equation,

Heat lost by the condenser = heat gained by the cooling water
$$m_s(h_2 - h_3) = m_w c(\Delta T)_w$$
$$41.46(2316.91 - 191.83) = 2000 \times 4.2 \times (\Delta T)_w$$

or

$$(\Delta T)_w = \mathbf{10.48\,°C\,or\,K}$$

Problem 2.7: Consider a coal-fired steam power plant that produces 300 MW of electric power. The power plant operates on a simple ideal Rankine cycle with turbine inlet conditions of 5 MPa and 450 °C and a condenser pressure of 25 kPa. The coal has a heating value of 29,300 kJ/kg. Assuming that 75% of this energy transferred to the steam in the boiler and that the electric generator has an efficiency of 96%, determine.

(a)   the overall plant efficiency,
(b)   the required rate of coal supply.

Solution: Given data:
    Electric power:

$$P_E = 300\,MW = 300 \times 10^3\,kW$$

At turbine inlet:

$$p_1 = 5\,MPa = 50\,bar$$
$$T_1 = 450\,°C$$

Condenser pressure:

$$p_2 = 25\,kPa$$

Heating value of the coal:

$$CV = 29300 \, \text{kJ/kg}$$

Heat supplied to the boiler

$$Q_{4-1} = 0.75 \times m_f CV = 0.75 \times m_f \times 29300 = 21975 \, m_f$$

Generator efficiency

$$\eta_G = 96\% = 0.96$$

also

$$\eta_G = \frac{\text{electric power}}{\text{net power output of the cycle}}$$

$$0.96 = \frac{300 \times 10^3}{P}$$

or

$$P = 312500 \, \text{kW}$$

From saturated steam table (pressure based),
At

$$p_1 = 50 \, \text{bar}$$
$$T_{sat} = 264 \, ^\circ\text{C}$$

As $T_1 > T_{sat}$, it means the steam is superheated.
From superheated steam table (Fig. 2.29),

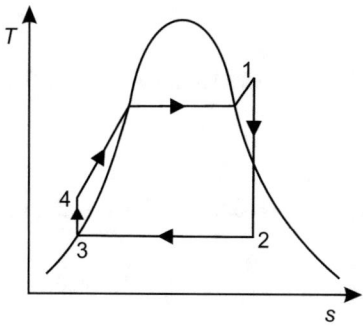

Fig. 2.29 T-s diagram for Problem 2.7

At

$$p_1 = 50\,\text{bar}$$
$$T_1 = 450\,°C$$
$$h_1 = 3316.2\,\text{kJ/kg}$$
$$s_1 = 6.818\,\text{kJ/kgK}$$

Entropy at state $1 =$ entropy at state 2

i.e.,

$$s_1 = s_2 = 6.81\,\text{kJ/kgK}$$

From saturated steam table (pressure based),
At

$$p_2 = 25\,\text{kPa} = 0.25\,\text{bar}$$
$$s_g = 7.831\,\text{kJ/kgK}$$

As $s_g > s_2$, it means the condition of steam is wet at turbine exit.
Other properties:

$$v_f = v_3 = 0.00102\,\text{m}^3/\text{kg}$$
$$h_f = h_3 = 271.9\,\text{kJ/kg},\, h_{fg} = 2346.3\,\text{kJ/kg}$$
$$s_f = 0.893\,\text{kJ/kgK},\, s_{fg} = 6.938\,\text{kJ/kgK}$$
$$s_2 = s_f + x_2 s_{fg}$$
$$6.818 = 0.893 + x_2 \times 6.938$$

or

$$x_2 = 0.8539$$
$$h_2 = h_f + x_2 h_{fg}$$
$$= 271.9 + 0.8539 \times 2346.3 = 2275.40\,\text{kJ/kg}$$

Pump work:

$$w_P = v_3(p_1 - p_2) = 0.00102(5000 - 25) = 5.07\,\text{kJ/kg}$$

also

$$w_P = h_4 - h_3$$

∴

$$5.07 = h_4 - 271.9$$

or

$$h_4 = 276.97 \, \text{kJ/kg}$$

Turbine work:

$$w_T = h_1 - h_2 = 3316.2 - 2276.40 = 1039.8 \, \text{kJ/kg}$$

Net work output:

$$w_{net} = w_T - w_P$$
$$= 1039.8 - 5.07 = 1034.73 \, \text{kJ/kg}$$

Net power output of the cycle:

$$P = m_s w_{net}$$
$$312500 = m_s \times 1034.73$$

or

$$m_s = 302.01 \, \text{kg/s}$$

Heat supplied to a boiler:

$$Q_{4-1} = m_s(h_1 - h_4)$$
$$= 302.01(3316.2 - 276.97) = 917877.85 \, \text{kW}$$

also

$$Q_{4-1} = 21973 \, m_f$$

∴

$$917877.85 = 21973 \, m_f$$

or

$$m_f = \frac{917877.85}{21973} = \mathbf{41.77 \, kg/s}$$

(a)   The overall plant efficiency:

$$\eta_0 = \frac{\text{electric power}}{\text{heat supplied by the coal}}$$

$$= \frac{300 \times 10^3}{41.77 \times 29300} = 0.2451 = \mathbf{24.51\%}$$

(b)   The rate of coal supplied:

$$m_f = \mathbf{41.77\,kg/s}$$

Problem 2.8: Consider a reheat cycle utilization steam as the working fluid steam enters the high-pressure turbine at 35 bar, 350 °C and expands to 8 bar. It is then reheated 350 °C and then expand to 10 kPa in the low-pressure turbine. Determine the thermal efficiency of the cycle.

Solution: Given data: (refer Fig. 2.30)
    At high-pressure turbine (HPT) inlet:

$$p_1 = 35\,\text{bar}$$
$$T_1 = 350\,^{\circ}\text{C}$$

From saturated steam table (pressure based),
At

$$p_1 = 35\,\text{bar}$$
$$T_{\text{sat}} = 242.6\,^{\circ}\text{C}$$

As $T_1 > T_{sat}$, it means that the steam is superheated.

From superheated steam table,
At

$$p_1 = 35\,\text{bar}$$
$$T_1 = 350\,^{\circ}\text{C}$$
$$h_1 = 3104\,\text{kJ/kg}$$
$$s_1 = 6.658\,\text{kJ/kgK}$$

At high-pressure turbine exit:

$$p_2 = 8\,\text{bar}$$
$$s_2 = s_1 = 6.658\,\text{kJ/kgK}$$

From saturated steam table (pressure based),

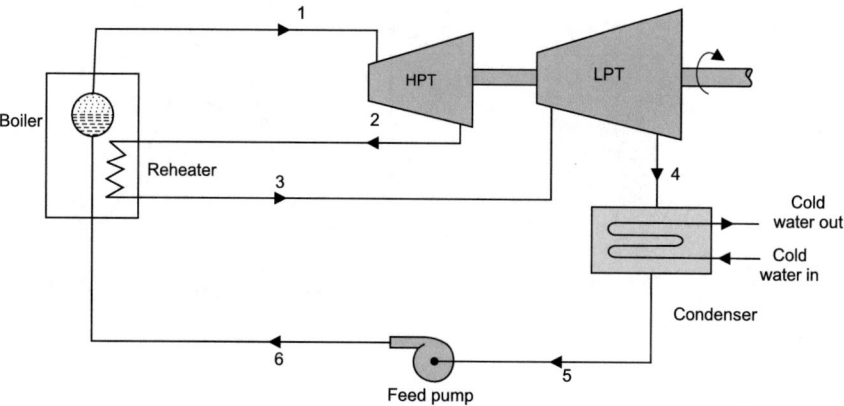

Fig. 2.30 Reheat cycle

At

$$p_2 = 8\,\text{bar}$$
$$s_g = 6.663\,\text{kJ/kgK}$$

As $s_2 \approx s_g$, it means that the steam is dry saturated.
Other properties at

$$p_2 = 8\,\text{bar}$$
$$h_g = h_2 = 2769.1\,\text{kJ/kg}$$

At exit of the reheater/low-pressure turbine (LPT) inlet

$$p_3 = p_2 = 8\,\text{bar}$$
$$T_3 = 350\,^\circ\text{C}$$

From superheated steam table,
At

$$p_3 = 8\,\text{bar},\ T_3 = 350\,^\circ\text{C}$$
$$h_3 = 3161.7\,\text{kJ/kg}$$
$$s_3 = 7.409\,\text{kJ/kgK}$$

At low-pressure turbine exit:

$$p_4 = 10\,\text{kPa}$$
$$s_4 = s_3 = 7.409\,\text{kJ/kgK}$$

Fig. 2.31 T-s diagram
for Problem 2.8

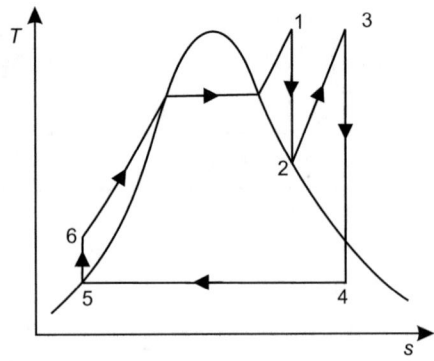

From saturated steam table (pressure based),
At

$$p_4 = 10\,\text{kPa}$$
$$s_g = 8.147\,\text{kJ/kgK}$$

As $s_4 > s_g$, it means that the steam is wet (Fig. 2.31).
Other properties at

$$p_4 = 10\,\text{kPa}$$
$$v_f = v_5 = 01.00101\,\text{m}^3/\text{kg}$$
$$h_f = h_5 = 192.6\ \text{kJ/kg}$$
$$h_{fg} = 2392.4\,\text{kJ/kg}$$
$$s_f = 0.625\,\text{kJ/kgK}$$
$$s_{fg} = 7.495\,\text{kJ/kgK}$$
$$s_4 = s_f + x_4 s_{fg}$$
$$7.409 = 0.652 + x_4 \times 7.495$$

or

$$7.495\,x_4 = 6.757$$
$$x_4 = 0.9015$$
$$h_4 = h_f + x_4 h_{fg}$$
$$= 192.6 + 0.9015 \times 2392.4 = 2349.34\,\text{kJ/kg}$$

Pump input work:

$$w_P = v_5(p_1 - p_4)\text{kJ/kg}$$

where v_5 in m³/kg p_1 and p_4 in kPa

$\therefore$

$$w_P = 0.00101(3500 - 10) = 3.525 \, \text{kJ/kg}$$

also

$$w_P = h_6 - h_5$$

$\therefore$

$$3.525 = h_6 - 192.6$$

or

$$h_6 = 196.125 \, \text{kJ/kg}$$

Net heat supplied:

q_{net} = heat supplied in a boiler + heat supplied in a reheater
$$= q_{6-1} + q_{2-3}$$
$$= (h_1 - h_6) + (h_3 - h_2)$$
$$= (3104 - 196.125) + (3161.7 - 2769.1) = 3300.475 \, \text{kJ/kg}$$

Net work output $= w_{1-2} + w_{3-4} - w_p$
$$= (h_1 - h_2) + (h_3 - h_4) - 3.525$$
$$= (3104 - 2769.1) + (3161.7 - 2349.34) - 3.525$$
$$= 1143.735 \, \text{kJ/kg}$$

Thermal efficiency of the cycle:

$$\eta_{th} = \frac{\text{net work output}}{\text{net heat supplied}} = \frac{1143.735}{3300.475} = 0.3465 = \textbf{34.65\%}$$

Problem 2.9: A steam power plant operates between a boiler pressure of 43 bar and condenser pressure of 0.035 bar. The steam is superheated to 500 °C. Neglecting feed pump work, compare the performance of the plant if reheat is included. The steam condition at inlet to the turbine is same in reheat cycle and also condenser pressure. Assume that the steam is just dry saturated on leaving the first turbine and is reheated to the inlet temperature of the high-pressure turbine.

Solution: Given data:
At inlet of the turbine:

$$p_1 = 42\,\text{bar},\ T_1 = 500\,°\text{C}$$

Condenser pressure:

$$p_c = 0.035\,\text{bar}$$

Case-I: Simple Rankine cycle (without reheat).
From superheated steam table (Fig. 2.32).
At

$$p_1 = 43\,\text{bar},\ T_1 = 500\,°\text{C}$$

T_1 500 °C	$p_a = 40$ bar		$p_1 = 43$ bar		$p_b = 45$ bar	
	h_a	s_a	h_1	s_1	h_b	s_b
	kJ/kg	kJ/kgK			kJ/kg	kJ/kgK
	3445.3	7.090	?	?	3439.6	7.030

By interpolation, we get

$$h_1 = h_a + (h_b - h_a)\frac{(p_1 - p_a)}{(p_b - p_a)}$$

$$= 3445.3 + (3439.6 - 3445.3) \times \frac{(43 - 40)}{(45 - 40)}$$

$$= 3445.3 - 3.24 = 3442.06\,\text{kJ/kg}$$

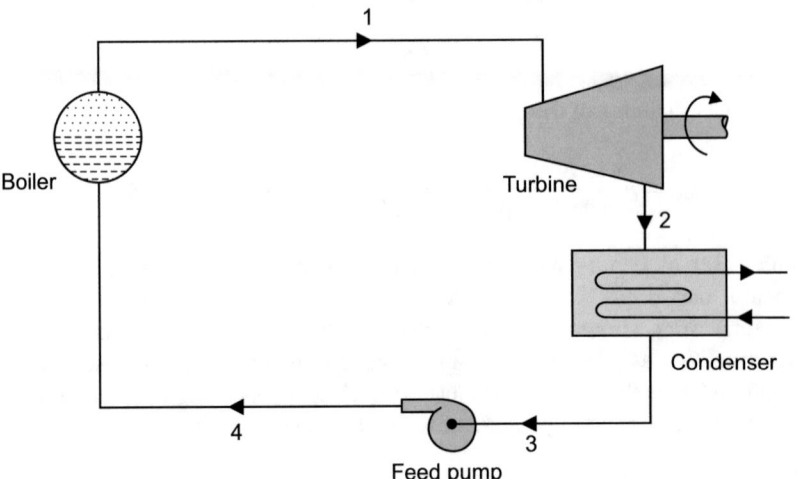

Fig. 2.32 Simple Rankine cycle

and

$$s_1 = s_a + (s_b - s_a)\frac{(p_1 - p_a)}{(p_b - p_a)}$$
$$= 7.09 + (7.03 - 7.07) \times \frac{(43 - 40)}{(45 - 40)}$$
$$= 7.09 - 0.024 = 7.066\,\text{kJ/kgK}$$
$$= s_2$$

From saturated steam table (pressure based) (Fig. 2.33),
At

$$p_c = p_2 = 0.035\,\text{bar}$$
$$s_g = 8.523\,\text{kJ/kgK}$$

As $s_2 < s_g$, it means that the steam is wet.
Other properties are

$$h_f = h_3 = 111.9\,\text{kJ/kg}$$
$$h_{fg} = 2438.4\,\text{kJ/kg}$$
$$s_f = 0.391\,\text{kJ/kgK}$$
$$s_{fg} = 8.132\,\text{kJ/kgK}$$
$$s_2 = s_f + x_2 s_{fg}$$

$$7.066 = .0.391 + x_2 + 8.132$$

or

Fig. 2.33 $T-s$ diagram for a simple Rankine cycle

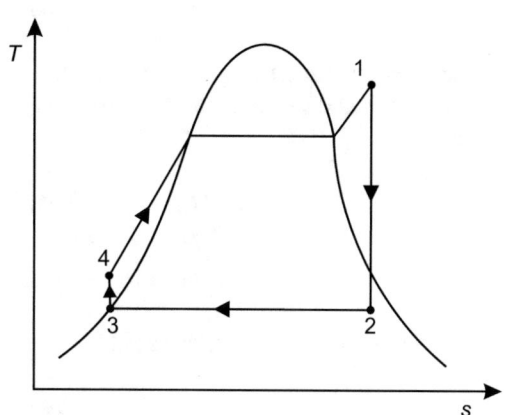

$$x_2 = 0.8208$$
$$h_2 = h_f + x_2 h_{fg}$$
$$= 111.9 + 0.8208 \times 2438.4 = 2113.33 \, \text{kJ/kg}$$
$$h_3 = h_4 = 111.9 \, \text{kJ/kg (pump work neglected)}$$

Heat supplied:

$$q_s = h_1 - h_4 = 3442.06 - 111.9 = \mathbf{3330.16 \, kJ/kg}$$

Net work output:

$$w_{net} = \text{turbine output work}$$
$$= w_T \quad \text{(pump work neglected)}$$
$$= h_1 - h_2$$
$$= 3442.06 - 2113.33 = \mathbf{1328.73 \, kJ/kg}$$

Thermal efficiency:

$$\eta_{th} = \frac{w_{net}}{q_s} = \frac{1328.73}{3330.16} = 0.3989 = \mathbf{39.89\%}$$

Case-II: Reheat cycle.
State 1 is same as Case-I

$$h_1 = 3442.06 \, \text{kJ/kg}$$
$$s_1 = 7.066 \, \text{kJ/kgK}$$

At high-pressure turbine (HPT) exit (Fig. 2.34):

$$s_2 = s_1 = 7.066 \, \text{kJ/kgK}$$

Steam is dry saturated (given condition).
From saturated steam table (pressure based),
At

$$s_2 = 7.066 \, \text{kJ/kgK}$$
$$p_2 = 2.3 \, \text{bar}$$
$$h_2 = h_g = 2713.1 \, \text{kJ/kg}$$

At inlet of low-pressure turbine (LPT):

$$p_3 = p_2 = 2.3 \, \text{bar}, \, T_3 = T_1 = 500 \, ^\circ\text{C}$$

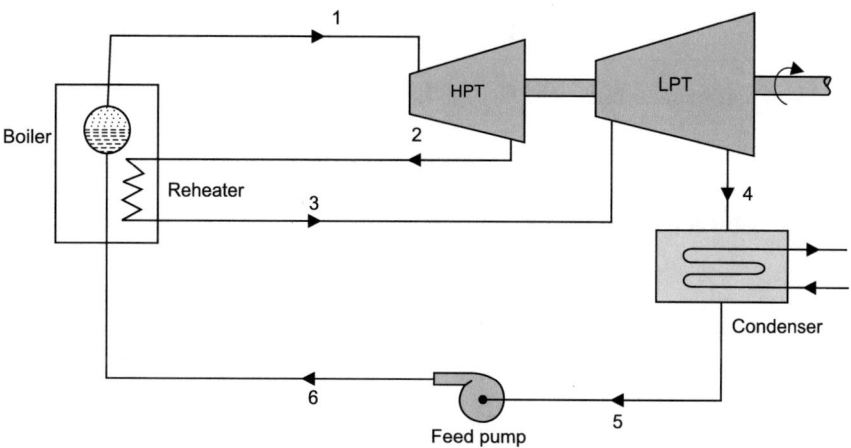

Fig. 2.34 Layout of the reheat cycle

From superheated steam table,
At

$$p_3 = 2.3 \, \text{bar}, \, T_1 = 500\,^{\circ}\text{C}$$

T_3 500 °C	$p_a = 2$ bar		$p_3 = 2.3$ bar		$p_b = 3$ bar	
	h_a	s_a	$h3$	$s3$	h_b	s_b
	kJ/kg	kJ/kgK			kJ/kg	kJ/kgK
	3487.1	8.513	?	?	3486	8.325

By interpolation, we get,

$$h_3 = h_a + (h_b - h_a)\frac{(p_3 - p_a)}{(p_b - p_a)}$$
$$= 3487.1 + (3486 - 3487.1) \times \frac{(2.3 - 2)}{(3 - 2)}$$
$$= 3487.1 - 0.33 = 3486.77 \, \text{kJ/kg}$$

and

$$s_3 = s_a + (s_b - s_a)\frac{(p_3 - p_a)}{(p_b - p_a)}$$
$$= 8.513 + (8.325 - 8.513) \times \frac{(2.3 - 2)}{(3 - 2)}$$
$$= 8.513 \text{ - } 0.056$$

$$= 8.457 \, \text{kJ/kgK} = s_4$$

At low-pressure turbine exit (Fig. 2.35):

$$p_4 = p_c = 0.035 \, \text{bar}$$

From saturated steam table (pressure based),
At

$$p_4 = 0.035 \, \text{bar}$$
$$s_g = 8.523 \, \text{kJ/kgK}$$

As $s_4 < s_g$, it means that the steam is wet.
Other properties are

$$h_f = h_5 = 111.9 \, \text{kJ/kg}$$
$$h_{fg} = 2438.4 \, \text{kJ/kg}$$
$$s_f = 0.391 \, \text{kJ/kgK}$$
$$s_{fg} = 8.132 \, \text{kJ/kgK}$$
$$s_4 = s_f + x_4 \, s_{fg}$$
$$8.457 = 0.391 + x_4 \times 8.132$$

or

$$x_4 = 0.9918$$
$$h_4 = h_f + x_4 \, h_{fg}$$
$$= 111.9 + 0.9918 \times 2438.4 = 2530.30 \, \text{kJ/kg}$$
$$h_6 = h_5 = 111.9 \, \text{kJ/kg}$$

Fig. 2.35 *T–s* diagram for a reheat cycle

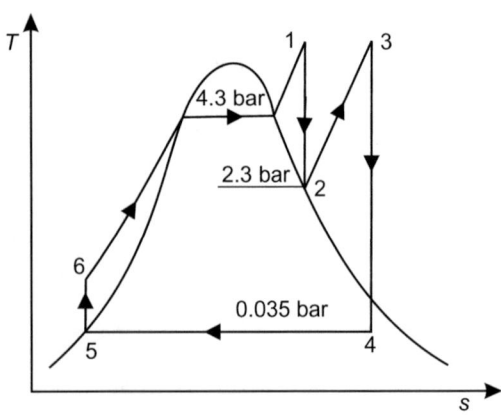

Net heat supplied:

$$q'_s = (h_1 - h_6) + (h_3 - h_2)$$
$$= (344.2.06 - 111.9) + (3486.77 - 2713.1)$$
$$= \mathbf{4103.83\,kJ/kg}$$

Net work output:

$$w'_{net} = w_{1-2} + w_{3-4} \quad \text{(pump work neglected)}$$
$$= (h_1 - h_2) + (h_3 - h_4)$$
$$= (3442.06 - 2713.1) + (3486.77 - 2530.30)$$
$$= \mathbf{1685.43\,kJ/kg}$$

Thermal efficiency:

$$\eta'_{th} = \frac{w'_{net}}{q'_s} = \frac{1685.43}{4103.83} = 0.4107 = \mathbf{41.07\%}$$

Increase in work output $= \dfrac{w'_{net} - w_{net}}{w_{net}} = \dfrac{1685.43 - 1328.73}{1328.73}$

$$= .2684 = \mathbf{26.84\%}$$

Increase in thermal efficiency $= \dfrac{\eta'_{th} - \eta_{th}}{\eta_{th}} = \dfrac{0.4107 - 0.3989}{0.3989} = 0.0295 = \mathbf{2.65\%}$

Problem 2.10: A steam power plant operates on an ideal regenerative Rankine cycle. Steam enters the turbine at 6 MPa and 450 °C and is condensed in the condenser at 20 kPa. Steam is extracted from the turbine at 0.4 MPa to heat the feedwater heater as a saturated liquid. Show the cycle on a $T-s$ diagram, and determine.

(a) the net work output per kg of steam flowing through the boiler.
(b) the thermal efficiency of the cycle.

Solution: Given data for an ideal regenerative Rankine cycle.
At turbine inlet:

$$p_1 = 6\,\text{MPa}$$
$$T_1 = 450\,°\text{C}$$

From saturated steam table (pressure based),
At

$$p_1 = 6\,\text{MPa} = 60\,\text{bar}$$
$$T_{sat} = 275.6\,°\text{C}$$

As $T_1 > T_{sat}$, it means that the steam is superheated.
From superheated steam table,
At

$$p_1 = 6\,\text{MPa} = 60\,\text{bar}, T_1 = 450\,°\text{C}$$
$$h_1 = 3301.8\,\text{kJ/kg}$$
$$s_1 = 6.719\,\text{kJ/kgK}$$
$$= s_2$$

At state 2, inlet to feedwater heater (FWH) (Fig. 2.36)

$$p_2 = 0.4\,\text{MPa} = 4\,\text{bar}$$

From saturated steam table (pressure based),
At

$$p_2 = 4\,\text{bar}$$
$$s_g = 6.896\,\text{kJ/kg}$$

As $s_1 > s_g$, it means the steam is wet at state 2.

$$v_f = 0.001084\,\text{m}^3/\text{kg}$$
$$= v_6$$

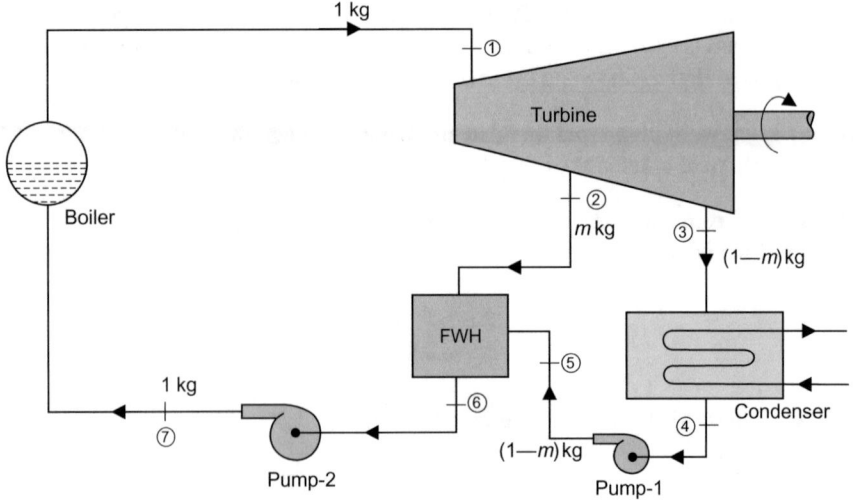

Fig. 2.36 Schematic for Problem 2.10

$$h_f = 604.7 \, \text{kJ/kg}, \, h_{fg} = 2133.8 \, \text{kJ/kg}$$
$$= h_6$$
$$s_f = 1.777 \, \text{kJ/kgK}, \, s_{fg} = 5.119 \, \text{kJ/kgK}$$
$$s_2 = s_1 = s_f + x_2 s_{fg}$$
$$6.719 = 1.777 + x_2 \times 5.119$$

or

$$x_2 = 0.9654$$
$$h_2 = h_f + x_2 h_{fg}$$
$$= 604.7 + 0.9654 \times 2133.8 = 2664.67 \, \text{kJ/kg}$$

From saturated steam table (pressure based),
At

$$p_3 = 20 \, \text{kPa} = 0.2 \, \text{bar}, \, s_3 = s_2 = 6.719 \, \text{kJ/kgK}$$

(Note that if steam is wet at state 2, it must be wet at state 3 also)

$$v_f = 0.001017 \, \text{m}^3/\text{kg}$$
$$= v_4$$
$$h_f = 251.4 \, \text{kJ/kg}, \, h_{fg} = 2358.3 \, \text{kJ/kg}$$
$$= h_4$$
$$s_f = 0.832 \, \text{kJ/kgK}, \, s_{fg} = 7.077 \, \text{kJ/kgK}$$
$$s_3 = s_f + x_3 s_{fg}$$
$$6.719 = 0.832 + x_3 \times 7.077$$

or

$$x_3 = 0.8318$$
$$h_3 = h_f + x_3 h_{fg}$$
$$= 251.4 + 0.8318 \times 2358.3 = 2213.03 \, \text{kJ/kg}$$

For pump-1 (Fig. 2.37):
Pump work:

$$w_{P1} = v_4(p_2 - p_3)\text{kJ/kg}$$

where p_2, p_3 are in kPa and v_4 is in m^3/kg

$\therefore$

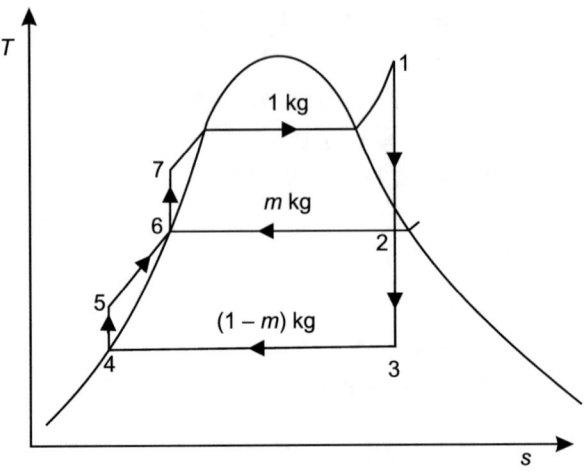

Fig. 2.37 T-s diagram for 2.10

$$w_{P1} = 0.001017(400 - 20)$$
$$= 0.3864 \text{ kJ/kg}$$

also

$$w_{P1} = h_5 - h_4$$

$\therefore$

$$0.3864 = h_5 - 251.4$$

or

$$h_5 = 251.78 \text{ kJ/kg}$$

For pump-2:
Pump work:

$$w_{P2} = v_6(p_1 - p_2) \text{ kJ/kg}$$

where v_6 are in m^3/kg and p_1, p_2 is in kPa.

$$= 0.001084 \, (6000 - 400)$$
$$= 6.07 \text{ kJ/kg}$$

also

$$w_{p2} = h_7 - h_6$$
$$6.07 = h_7 - 604.7$$

or

$$h_7 = 610.77 \, \text{kJ/kg}$$

Applying the energy balance equation on FWH

$$m\,h_2 + (1 - m)h_5 = 1\,h_6$$
$$m \times 2664.67 + (1 - m)251.78 = 604.7$$
$$2664.67\,m + 251.78 - 251.78\,m = 604.7$$
$$2412.89\,m = 352.92$$

or

$$m = 0.1462$$

Turbine work:

$$w_T = (h_1 - h_2) + (1 - m)(h_2 - h_3)$$
$$= (3301.8 - 2664.67) + (1 - 0.1462)(2664.67 - 2213.03)$$
$$= 637.13 + 385.61 = 1022.74 \, \text{kJ/kg}$$

Heat supplied:

$$q_{7-1} = h_1 - h_7 = 3301.8 - 610.77 = 2691.03 \, \text{kJ/kg}$$

(a) Net work output per kg of steam: w_{net}

$$w_{net} = w_T - w_{P1} - w_{P2}$$
$$= 1022.74 - 0.3864 - 6.07 = \mathbf{1016.28 \, kJ/kg}$$

(b) Thermal efficiency: η_{th}

$$\eta_{th} = \frac{w_{net}}{q_{7-1}} = \frac{1016.28}{2961.03} = 0.3776 = \mathbf{37.76\%}$$

Summary

1. The cycles which are used to convert heat into work is called the power cycle. If the working fluid in the cycle is gas, it is called gas power cycle. When the working fluid in the cycle is vapour, it is called vapour power cycle.
2. The main components of simple vapour power cycle are as follows:

    (i)   Turbine,
    (ii)   Condenser,
    (iii) Feed pump,
    (iv)   Boiler.

3. The simplest vapour power cycle is called the Rankine cycle.
4. The thermal efficiency of the Rankine cycle:

$$\eta_{th} = \frac{net\, work\, output}{heat\, supplied} = \frac{w_{net}}{q_s}$$

    where

$$w_{net} = w_T - w_P$$

5. **Work Ratio.** It is defined as the ratio of the net work output to the turbine work output.

$$Work\, ratio = \frac{w_{net}}{q_s}$$

6. The efficiency of the Rankine cycle can be improved by.

    (i)   increasing the boiler pressure,
    (ii)   increasing temperature of the steam at inlet of the turbine,
    (iii) decreasing the condenser pressure.

7. **Reheat Cycle.** In the reheat cycle, steam is reheated after expanding partially in the high-pressure turbine. This is done by the steam partial expansion in the high-pressure turbine and sent back to the boiler where it is reheated at constant pressure, usually to the inlet temperature of the high-pressure turbine.
    Advantages of a reheat cycle:

    (i)   It increases the thermal efficiency of the cycle when the cycle works at optimum values of the boiler pressure and the reheat pressure.
    (ii)   It increases the net work output.
    (iii) It decreases the moisture content of the steam at the low-pressure turbine exit and reducing the turbine blade erosion.

8. **Regenerative Cycle**. In this cycle, the feed water is preheated before it enters the boiler by extracting some of the steam as it expands in the turbine and mixing it with the water as it exits the condensate pump.

   Advantages of a regenerative cycle:

   (i)   It increases the thermal efficiency of the cycle.
   (ii)   Water particles escape with the bled steam through extraction belts. This reduces the moisture content in the steam, thus decreasing turbine blade erosion.
   (iii) Feed heating by bled steam decreases the quantity of steam reaching the condenser. Thus, in turn, reduces the size of the condenser and the cooling water requirements.
   (iv)   Steam extraction for feed heating reduces the flow rate significantly in the low-pressure stages, thus allowing the blades to be shorter. This is a great advantage in large units where designing long turbine blades is very critical.

9. **Combined Gas-vapour Power Cycle**. The combined cycle means that both gas turbine cycle and vapour power cycle are combined together. The exhaust of the gas turbine at high temperature is supplied to the boiler to raise steam which can be utilized for power generation in a steam turbine.

   Advantages of the combined cycle plants:

   (i)   High thermal efficiency.
   (ii)   High total power.
   (iii) Lower cooling water requirements. Gas turbine needs no cooling. Only cooling water required is for the condenser.
   (iv)   Lower specific investment cost.

Assignment-1

1. What are the basic components of a vapour power plant?
2. What is Rankine cycle?
3. Sketch a Carnot cycle for the steam power plant. Why is Carnot cycle not used as thermodynamic cycle for the steam power plant?
4. What is the mean temperature of heat supplied?
5. Discuss the limitations of maximum and minimum temperatures in the vapour power cycle.
6. Explain the effect of the following parameters on the performance of Rankine cycle:

    (i)   temperature of steam at inlet to the turbine,
    (ii)   pressure of steam at inlet to the turbine,
    (iii) pressure of steam at exit from the turbine.

7.   Explain reheat cycle and compare it with simple Rankine cycle.
8.   Explain how the quality of steam at exit from the turbine gets restricted.
9.   Explain regenerative cycle and compare it with simple Rankine cycle.
10.   Show that the Rankine cycle efficiency measures with increase in temperature and pressure of steam in turbine inlet, decrease in condenser pressure, reheating and regeneration.
11.   Explain regenerative cycle using layout diagrams and $T-s$ chart. How does the regeneration improve the cycle efficiency over a sample Rankine cycle? With the same initial conditions and condenser pressure?
12.   What is the effect of reheat on

(i)   the specific output, and
(ii)   the cycle efficiency of a steam power plant?

13.   Draw the $T-s$ diagram of Rankine cycle using dry saturated steam and develop the equation for Rankine cycle efficiency?
14.   Explain with help of neat diagram a regeneration steam power cycle and state advantages of this cycle.
15.   Why is the employment of more than one reheaters not popular?
16.   In combined gas-vapour cycles, what is the energy source for the steam?
17.   Why is the combined gas-vapour cycle more efficient than either of the cycle operated alone?
18.   For improvement in thermal efficiency in steam power plants, regenerative cycle is preferred over reheat cycle. Give the thermodynamic reasoning for this.

Assignment-2

1.   For a vapour power plant, the following observation was made.
          Steam supply condition: 60 bar, 450 °C.
          Condenser pressure: 0.1 bar.
          Steam flow rate: 5000 kg/hr.
          Determine the following:

(i)   Turbine work,
(ii)   Percentage of pump work compared to turbine work,
(iii) Heat addition in the boiler,
(iv)   Heat rejection in the condenser,
(v)   Thermal efficiency.

[**Ans.** (i) 1630.20 kW, (ii) 0.51%, (iii) 4307.77 kW, (iv) 2689.23 kW, (v) 37.64%].

2.   A steam turbine plant operates on Rankine cycle with steam entering turbine at 40 bar, 350 °C and leaving at 0.05 bar. Determine the net work per kg of steam and the thermal efficiency of the cycle.

[**Ans.** 1081.74 kJ/kg, 36.66%].

3. In a steam turbine installation running on ideal Rankine cycle steam leaves the boiler at 10 MPa and 700 °C and leaves turbine at 0.005 MPa. For the 50 MW output of the plant and cooling water entering and leaving condenser at 15 °C and 30 °C, respectively, determine.

    (i)   the mass flow rate of steam in kg/s,
    (ii)   the mass flow rate of condenser cooling water in kg/s,
    (iii)   the thermal efficiency of cycle.

[**Ans.** (i) 29.69 kg/s, (ii) 969.80 kg/s, (iii) 45.12%].

4. A vapour power plant operates on Rankine cycle with 30 bar and 400 °C at the turbine inlet and 10 kPa at the turbine exit. The isentropic efficiency of the turbine is 85% and the pump is 80%. Determine.

    (i)   the thermal efficiency of the cycle, and
    (ii)   the mass flow rate of the steam if the power output of the cycle is 20 MW.

[**Ans.** (i) 28.96%, (ii) 22.75 kg/s].

5. Steam enters the turbine of a 250 MW vapour power plant operating on a simple ideal Rankine cycle at 15 MPa and 400 °C and is cooled in the condense at 20 kPa. Determine the quality of the steam at the turbine exit, the thermal efficiency of the cycle, and the mass flow rate of the steam.
    [**Ans.** 71.34%, 37.89%, 243.55 kg/s].

6. Consider a steam power plant operating on a Rankine cycle. The steam enters the turbine at 2 MPa and 350° and leaves at 5 kPa. The turbine and pump each have an efficiency of 80%. Determine the net work output per kg of steam and the cycle efficiency. Also, find the percentage increase in net work output and the cycle efficiency if the cycle is assumed to be ideal.
    [**Ans.** 810.91 kJ/kg, 27.12%, 25.06%, 24.74%].

7. A steam power plant operating on a Rankine cycle has a pressure of 10 MPa and a temperature of 500 °C at the boiler exit and a temperature of 35 °C at the turbine exit. The isentropic efficiency of the turbine is 90% and the isentropic efficiency of the pump is 60%. If the power plant generates 100 MW, determine the cycle efficiency and the steam flow rate.
    [**Ans.** 37.4%, 83.44 kg/s].

8. A regenerative steam cycle operates between pressure limits 3.5 MPa and 15 kPa. The temperature of the steam entering the turbine is 400 °C. Steam extraction occurs 800 kPa and an open feed water heater is employed. Calculate the thermal efficiency of the cycle assuming isentropic expansion in the turbine.
    [**Ans.** 35.28%].

9. A regenerative Rankine cycle has steam entering turbine at 200 bar, 650 °C and leaving at 0.05 bar. Considering feed water heaters to be of open type determine thermal efficiency for the following conditions:

      (a)   there is no feed water heater, and

      (b)   there is only one feed water heater working at 8 bar.

      [**Ans**. (a) 46.18% (b) 49.76%].

10.   A reheat cycle has steam generated at 50 bar, 500 °C for being sent to high-pressure turbine and expanded up to 5 bar before supplied to low-pressure turbine. Steam enters at 5 bar, 400 °C into low-pressure turbine after being reheated in boiler. Steam finally enters condenser at 0.05 bar. Determine the thermal efficiency and work ratio.

      [**Ans**. 45.74%, 0.9967].

11.   Steam expands from 150 bar, 500 °C to 0.06 bar in a steam turbine plant. Compare the thermal efficiencies, heat supplied, and dryness fractions of the exhaust steam for reheat pressure of 70 bar and 25 bar. The steam is reheated to the inlet temperature of the high-pressure turbine. Ignore the feed pump work.

      [**Ans**. (i) 42.94%, 0.7456 (without reheat), (ii) 44.13%, 0.8038 (with reheat at 70 bar), (iii) 44.20%, 0.8711 (with reheat at 25 bar)].

Chapter 3
Introduction to Working of IC Engines

Nomenclature

The following is a list of the nomenclature in this chapter:

IC	–	Internal combustion
EC	–	External combustion
TDC	–	Top dead centre
BDC	–	Bottom dead centre
l	m	Stroke
r	m	Radius of crank
A	m²	Piston area
V_s	m³	Swept or displacement volume
V	m³	Cylinder volume
V_c	m³	Clearance volume
r	–	Compression ratio
SI	–	Spark ignition
CI	–	Compression ignition
ip	kW	Indicated power
bp	kW	Brake power
fp	kW	Friction power
mep	kPa	Mean effective pressure
η_{ith}	–	Indicated thermal efficiency
η_{bth}	–	Brake thermal efficiency
η_{ret}	–	Relative efficiency
η_v	–	Volumetric efficiency
sfc	kg/kWh	Specific fuel consumption

(continued)

© The Author(s) 2022
S. Kumar, *Thermal Engineering Volume 2*,
https://doi.org/10.1007/978-3-030-89216-6_3

n	rpm	Speed
N	rpm	Speed of the crank shaft
x	–	Number of cylinder
T	Nm	Torque
S	N	Spring balance reading
R	m	Radius of brake drum
r_p	m	Radius of rope
R_e	m	Efflective radius of brake drum
η_m	–	Mechanical efficiency
CV	kJ/kg	Calorific value of fuel
m_f	kg/s	Mass flow rate of fuel
W_{net}	kJ	Net work produced per cycle
a	m^2	Area of the actual indicator diagram
l	m	Base length of the indicator diagram
k	N/m^3	Spring constant
η_{Otto}	–	Air-standard efficiency of Otto cycle
η_{Diesel}	–	Air-standard efficiency of Diesel cycle
γ	–	Adiabatic index
r_c	–	Cut-off ratio
c_{pw}	kJ/kgK	Specific heat of water
c_{pg}	kJ/kgK	Specific heat of exhaust gases.
A/F	–	Air–fuel ratio

3.1   Heat Engine

Heat engine is defined as a device that converts heat energy into mechanical work. It can be broadly classified into two categories:

(i)   Internal combustion engines (IC engine),
(ii)   External combustion engines (EC engines) (Fig. 3.1).

(i)   **Internal Combustion Engine**: It is a heat engine that converts the chemical energy of fuel (liquid or gas) into heat energy due to the combustion of fuel that takes place inside the engine cylinder and further a part of this heat energy converted into useful work. In other words, it converts the chemical energy of the fuel into mechanical work due to the combustion of fuel that takes place inside to engine cylinder. Internal combustion engine is more popularly known as IC engine. Petrol engine, diesel engine, gas engine, wankel engine, and open-cycle gas turbine are examples of internal combustion engines.

(ii)   **External Combustion Engine**: It is a heat engine that converts the heat energy into mechanical work. The conversion of chemical energy of fuel (solid, liquid,

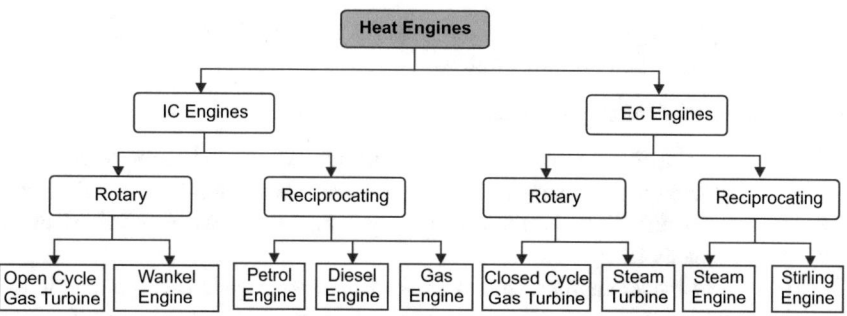

Fig. 3.1 Classification of heat engine

or gas) into heat energy takes place outside the engine. Hence it is called an external combustion engine. Steam engine, steam turbine, stirling engine, and closed-cycle gas turbine are examples of external combustion engines.

3.2 Classification of IC Engines

IC engines are classified based on the following:

1. **Working cycles**.

    (i)   Otto cycle engine (constant volume cycle). Most of the petrol and gas engines work on this cycle.
    (ii) Diesel cycle engine. Most of the low-speed diesel engines work on this cycle.
    (iii) Dual combustion cycle engine. Most of the high-speed diesel engines work on this cycle.

2. **Types of fuel used**.

    (i)   Petrol (or gasoline) engine,
    (ii) Diesel engine,
    (iii) Gas engine,
    (iv) Bi-fuel engine.

3. **Ignition system**.

    (i)   Spark ignition (SI) engines. In this type of engines, electric spark is used to ignite the fuel–air mixture. Most of the engines using petrol and gas as fuel belong to this category.
    (ii) Compression ignition (CI) engines. In this type of engines, air is compressed to a very high temperature and pressure and then fuel is injected into it in the form of a spray. The fuel gets ignited due to the

high temperature of the compressed air. The diesel engines belong to compression ignition engines.

4. **Number of strokes per cycle**.

(i)   Four-stroke engine. In four-stroke engine, one cycle is completed in four strokes of the piston in the cylinder. In other words, one power stroke is obtained in four strokes of the piston, i.e., in two revolutions of the crankshaft.

(ii) Two-stroke engines. In two-stroke engine, one cycle is completed in two strokes of the piston in the cylinder. In other words, one power stroke is obtained in two strokes of the piston, i.e., in one revolution of the crankshaft.

5. **Speed of engines**.

(i)   Low speed engines (up to 400 rpm),
(ii)   Medium speed engines (400–1200 rpm),
(iii) High speed engines (above 1200 rpm).

6. **Application of engines**.

(i)   Automotive. Bus, truck, car, bike, etc.
(ii)   Locomotive,
(iii) Light aircraft,
(iv)   Marine. Ship, boat, etc.,
(v)   Power generation,
(vi)   Agricultural. Tractor, pump etc.,
(vii) Earthmoving. Dumpers, mining equipment, etc.

7. **Number of cylinder**.

(i)   Single-cylinder engine,
(ii) Multi-cylinder engine.

8. **Cylinder arrangement**.

(i)   Vertical engine,
(ii)   Horizontal engine,
(iii) V-engine,
(iv)   In-line engine,
(v)   Opposed cylinder engine,
(vi)   Radial engine.

9. **Cooling system**.

(i)   Air-cooled engine,
(ii) Water-cooled engine.

3.3 Nomenclature for Reciprocating Engines

1. **Bore**. The inner diameter of the cylinder is called the bore.
2. **Dead Centres**. The piston reciprocates in the cylinder between two fixed positions called dead centres. The name 'dead' means the speed of the piston is zero at these positions. At dead centre piston changes its direction.

    (i)   *Top Dead Centre* (*TDC*). The position of the piston towards the cylinder head when it forms the smallest volume in the cylinder is called the top dead centre. For horizontal engines, this position is called the inner dead centre (IDC).

    (ii) *Bottom Dead Centre* (*BDC*). The position of the piston towards the crank end side when it forms the largest volume in the cylinder is called the bottom dead centre (BDC). For horizontal engines, this position is called the outer dead centre (ODC) (Figs. 3.2 and 3.3).

Fig. 3.2 Nomenclature for reciprocating engine

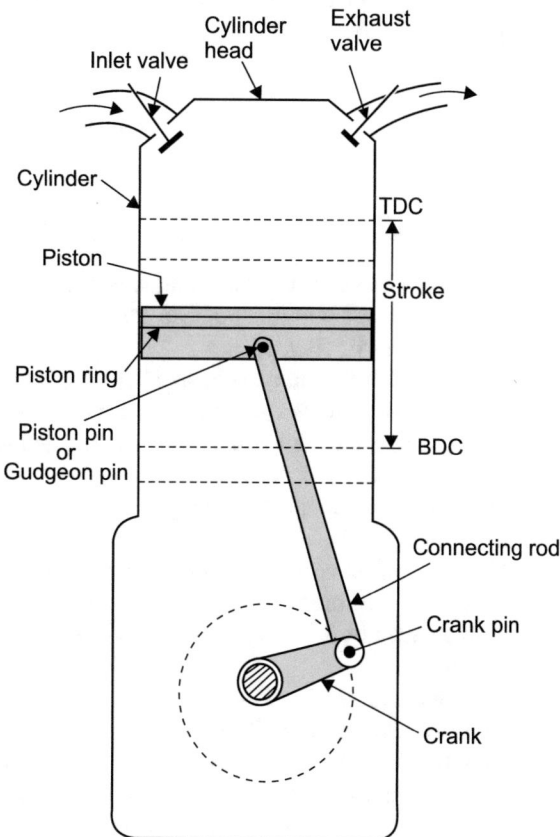

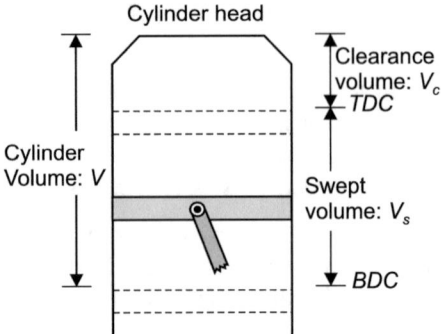

Fig. 3.3 Clearance volume and swept or displacement volume

3. **Stroke.** The distance travelled by the piston from top dead centre to bottom dead centre or bottom dead centre to top dead centre is called the stroke of the engine. It is equal to twice the radius of the crank. It is denoted by l.
    Mathematically,
    Stroke:

$$l = 2r,$$

    where

    r = radius of the crank.

4. **Clearance volume.** V_c. The volume between the piston and the cylinder head when the piston is at the top dead centre (TDC) is called clearance volume. It is denoted by V_c (Fig. 3.4).

5. **Swept or displacement volume (V_s).** The volume displaced by the piston as it moves between the top dead centre and the bottom dead centre is called the swept or displacement volume. It is simply defined as the volume between the TDC and BDC. It is denoted by V_s. Mathematically, it is equal to the product of the piston area and stroke length.
    Swept volume: $V_s = Al$.

6. **Cylinder volume:** V. The volume between the BDC and cylinder head when the piston is at the bottom dead centre (BDC) is called the cylinder volume. Mathematically, it is equal to the sum of the clearance volume (V_c) and swept volume (V_s).
    Cylinder volume: $V = V_c + V_s$.

7. **Compression ratio:** r. The compression ratio is a volume ratio. It is defined as the ratio of the volume in the cylinder when the piston is at BDC to the volume in the cylinder when the piston is at TDC. In other words, it is defined as the ratio of the maximum volume in the cylinder to the clearance volume. It is denoted by r.
    Mathematically,

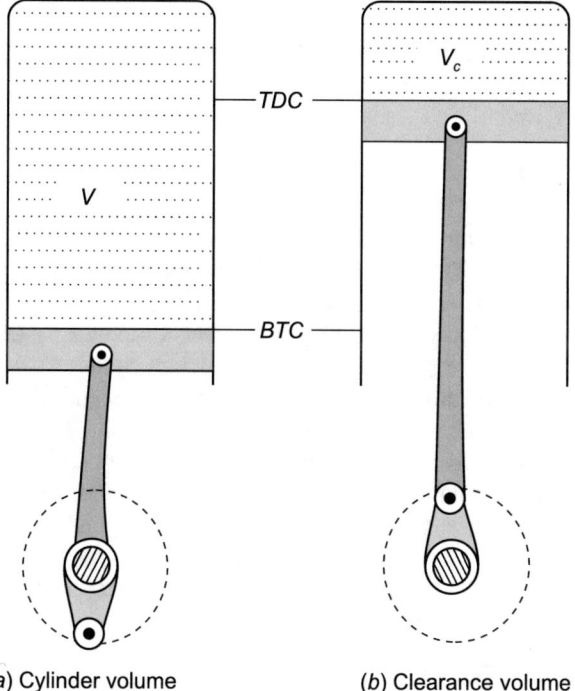

(a) Cylinder volume (b) Clearance volume

Fig. 3.4 Compression ratio

Compression ratio:

$$r = \frac{V}{V_c}$$

where

$$V = V_c + V_s$$
$$r = \frac{V_c + V_s}{V_c}$$
$$r = 1 + \frac{V_s}{V_c}.$$

3.4 Working of Four-Stroke Petrol (SI) Engine

Petrol engine works on the principle of Otto cycle, also known as constant volume cycle. In this type of engine, one cycle is completed in four strokes of the piston or two revolutions of the crankshaft. The cycle of the four-stroke spark-ignition (SI) engine consists of the following four strokes:

(i)   Suction stroke,
(ii)   Compression stroke,
(iii)   Expansion or power stroke,
(iv)   Exhaust stroke.

(i)   Suction Stroke

During suction stroke, the inlet valve is opened and the exhaust valve is closed. The piston moves from the TDC to the BDC. The crankshaft revolves by half rotation. The energy required to perform this stroke is supplied by 'cranking' in the first cycle at the time of starting but while running the flywheel supplies the energy stored by it during the expansion or power stroke of the previous cycle.

At the beginning of this stroke, the inlet valve is open and the pressure in the cylinder will be atmospheric. As the piston moves from the TDC to the BDC, the volume in the cylinder increases and the pressure decreases. This sets up a pressure difference between the atmosphere and inside of the cylinder. Due to this pressure difference, the petrol–air mixture will be drawn into the cylinder through the carburetor. At the end of this stroke, the cylinder will be filled completely with petrol–air mixture and the inlet valve is closed. The suction stroke is represented by 0–1 on the p–v diagram as shown in Fig. 3.5e.

(ii)   Compression Stroke

During this stroke, both the inlet and exhaust valves are closed. The piston moves from the BDC to the TDC. The crankshaft revolves by half rotation. The energy required to perform this stroke is supplied by 'cranking' in the first cycle at the time of starting but while running the flywheel supplies the energy stored by it during the expansion or power stroke of the previous cycle.

As this stroke is performed, the petrol-air mixture contained in the cylinder will be compressed. The compression ratio ranges from 6: 1 to 12: 1. This process of compression is a reversible adiabatic represented by the curve 1–2 on the p–V diagram as shown in Fig. 3.5e.

At the end of the stroke at state 2, the petrol–air mixture is ignited by the electric spark given out by the spark plug. The temperature of the spark is more than 40,000 °C and the ignition temperature of the petrol lies in the range of 246–280 °C. Since the ignition of the petrol is by the spark given out by the spark plug, this type of engine is called spark ignition engine abbreviated as SI engine. The combustion of petrol release hot gases, which increase the temperature (about 2000 °C) and pressure at constant volume. This constant volume combustion process is represented by process 2–3 on the p–v diagram as shown in Fig. 3.5e.

(iii) Expansion or Power Stroke

During this stroke, both the inlet and exhaust valves are closed. The piston moves from the TDC to the BDC. The crank shaft revolves by half rotation. The high-pressure

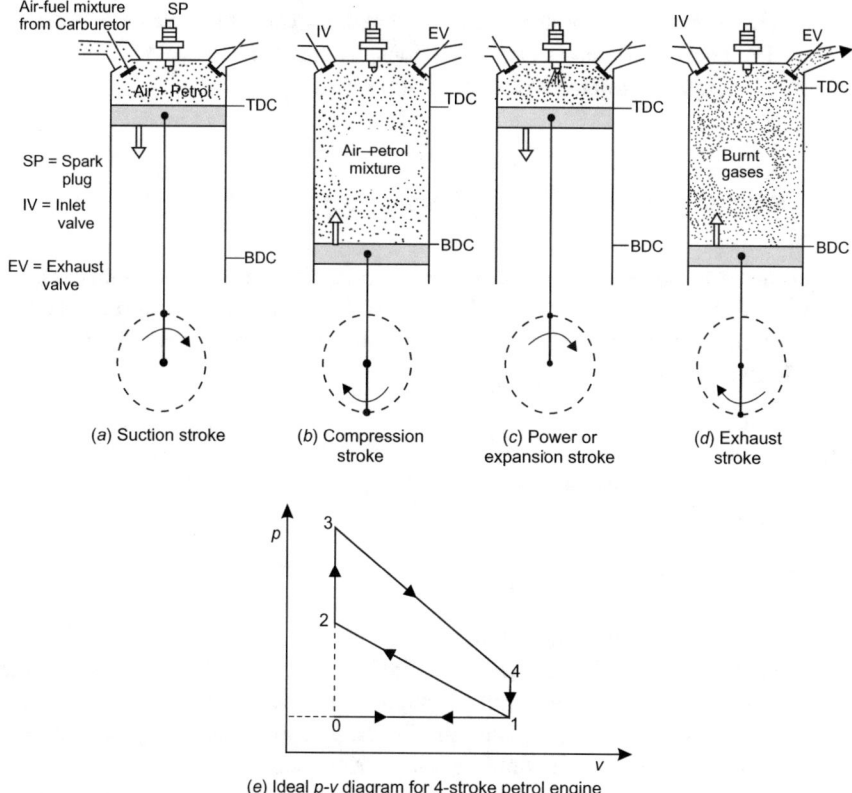

(a) Suction stroke   (b) Compression   (c) Power or   (d) Exhaust
                              stroke   expansion stroke   stroke

(e) Ideal p-v diagram for 4-stroke petrol engine

Fig. 3.5 Working of 4-stroke petrol (SI) engine

burnt gases force the piston to perform this stroke, called expansion or working or power stroke. As the piston moves, the pressure of the hot gases gradually decreases. The process of expansion is a reversible adiabatic represented by the curve 3–4 on the p–v diagram as shown in Fig. 3.5e.

At the end of this stroke, the exhaust valve opens which will release the burnt gases to the atmosphere. This will suddenly bring down the pressure in the cylinder to that of atmosphere. This drop in pressure at constant volume is represented by process 4–1 on the p–v diagram as shown in Fig. 3.5e.

(iv)   Exhaust Stroke

During this stroke, the exhaust valve is opened and the inlet valve is closed. The piston moves from the BDC to the TDC. The crank shaft revolves by half rotation. The energy required to perform this stroke is supplied by the flywheel from the energy stored by it during the previous stroke.

As the piston moves from the BDC to the TDC, the burnt gases will be expelled out of the cylinder at atmospheric pressure. This process is represented by line 1–0 on p–v diagram as shown in Fig. 3.5e.

This engine requires four strokes to complete one cycle, it is called a four-stroke engine. The crank shaft makes two revolutions to complete one cycle. The power is developed in every alternate revolution of the crank shaft.

3.5 Working of Four-Stroke Diesel (CI) Engine

Diesel engine works on the principle of diesel cycle, also known as constant pressure cycle. In this type of engine, one cycle is completed in four strokes of the piston or two revolutions of the crank shaft. The cycle of the four-stroke diesel engine consists of the following four strokes:

(i)   Suction stroke,
(ii)   Compression stroke,   ·
(iii)   Expansion or power stroke,
(iv)   Exhaust stroke.

(i)   **Suction Stroke**

During suction stroke, the inlet valve is opened and the exhaust valve is closed. The piston moves from the TDC to BDC. The crank shaft revolves by half rotation. The energy required to perform this stroke is supplied by 'cranking' in the first cycle at the time of starting but while running the flywheel supplies the energy stored by it during the expansion airpower stroke of the previous cycle.

At the beginning of this stroke, the inlet valve is open and the pressure in the cylinder will be atmospheric. As the piston moves from the TDC to the BDC, the volume in the cylinder increases and the pressure decreases. This sets up a pressure difference between the atmosphere and inside the cylinder. Due to this pressure difference, the atmospheric air will be drawn into the cylinder through the air filter. At the end of this stroke (i.e., at BDC), the cylinder will be filled completely with air and the inlet valve is closed. The suction stroke is represented by 0–1 on the p–v diagram as shown in Fig. 3.6e.

(ii)   **Compression Stroke**

During this stroke, both the inlet and exhaust valves are closed. The piston moves from the BDC to TDC. The crank shaft revolves by half rotation (i.e., 180°). The energy required to perform this stroke is supplied by 'cranking' in the first cycle at the time of starting but while running. The flywheel supplies the energy stored by it during the expansion or power stroke of the previous cycle.

As this stroke is performed, the air contained in the cylinder will be compressed. The compression ratio ranges from 14:1 to 22:1. This process of compression is a reversible adiabatic represented by the curve 1–2 on the p–V diagram as shown in Fig. 3.6e.

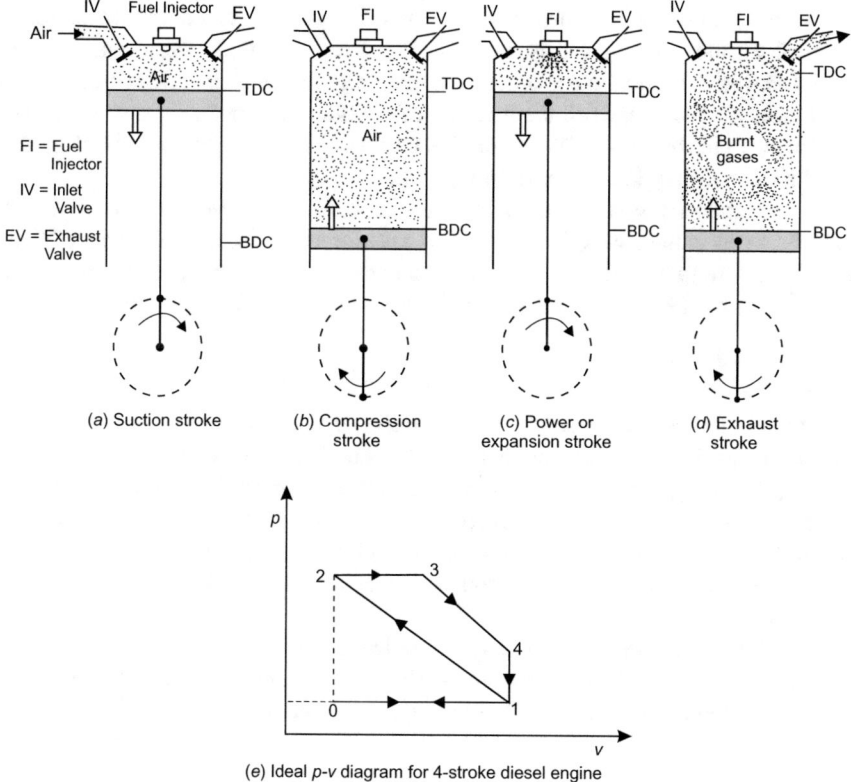

(a) Suction stroke   (b) Compression   (c) Power or   (d) Exhaust
                            stroke   expansion stroke   stroke

(e) Ideal p-v diagram for 4-stroke diesel engine

Fig. 3.6 Working of 4-stroke diesel (CI) engine

The compression ratio in the diesel engine is higher than the petrol engine. The high compression of the air will increase its pressure (about 50–60 bar) and temperature (about 500–700 °C). At the end of this state at state 2, the air will have attained a temperature greater than the ignition temperature (210 °C) of the diesel. The diesel oil is sprayed into the cylinder by the injector. The high temperature of the air ignites the diesel oil as soon as it is sprayed.

Since the compressed air ignites the diesel oil, this type of engine is called compression ignition engine abbreviated as CI engine.

(iii) Expansion or Power Stroke

During this stroke, both the inlet valve and the exhaust valve are closed. The piston moves from the TDC to the BDC. The crank shaft revolves by half rotation (180°). The self-ignition of the diesel oil starts the combustion as a result, the hot gases are released. The burnt gases released by the combustion of the diesel oil which is continuously inserted into the cylinder, force the piston to perform the earlier part of this stroke at constant pressure till the injection of the diesel oil stops. The constant pressure expansion with simultaneous combustion is represented by the horizontal line 2–3 on the p–v diagram as shown in Fig. 3.6e. The high-pressure burnt gases exert force on the piston to perform this stroke, called expansion or working or power stroke. As the piston moves, the pressure of the hot gases gradually decreases. The expansion of the burnt gases is a reversible adiabatic process represented by the curve 3–4 on the p–v diagram as shown in Fig. 3.6e.

At the end of the stroke at state 4, the exhaust valve opens and releases the exhaust gases to the atmosphere. This will affect the sudden drop of pressure inside the cylinder to that of the atmosphere. The constant volume release of hot gases is represented by the vertical line 4–1 on the p–v diagram as shown in Fig. 3.6e.

(iv)   **Exhaust Stroke**

During this stroke, the exhaust valve is opened and the inlet valve is closed. The piston moves from the BDC to the TDC. The crank shaft revolves by half rotation (180°). The energy required to perform this stroke is supplied by the flywheel from the energy stored by it during the previous stroke. As the piston performs this stroke, the exhaust gases will be expelled out of the cylinder at atmospheric pressure. This process is represented by the horizontal line 1–0 on the p–v diagram as shown in Fig. 3.6e.

This engine requires four strokes to complete one cycle, it is called a four-stroke engine. The crank shaft makes two revolutions to complete one cycle. The power is developed in every alternate revolution of the crank shaft.

3.6   Working of Two-Stroke Petrol Engine

Two-stroke engine performs only two strokes to complete one working cycle. That is, two strokes out of the four strokes of a working cycle have to be eliminated. The two strokes that are eliminated, the suction stroke, and the exhaust stroke. The suction and the exhaust strokes are combined with compression and the working strokes, respectively. Two-stroke petrol engine also works on the principle of the Otto cycle. Figure 3.7 shows the working of the two-stroke petrol engine.

(i)   First Stroke: Working and exhaust stroke
(ii)   Second stroke: Compression and suction stroke.

(i)   **First Stroke**

At the beginning of the first stroke, the piston is in the highest position as shown in Fig. 3.7a. The spark plug ignites the compressed petrol-air mixture. The combustion of petrol will release hot gases which increases the pressure in the cylinder. The high-pressure burned gases thrust the piston downwards. The piston performs the power stroke till it uncovers the exhaust port as shown in Fig. 3.7b. The burned gases which are still at high pressure escape through the exhaust port. As soon as the top edge of

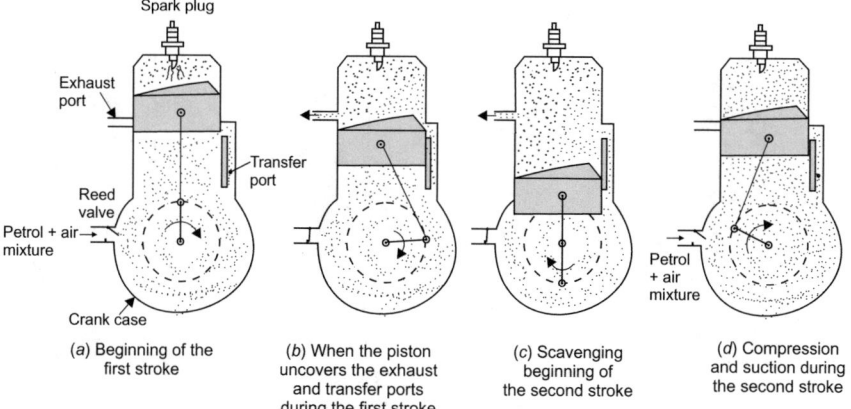

Fig. 3.7 Working of a two-stroke petrol engine

the piston uncovers the transfer port, the petrol-air mixture from the crankcase enters into the cylinder. The petrol-air mixture enters the cylinder and drives most of the remaining exhaust gases out of the cylinder. This process is called scavenging.

(ii) Second Stroke

In this stroke, the piston moves upward and the pressure decreases in the crankcase. This sets up a pressure difference between the atmosphere and inside the crankcase. Due to this pressure difference, the reed valve (check valve or one-way valve) is opened and the fresh petrol-air mixture enters into the crankcase through the carburetor. On the other hand, the petrol-air mixture will continue to drive out the burned gases till the piston cover both the exhaust and the transfer ports. The piston compresses the petrol-air mixture in the cylinder as shown in Fig. 3.7d. The compression ratio ranges from 6:1 to 12:1. After the piston reaches the top, the cycle is completed and the first strokes repeat.

This engine requires only two strokes to complete one cycle, hence it is called a two-stroke engine. The crankshaft makes only one revolution to complete the cycle. The power is developed in every revolution of the crankshaft.

3.7 Working of Two-Stroke Diesel Engine

Two-stroke diesel engines also work on the principle of diesel cycle, Fig. 3.8 shows the working of the two-stroke diesel engine.

(i) First Stroke

At the beginning of the first stroke, the piston is in the highest position (at TDC) as shown in Fig. 3.8a. The injector injects the diesel oil into the cylinder with high

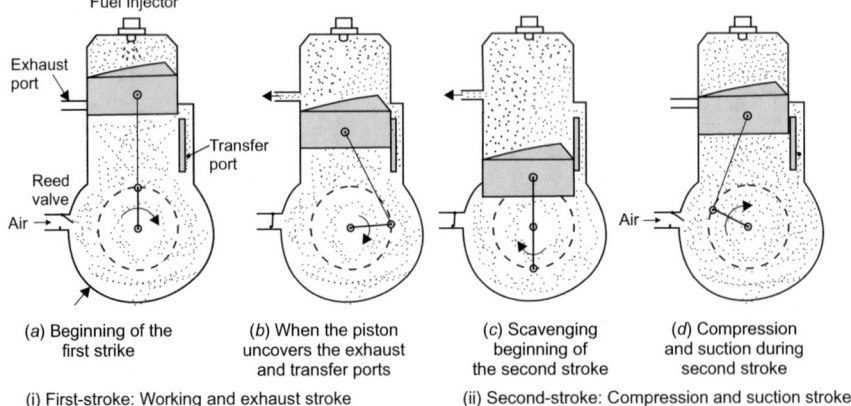

(a) Beginning of the   (b) When the piston   (c) Scavenging   (d) Compression
    first strike   uncovers the exhaust   beginning of   and suction during
               and transfer ports   the second stroke   second stroke

(i) First-stroke: Working and exhaust stroke   (ii) Second-stroke: Compression and suction stroke

Fig. 3.8 Working of two-stroke diesel engine

pressure as fine sprays. High temperature of air inside the cylinder is sufficient to ignite the diesel. The combustion of the diesel oil will release hot gases which increases the pressure in the cylinder. The high-pressure burned gases thrust the piston downwards. The piston performs the power stroke till it uncovers the exhaust port as shown in Fig. 3.8b. The burned gases which are still at high pressure escape through the exhaust port. As soon as the top edge of the piston uncovers the transfer port, the airflow from the crankcase into the cylinder. The air entering the cylinder through the transfer port drives out the burned gases as shown in Fig. 3.8c. This driving out of the burned gases is called scavenging.

(ii)   **Second Stroke**

In this stroke, the piston moves upward. When it covers the transfer and exhaust ports, the air supply from the crankcase and the scavenging will stop. The piston compresses the air in the cylinder as shown in Fig. 3.8d. The compression ratio ranges from 14:1 to 22:1. In the crankcase, the pressure decreases as soon as the piston moves upward. This sets up a pressure difference between the atmosphere and inside the crankcase. Due to this pressure difference, the reed valve is opened and the fresh air enters into the crankcase through the air filter. After the piston reaches the top dead centre and the engine is ready to repeat the cycle of operations.

3.8   Difference Between Petrol Engine and Diesel Engine

The following are the main difference between the petrol engine and diesel engine:

S. no.	Petrol engine	Diesel engine
1.	Petrol engine works on the Otto cycle	Diesel engine works on Diesel or Dual cycle
2.	The carburetor is used to supply the homogeneous mixture of the petrol and air during the suction stroke	No need for the carburetor in a diesel engine. In diesel engine, the fuel injector is used to supply the diesel oil into the cylinder
3.	The petrol-air mixture enters into the cylinder during the suction stroke	Only the air enters into the cylinder during the suction stroke
4.	Petrol engine works in low-compression ratio of the range of 6:1 to 12:1	Diesel engine works in high-compression ratio of the range of 14:1 to 22:1
5.	Lower thermal efficiency (about 25%) because it works in low-compression ratio	Higher thermal efficiency (about 35–40%) because it works in high-compression ratio
6.	Lower weight (about 0.5–3 kg/kW) per unit power output. Because of low-compression ratio, the engine cylinder undergoes less pressure	Higher weight (about 2–10 kg/kW) per unit power output. Because of high-compression ratio, the engine cylinder undergoes high pressure
7.	Petrol engine is designed for high speed	Diesel engine is designed for low speed
8.	Supercharging is limited because of detonation	Supercharging to increase the power output
9.	Petrol-air mixture will be ignited by spark plug	Temperature of the compressed air ignites the diesel
10.	Petrol engine starts easily in cold conditions because of low compression ratio and petrol is highly volatile	Greater cranking effort is required to overcome the high compression ratio and due to the cold air in the combustion chamber
11.	Noise and vibration is less because of low compression ratio	More noise and vibration because of high compression ratio
12.	Running cost is more as petrol is costly	Running cost is less as diesel is cheaper

3.9   Difference Between Two-Stroke and Four-Stroke Engines

S. no.	Two-stroke engine	Four-stroke engine
1.	Two-stroke engine occupies less space for the same power output	Four-stroke engine is large in size for the same power output
2.	The torque is more uniform than a four-stroke engine, hence a lighter flywheel is required	The torque is not uniform, hence requires a heavier flywheel

(continued)

(continued)

S. no.	Two-stroke engine	Four-stroke engine
3.	This engine requires two strokes to complete one cycle OR The crank shaft makes one revolution to complete one cycle	This engine requires four strokes to complete one cycle OR The crank shaft makes two revolutions to complete one cycle
4.	Two-stroke engine produces more noise than four-stroke engine	Four-stroke engine produces less noise
5.	The thermal efficiency is less	The thermal efficiency is more than that of two-stroke engine of the same dimension
6.	The mechanical efficiency is higher about 90–95%	The mechanical efficiency is less about 85–90%
7.	Power developed in every revolution of the crank shaft. Hence, theoretically, the power developed for the same engine speed and cylinder volume is twice that of is a four-stroke engine	Power developed in two revolutions of the crank shaft
8.	Three ports like inlet, exhaust, and transfer ports are used in two-stroke engine. But nowadays, reed valve (one-way valve) is used instead of inlet port	Two valves like inlet and exhaust valves are used in a four-stroke engine
9.	More engine cooling is required in two-stroke engine. Since the fuel burns in every revolution of the crank shaft or in alternate stroke of the piston. That is the main cause, two-stroke engine is not used in heavy vehicles such as trucks, buses, tractor etc.	Less engine cooling is required in four-stroke engine. Since the fuel burns in an alternate revolution of the crank shaft or after three strokes of the piston in the cylinder

3.10 Advantages of a Two-Stroke Engine Over a Four-Stroke Engine

The following are the advantages of a two-stroke engine over a four-stroke engine.

1. Two-stroke engine has twice the number of power strokes than the four-stroke engine at the same speed. Hence, theoretically, a two-stroke engine develops double the power per cc (cubic metre) of the displacement volume than the four-stroke engine.
2. The weight of the two-stroke engine is less than the four-stroke engine because of the lighter flywheel due to uniform torque on the crank shaft.
3. The scavenging is more complete in low-speed two-stroke engines, since gases does not leave the clearance volume as in the four-stroke engine.

4. There are no mechanical valves, valve gears, cams, rocker, the construction of the two-stroke engine is simple which reduces its initial cost.
5. Due to the absence of cams, valves gears, rockers, etc., and to complete one cycle in two strokes of the piston, the mechanical efficiency is higher.
6. Two-stroke engine occupies less space.
7. Two-stroke engine has less maintenance cost since it requires only a few parts.
8. Two-stroke engine can be easily started than a four-stroke engine.

3.11 Disadvantages of a Two-Stroke Engine

The following are the disadvantages of two-stroke engines when compared with a four-stroke engine.

1. The fuel combustion takes place in every revolution, the time available for cooling will be less than four-stroke engine, which results in overheating of the cylinder and the piston. This is the main reason, two-stroke engine is not used in buses, tractors, and trucks (heavy automobile vehicles).
2. Both the exhaust and transfer ports are kept open for the same period, there is a possibility of escaping of the fresh charge through the exhaust port which will reduce the thermal efficiency.
3. It gives greater noise due to the sudden release of the burnt gases.
4. For the given stroke and the clearance volume, the effective compression stroke is less in a two-stroke engine than four-stroke engine.
5. Two-stroke engine consumes more lubricating oil.
6. In a crankcase compressed type of two-stroke engines, the volume of charge drawn into the crankcase is less due to the reduction in the crankcase volume because the rotating parts like crank, crank shaft, and connecting rod occupy some volume.

3.12 Performance Parameter of IC Engines

The performance of IC engines are calculated on basis of the following parameters:

1. Indicated power (ip),
2. Brake power (bp),
3. Friction power (fp),
4. Mean effective pressure (mep),
5. Indicated thermal efficiency (η_{ith}),
6. Brake thermal efficiency (η_{bth}),
7. Mechanical efficiency (η_m),
8. Relative efficiency or efficiency ratio (η_{ret}),
9. Volumetric efficiency (η_v), and
10. Specific fuel consumption (sfc).

3.12.1 Indicated Power: ip. The Power Produced Inside the Cylinder is Called the Indicated Power

In IC engine, fuel and air is supplied and burned in the cylinder which converts the chemical energy of the fuel into heat energy. The total heat generated cannot be converted into work, and a part of that heat is lost to the exhaust, coolant. The remaining part of the heat energy in the gases is used to drive the piston and converted into the power is called **indicated power (*ip*)**. It is calculated as (Fig. 3.9)

Indicated power:

$$ip = \frac{p_m \, Al \, nx}{60} \text{ kW}$$

where

p_m = mean effective pressure in kPa,
l = stroke length in m,
A = cross-sectional of the cylinder in m^2,
n = speed in rpm,

$= N$ for two-stroke engine,

$= \dfrac{N}{2}$ for four-stroke engine,

where

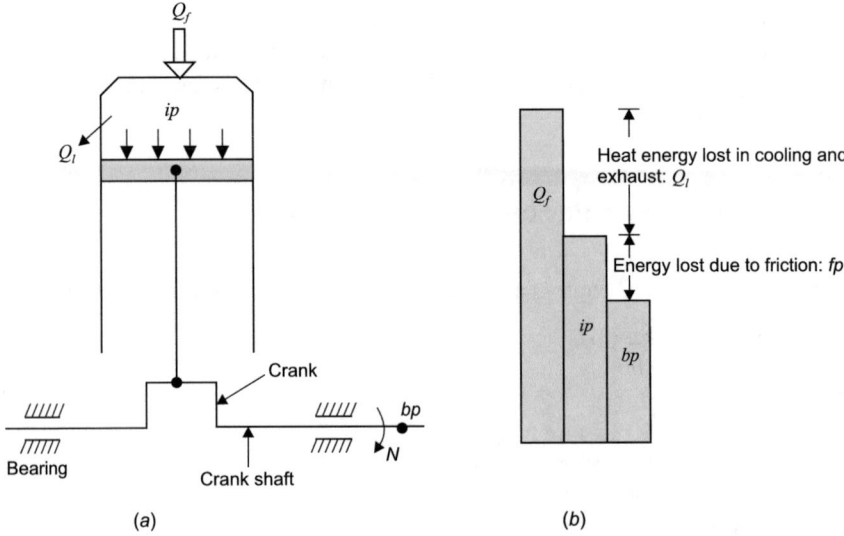

(a)   (b)

Fig. 3.9 Representation of heat supplied by the fuel (Q_f), indicated power (*ip*)and brake power (*bp*)

Fig. 3.10 Rope brake
dynamometer

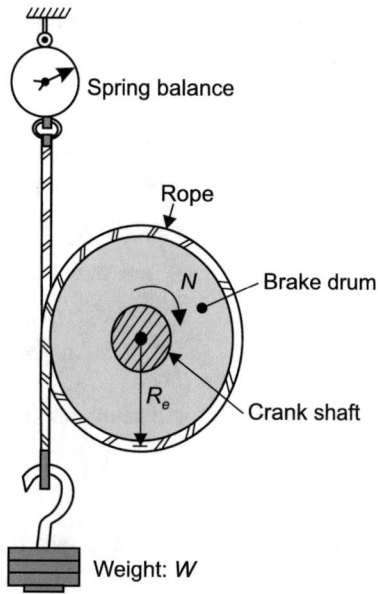

N = speed of the crank shaft in rpm,
x = number of cylinder.

3.12.2 Brake Power: bp. The Power Available at the Crank Shaft is Called Brake Power

The indicated power produced inside the engine cylinder will be transmitted to the crank shaft through the piston, connecting rod and crank. Since these parts are moving relative to each other, they will have to encounter resistance due to friction. Therefore, a certain portion of the indicated power produced will be lost due to friction of the moving parts of the engine. Therefore, the power available at the crank shaft is equal to the difference between the indicated power and the power lost due to friction (i.e., $bp = ip - fp$). The dynamometers are used to measure the brake power. The brake power is calculated as follows (Fig. 3.10):

Brake power:

$$bp = \frac{2\pi NT}{60} \text{ kW}$$

where

N = speed of the crank shaft in rpm,
T = torque applied on the crank shaft in kNm.

(i)   **Rope Brake Dynamometer**

Let

$$W = \text{external weight applied; N}$$
$$S = \text{spring balance reading; N}$$
$$R_e = \text{effective radius of brake drum}$$
$$= \text{radius of brake drum} + \text{radius of rope}$$
$$= R + r_p$$
$$R_e = R, \text{ neglecting the radius of rope.}$$

Net resisting force acting on the brake drum:

$$= W - S$$

The effective radius R_e at which the net resisting force acts:

$$R_e = R + r_p$$

$\therefore$ The resisting torque acting on the brake wheel is given by

$$\boldsymbol{T = (W - S)R_e}$$

$\therefore$ Brake power:

$$\boldsymbol{bp = \frac{2\pi NT}{60} \text{ kW.}}$$

3.12.3   *Friction Power:* **fp**

A certain part of the indicated power produced inside the cylinder is lost due to friction of the moving parts of the engine. The power lost due to friction is called **friction power (*fp*)**. The friction power is the difference between the *ip* and the *bp*.
That is, friction power:

$$\boldsymbol{fp = ip - bp.}$$

3.12.4 Mean Effective Pressure: mep

The mean effective pressure is the constant hypothetical pressure that, if it acting on the piston during the entire expansion (or working) stroke, would have produced the same amount of net work as that produced during the actual cycle.

Net work produced per cycle = mean effective pressure × displacement volume
$$W_{net} = mep \times V_s$$

or

$$mep = \frac{W_{net}}{V_s} \quad \frac{kJ}{m^3} \quad \text{or} \quad \frac{kN \cdot m}{m^3}$$
$$= \frac{W_{net}}{V_s} \quad kN/m^2 \text{ or } kPa.$$

or

$$mep = \frac{w_{net}}{w_s} \quad kPa$$

where

w_{net} = net specific work output; kJ/kg,
v_s = specific displacement volume; m^3/kg.

$\therefore$

$$mep = \frac{W_{net}}{V_s} = \frac{w_{net}}{v_s}$$

The actual mean effective is found by using the indicator diagram as follows:
Let

$$a = \text{area of the actual indicator diagram in m}^2$$
$$l = \text{base length of the indicator diagram in m}^3$$
$$k = \text{spring constant in N/m}^3$$
$$mep = \frac{\text{spring value} \times \text{area}}{\text{base length}}$$
$$mep = \frac{ka}{l} \quad N/m^2.$$

3.12.5 Indicated Thermal Efficiency: η_{ith}

It is defined as the ratio of the indicated power (ip) to the energy supplied by the fuel. Indicated thermal efficiency:

$$\eta_{ith} = \frac{\text{indicated power: } ip}{\text{energy supplied by the fuel per second: } m_f \times CV}$$

$$h_{ith} = \frac{ip}{m_f CV}$$

where

m_f = mass flow rate of fuel in kg/s,
CV = calorific value of fuel in kJ/kg.

☞**Note**: The units of both the ip and $m_f \times CV$ must be the same.

3.12.6 Brake Thermal Efficiency: η_{bth}

It is defined as the ratio of the brake power (bp) to the energy supplied by the fuel. Brake thermal efficiency:

$$\eta_{bth} = \frac{\text{brake power: } bp}{\text{energy supplied by the fuel per second: } m_f \times CV}$$

$$\eta_{bth} = \frac{bp}{m_f CV}.$$

3.12.7 Mechanical Efficiency: η_m

It is defined as the ratio of the brake power to the indicated power. Mechanical efficiency:

$$\eta_m = \frac{\text{brake power: } bp}{\text{indicated power: } ip}$$

$$\eta_m = \frac{bp}{ip}$$

$$\eta_m = \frac{bp/m_f \cdot CV}{ip/m_f \cdot CV}$$

$$\eta_m = \frac{\eta_{bth}}{\eta_{ith}} \tag{3.1}$$

The mechanical efficiency is also defined as the ratio of the brake thermal efficiency (η_{bth}) to the indicated thermal efficiency (η_{ith}).

We know that

$$ip = bp + fp$$

or

$$bp = ip - fp$$

Substituting.
$bp = ip - fp$ in Eq. (3.1), we get

$$\eta_m = \frac{ip - fp}{ip} = 1 - \frac{fp}{ip}.$$

3.12.8 Relative Efficiency or Efficiency Ratio: η_{ret}

It is defined as the ratio of actual thermal efficiency to the ideal thermal efficiency (i.e., air-standard efficiency).

Relative efficiency:

$$\eta_{ret} = \frac{\text{actual thermal efficiency}}{\text{air-standard efficiency}}$$

Air-standard efficiency of Otto cycle:

$$\eta_{Otto} = 1 - \frac{1}{r^{\gamma-1}}$$

where

$r =$ compression ratio,
$\gamma =$ adiabatic index of the air,
  $=1.4$

Air-standard efficiency of Diesel cycle:

$$\eta_{Diesel} = 1 - \frac{1}{r^{\gamma-1}} \frac{\left(r_c^{\gamma} - 1\right)}{\gamma(r_c - 1)}$$

where

> $r = $ compression ratio,
> $r_c = $ cut-off ratio,
> $\gamma = $ adiabatic index of the air.

3.12.9 Volumetric Efficiency: η_V

The engine output depends upon the amount of maximum air taken in during the suction stroke because the amount of fuel burned depends upon the amount of air available. Volumetric efficiency is an indication of the breathing ability of the engine and it is defined as the ratio of the actual volume sucked at the ambient condition to the displacement volume of the engine. It is denoted by η_v.

Volumetric efficiency:

$$\eta_v = \frac{\text{actual volume sucked at ambient condition}}{\text{displacement volume}}$$

$$= \frac{V_a}{V_s}.$$

3.12.10 Specific Fuel Consumption: sfc

It is the ratio of the mass of fuel consumed per hour to the power developed in kW. Thus specific fuel consumption:

$$sfc = \frac{\text{fuel consumed in kg/h}}{\text{power developed in kW}}$$

The units of *sfc* is kg/kWh.

It is reciprocal of the thermal efficiency. The lower value of *sfc* reflects the good performance of the engine.

It can be indicated on the basis of indicated power and brake power as
Indicated specific fuel consumption: *isfc*

$$isfc = \frac{\text{fuel consumed in kg/h}}{\text{indicated power in kW}}$$

$$= \frac{m_f}{ip} \quad \text{kg/kWh}$$

Brake specific fuel consumption: *bsfc*

$$bsfc = \frac{\text{fuel consumed in kg/h}}{\text{brake power in kW}}$$

$$= \frac{m_f}{bp} \quad \text{kg/kWh.}$$

3.12.11 Piston Speed

The average distance travelled by the piston per minute in an engine is known as **piston speed**. Thus

$$\text{Piston speed} = 21N \quad \text{m/min}$$

where

$l =$ stroke length in m,
$N =$ speed of the crank shaft in rpm.

3.13 Heat Balance Sheet

Heat balance sheet for IC engine given the information about a part of heat supplied is converted into work and the rest part of heat supplied is carried away by the exhaust gases and the cooling medium. The heat balance is generally done on a minute or hourly basis. For the heat balance sheet, the following data noted when the engine is run on constant load and at constant speed:

(i)   The quantity of fuel consumed during a given time,
(ii)   Inlet and outlet temperatures of the cooling water,
(iii)   Mass flow rate of the cooling water,
(iv)   Mass flow rate of the exhaust gases,
(v)   Temperature of exhaust gases and ambient temperature, and
(vi)   Brake power (Table 3.1).

Now the parameters are calculated for the heat balance sheet as

(a)   Heat supplied:

$$Q_s = m_f \times CV \text{ kJ/min for oil engine}$$

where

$m_f =$ mass of fuel supplied; kg/min,
$CV =$ caloric value of fuel; kJ/kg,

$$Q_s = V \times CV \text{ kJ/min for gas engines}$$

where

V = volume of gas supplied; m^3/min, and
CV = calorific value of fuel; kJ/m^3.

(b)   Heat equivalent to brake power = bp kJ/min
(c)   Heat lost to cooling water:

$$Q_w = m_w c_{pw}(T_2 - T_1) \text{ kJ/min}$$

where

m_w = mass flow rate of cooling water; kg/min,
c_{pw} = specific heat of water,

      = 4.18 kJ/kgK,
T_1 = temperature of cooling water at inlet in °C, and
T_2 = temperature of cooling water at outlet in °C.

(d)   Heat carried away by exhaust gases:

$$Q_e = m_g c_{pg}(T_g - T_o) \text{ kJ/min}$$

where

m_g = mass flow rate of the exhaust gas in kg/min,
c_{pg} = specific heat of exhaust gases in kJ/kgK,
T_g = temperature of exhaust gases in °C, and
T_o = ambient temperature in °C.

Problem 3.1: A single cylinder 4-s engine develops an indicated power of 150 kW and delivers a brake power of 120 kW.

Determine:

(i)   Frictional power,
(ii)   Mechanical efficiency of the engine.

Table 3.1 Heat balance sheet

Heat input per minute	kJ	%	Heat expenditure per minute	kJ	%
Heat supplied by fuel: Q_s	–	100%	1. Heat equivalent to bp	–	
			2. Heat lost to cooling water	–	–
			3. Heat carried away by exhaust gases	–	–
			4. Heat unaccounted $= Q_s - (bp + Q_w + Q_e)$	–	–
		Total	–		100%

Solution: Given data:
Indicated power:

$$ip = 150\,\text{kW}$$

Brake power:

$$bp = 120\,\text{kW}$$

also

$$bp = ip - fb$$

(i) Frictional power:

$$fp = ip - bp = 150 - 120 = 30\,\text{kW}$$

(ii) Mechanical efficiency:

$$\eta_m = \frac{bp}{ip} = \frac{120}{150} = 0.8 = \mathbf{80\%}$$

Problem 3.2: A single cylinder, CI engine with a brake thermal efficiency of 32% use high-speed diesel oil having a calorific value of 42,000 kJ/kg. If its mechanical efficiency is 84%. Determine

(i)   brake specific fuel consumption,
(ii)   indicated specific fuel consumption, and
(iii) indicated thermal efficiency.

Solution: Given data:
Brake thermal efficiency:

$$\eta_{bth} = 32\% = 0.32$$

Calorific value of fuel:

$$CV = 42000\,\text{kJ/kg}$$

Mechanical efficiency:

$$\eta_m = 84\% = 0.84$$

Brake thermal efficiency:

$$\eta_{bth} = \frac{\text{brake power: } bp}{m_f \times CV}$$

$$\eta_{bth} = \frac{bp}{m_f CV}$$

$$0.32 = \frac{bp}{m_f \times 42000}$$

or

$$\frac{m_f}{bp} = \frac{1}{0.32 \times 42000} \quad \frac{\text{kg/s}}{\text{kW}}$$

$$= \frac{3600}{0.32 \times 42000} \quad \frac{\text{kg/h}}{\text{kW}} = 0.267\,\text{kg/kWh}$$

(i)   Brake specific fuel consumption:

$$bsf_c = \mathbf{0.267\,kg/kWh}$$

(ii)   Indicated specific fuel consumption: isfc

Mechanical efficiency:

$$\eta_m = \frac{bp}{ip}$$

$$0.84 = \frac{m_f \times 0.32 \times 42000}{ip}$$

or

$$\frac{m_f}{ip} = 6.25 \times 10^{-5} \quad \frac{\text{kg/s}}{\text{kW}}$$

$$= 6.23 \times 10^{-5} \times 3600 \quad \frac{\text{kg/h}}{\text{kW}} = 0.225\,\text{kg/kWh}$$

$$\therefore$$

$$isfc = \mathbf{0.225\,kg/kWh}$$

(iii)   Indicated thermal efficiency: η_{ith}

$$\eta_{ith} = \frac{ip}{m_f \times CV} = \frac{1}{6.25 \times 10^{-5} \times 42000}$$

$$= 0.3809 = \mathbf{38.09\%}$$

Problem 3.3: A two-stroke CI engine delivers a brake power of 400 kW while 75 kW is used to overcome the friction losses. It consumes 200 kg/h of fuel at an air–fuel ratio of 20: 1. The calorific value of the fuel is 42000 kJ/kg. Determine

(i)   indicated power,
(ii)   mechanical efficiency,
(iii)   air consumption,
(iv)   indicated thermal efficiency, and.
(v)   brake thermal efficiency.

Solution: Given data:
    Brake power:

$$bp = 400\,\text{kW}$$

Friction power:

$$fp = 75\,\text{kW}$$

∴ Indicated power:

$$ip = bp + fp = 400 + 75 = 475\,\text{kW}$$

Mass flow rate of the fuel:

$$m_f = 200\,\text{kg/h} = \frac{200}{3600}\,\text{kg/s}$$

Air–fuel ratio:

$$\frac{m_a}{m_f} = 20$$

or

$$m_a = 20 \times m_f = \frac{20 \times 200}{3600} = 1.11\,\text{kg/s}$$

(i)   Indicated power:

$$ip = \mathbf{475\,kW}$$

(ii)   Mechanical efficiency:

$$\eta_m = \frac{bp}{ip} = \frac{400}{475} = 0.8421 = \mathbf{84.21\%}$$

(iii)   Air consumption:

$$m_a = \mathbf{1.11\,kg/s}$$

(iv)   Indicated thermal efficiency:

$$\eta_{ith} = \frac{ip}{m_f \times CV}$$

$$\eta_{ith} = \frac{475}{\frac{200}{3600} \times 42000} = 0.2035 = \mathbf{20.35\%}$$

(v)   Brake thermal efficiency:

$$\eta_{ith} = \frac{bp}{m_f \times CV} = \frac{400}{\frac{200}{3600} \times 42000} = 0.1714 = \mathbf{17.14\%}$$

Problem 3.4: A two-cylinder, two–stroke IC engine develops an indicated power of 32 kW at a speed of 1000 rpm. If the mean effective pressure 6 bar, determine bore and the stroke. Assuming stroke 1.5 times the bore.

Solution: Given data:
   Number of cylinder:

$$x = 2$$

Indicated power:

$$ip = 32\,kW$$

Speed:

$$N = 1000\,rpm$$

Mean effective pressure:

$$p_m = 6\,bar = 600\,kPa$$

Let

$$d = \text{bore of the cylinder}$$

$\therefore$ Stroke:

$$l = 1.5d$$

Indicated power:

$$ip = \frac{p_m l \, An \, x}{60} \text{ kW}$$

where

p_m is in kPa
l is in m
A is in m^2
$n = N$ for 2-s engine
x = number of cylinder

$$32 = \frac{600 \times lA \times 1000 \times 2}{60}$$

or

$$lA = 1.6 \times 10^{-3}$$
$$1.5d \times \frac{\pi}{4} d^2 = 1.6 \times 10^{-3}$$
$$d^3 = \frac{1.6 \times 10^{-3} \times 4}{1.5 \times 3.14} = 0.001358$$

or

$$d = 0.11076 \, \text{m} = \mathbf{110.76 \, mm}$$

Stroke:

$$l = 1.5d = 1.5 \times 110.76 = \mathbf{166.14 \, mm}$$

Problem 3.5: A single cylinder, four-stroke SI engine has a cylinder 250 mm in diameter and the stroke of the piston is 450 mm. It works under the following condition:

$$\text{Speed} = 180 \, \text{rpm}$$
$$\text{Misfires per minute} = 8$$
$$\text{Mean effective pressure} = 6.5 \, \text{bar}$$
$$\text{Mechanical efficiency} = 80\%$$

Determine:

  (i)   Indicated power,
  (ii)   Brake power, and
  (iii)   Friction power.

Solution: Given data:
  Number of cylinder:

$$x = 1$$

  Diameter of cylinder:

$$d = 250\,\text{mm} = 0.25\,\text{m}$$

  Stroke:

$$l = 450\,\text{mm} = 0.45\,\text{m}$$
$$\text{Speed} = 180\,\text{rpm}$$
$$\text{Misfires per minute} = 8$$

  Speed of the crank shaft during 8 misfires

$$= 8 \times 2 = 16\,\text{rpm} \quad \because \text{ four-stroke engine}$$

  $\therefore$ Actual speed:

$$N = 180 - 16 = 164\,\text{rpm}$$

  Mean effective pressure:

$$p_m = 6.5\,\text{bar} \ = 650\,\text{kPa}$$

  Mechanical efficiency:

$$\eta_m = 80\% = 0.80$$

  (i)   Indicated power:

$$ip = \frac{p_m l\, Anx}{60}$$
$$ip = \frac{p_m l \times \frac{\pi}{4} d^2 \times nx}{60}$$

       where

$$n = \frac{N}{2} = \frac{164}{2} = 82 \, \text{rpm} \quad \text{for 4-s engine}$$

$\therefore$

$$ip = \frac{650 \times 0.45 \times 3.14 \times (0.5)^2 \times 82 \times 1}{60 \times 4} = \mathbf{19.61 \, kW}$$

(ii)   Brake power: bp
       Mechanical efficiency:

$$\eta_m = \frac{bp}{ip}$$

or

$$bp = ip \times \eta_m = 19.61 \times 0.8 = \mathbf{15.68 \, kW}$$

(iii)   Friction power: fp

$$fp = ip - bp = 19.61 - 15.68 = \mathbf{3.93 \, kW}$$

Problem 3.6: An Otto gas engine having compression ratio of 6 develops an indicated power of 16.7 kW. The gas consumptions is 13.64 m³/h measured at 25 °C and at a pressure of 125 mm of water above the atmospheric pressure 735 mm of mercury. The calorific value of the gas at 1 bar and 0 °C was 18,700 kJ/m3. Determine the relative efficiency. Take $\gamma = 1.4$.

Solution: Given data:
   Compression ratio:

$$r = 6$$

Indicated power:

$$ip = 16.7 \, \text{kW}$$

Gas consumption:

$$V_2 = 13.64 \, \text{m}^3/\text{h}$$

At

$$T_2 = 25 \, ^\circ\text{C} = (25 + 273) \, \text{K} = 298 \, \text{K}$$

Gauge pressure head:

$$h = 125 \text{ mm of } H_2O = 0.125 \text{ m of } H_2O$$

∴ Gauge pressure:

$$
\begin{aligned}
p_{g2} &= (\rho g h)_{H_2O} \\
&= 1000 \times 9.81 \times 0.125 \text{ N/m}^2 \\
&= 1226.25 \text{ N/m}^2 = 1.226 \text{ kPa} \\
\text{Atmospheric pressure head} &= 735 \text{ mm of Hg} = 0.735 \text{ m of Hg}
\end{aligned}
$$

∴ Atmospheric pressure:

$$
\begin{aligned}
p_{atm} &= (\rho g h)_{Hg} \\
&= 13600 \times 9.81 \times 0.735 = 98060.76 \text{ N/m}^2 \\
&= 98.06 \text{ kPa}
\end{aligned}
$$

∴ Absolute pressure:

$$p_2 = p_{g2} + p_{atm} = 1.226 + 98.06 = 99.286 \text{ kPa}$$

Calorific value of the gas:

$$CV = 18700 \text{ kJ/m}^3$$

At

$$p_1 = 1 \text{ bar} = 100 \text{ kPa}$$

and

$$T_1 = 0\,°C = 273 \text{ K}$$

On applying the gas equation,

$$
\frac{p_1 V_1}{T_1} = \frac{p_2 V_2}{T_2}
$$
$$
\frac{100 \times V_1}{273} = \frac{99.286 \times 15}{298}
$$

or

$$V_1 = 13.64 \text{ m}^3/\text{h} = \frac{13.64}{3600} \text{ m}^3/\text{s}$$

Heat supplied:

$$Q_s = V_1 \times CV$$
$$= \frac{13.64}{3600} \times 18700 \, \text{kW} = 70.85 \, \text{kW}$$

Indicated (or actual) thermal efficiency:

$$\eta_{ith} = \frac{ip}{Q_s} = \frac{16.7}{70.85} = 0.2357$$

Air-standard efficiency for Otto cycle:

$$\eta_{Otto} = 1 - \frac{1}{r^{\gamma-1}} = 1 - \frac{1}{(6)^{1.4-1}} = 0.5116$$

∴ Relative efficiency:

$$\eta_{rev} = \frac{\text{actual thermal efficiency}}{\text{air standard efficiency}}$$
$$= \frac{0.2357}{0.5116} = 0.4607 = \textbf{46.07\%}$$

Problem 3.7: The following data were recorded during the trial of a four–stroke gas engine:

Area of indicator diagram: 5.66 cm^2
Length of indicator diagram: 7.5 cm
Spring index: 9 bar/cm
Cylinder diameter: 22 cm
Stroke: 43 cm
Number of explosion per minute: 100

Determine:

(i)   Indicated mean effective pressure, and
(ii) Indicated power.

Solution: Given data:
Area of indicator diagram:

$$a = 5.66 \, \text{cm}^2$$

Length of indicator diagram:

$$l = 7.5 \, \text{cm}$$

Spring index:

$$k = 9 \, \text{bar/cm}$$

Cylinder diameter:

$$d = 22 \, \text{cm} = 0.22 \, \text{m}$$

$\therefore$ Area:

$$A = \frac{\pi}{4}d^2 = \frac{5.14}{4} \times (0.22)^2 = 0.03799 \, \text{m}^2$$

Stroke:

$$L = 43 \, \text{cm} = 0.43 \, \text{m}$$

Number of explosion per minute $= 100$
That is speed of the crank shaft:

$$N = 200 \, \text{rpm} \quad \because 4\text{-s engine}$$

(i)   Indicated mean effective pressure:

$$p_m = \frac{ka}{l} = \frac{9 \times 5.66}{7.5} = 6.792 \, \text{bar} = \mathbf{679.2 \, kPa}$$

(ii)   Indicated power:

$$ip = \frac{p_m \, AL \, nx}{60} \, \text{kW}$$

where

p_m is in kPa
A is in m^2,
L is in m,

$$n = \frac{N}{2} \, \text{rpm for 4} - \text{s engine}$$
$$= \frac{200}{2} = 100 \, \text{rpm}$$
$$x = 1, \quad \text{number of cylinder}$$

$\therefore$

$$ip = \frac{679.2 \times 0.03799 \times 0.43 \times 100 \times 1}{60} = 18.49\,\text{kW}$$

Problem 3.8: The following data relate to a test of a single-cylinder petrol engine working on the 4-s cycle:

Duration of test: 60 min
Calorific value of fuel: 40,000 kJ/kg
Fuel consumption: 1.6 kg
Mean effective pressure: 8.23 bar
Speed: 2500 rpm
Diameter of cylinder: 65 mm
Stroke: 100 mm
Diameter of brake pulley: 600 mm
Net load on brake: 60 N
Compression ratio: 5.5

Determine:

(i)   Indicated thermal efficiency,
(ii)   Brake thermal efficiency,
(iii) Air-standard efficiency, and
(iv)   Mechanical efficiency.

Solution: Given data:
Duration of test:

$$t = 60\,\text{min}$$

Calorific value of fuel:

$$CV = 40000\,\text{kJ/kg}$$

Fuel consumption:

$$M_f = 1.6\,\text{kg}$$

$\therefore$ Mass flow rate of fuel consumption:

$$m_f = \frac{M_f}{t} = \frac{1.6}{60}\,\text{kg/min} = \frac{1.6}{3600}\,\text{kg/s}$$

Mean effective pressure:

$$p_m = 8.23\,\text{bar} = 823\,\text{kPa}$$

Speed:

$$N = 2500\,\text{rpm}$$

Diameter of cylinder:

$$d = 65\,\text{nm} = 0.065\,\text{m}$$

∴ Area:

$$A = \frac{\pi}{4}d^2 = \frac{3.14}{4} \times (0.065)^2 = 0.00331\,\text{m}^2$$

Stroke:

$$l = 100\,\text{mm} = 0.1\,\text{mm}$$

Diameter of brake pulley:

$$D = 600\,\text{mm} = 0.6\,\text{m}$$

∴ Radius of brake pulley:

$$R = \frac{D}{2} = \frac{0.6}{2} = 0.3\,\text{m}$$

Net load on brake:

$$W_{net} = 60\,\text{N}$$

∴ Torque applied on crank shaft:

$$T = W_{net}R = 60 \times 0.3 = 18\,\text{Nm} = 0.018\,\text{kNm}$$

Compression ratio:

$$r = 5.5$$

Indicated power:

$$ip = \frac{p_m\,Al\,nx}{60}\,\text{kW}$$

where

p_m is in kPa

A is in m^2

l is in m

$$N = \frac{N}{2}\text{rpm} \quad \text{for 4 - s engine}$$
$$= \frac{2500}{2}\text{rpm} = 1250\,\text{rpm}$$
$$x = 1, \quad \text{number of cylinder}$$

$\therefore$

$$ip = \frac{823 \times 0.00331 \times 0.1 \times 1250 \times 1}{60} = 5.675\,\text{kW}$$

Brake power:

$$bp = \frac{2\pi NT}{60}\,\text{kW}$$

where

T is in kNm

N is in rpm

$\therefore$

$$bp = \frac{2 \times 3.14 \times 2500 \times 0.018}{60} = 4.71\,\text{kW}$$

(i)   Indicated thermal efficiency: η_{ith}

$$\eta_{ith} = \frac{ip}{m_f \times CV}$$
$$= \frac{5.675}{\frac{1.6}{3600} \times 40000} = \frac{5.675 \times 3600}{1.6 \times 40000}$$
$$= 0.3192 = \mathbf{31.92\%}$$

(ii)   Brake thermal efficiency: η_{bth}

$$\eta_{bth} = \frac{bp}{m_f \times CV}$$
$$= \frac{4.71}{\frac{1.6}{3600} \times 40000} = \frac{4.71 \times 3600}{1.6 \times 40000}$$

$$= 0.2649 = \mathbf{26.49\%}$$

(iii) Air-standard efficiency: η_{Otto}

$$\eta_{Otto} = 1 - \frac{1}{r^{\gamma-1}}$$

where

$r =$ compression ratio
$\gamma =$ adiabatic index
$\quad = 1.4$

$\therefore$

$$\eta_{Otto} = 1 - \frac{1}{(5.5)^{1.4-1}} = 1 - 0.5056$$
$$= 0.4944 = \mathbf{49.44\%}$$

(iv) Mechanical efficiency: η_m

$$\eta_m = \frac{bp}{ip} = \frac{4.71}{5.675} = 0.8299 = \mathbf{82.99\%}$$

Problem 3.9: The air flow to a four cylinder, 4-s oil engine is measured by means of a 50 mm diameter orifice having a coefficient of discharge of 0.6. During a test on the engine the following data were recorded:

Bore: 100 mm
Stroke: 120 mm
Speed: 1200 rpm
Brake torque: 120 Nm
Full consumption: 5 kg/h
Calorific value of fuel: 42,000 kJ/kg
Pressure drop across orifice: 46 mm of water
Ambient temperature: 17 °C
Ambient pressure: 1 bar

Determine:

(i)   Brake thermal efficiency
(ii)   Brake mean effective pressure
(iii) Volumetric efficiency

Solution: Given data:
Number of cylinder:

$$x = 4$$

Diameter of orifice:

$$d_0 = 50 \, \text{mm} = 0.05 \, \text{m}$$

∴ Cross-sectional area of orifice:

$$a = \frac{\pi}{4} d_0^2 = \frac{3.14}{4} \times (0.05)^2 = 1.96 \times 10^{-3} \, \text{m}^2$$

Coefficient of discharge:

$$C_d = 0.6$$

Bore:

$$d = 100 \, \text{mm} = 0.1 \, \text{m}$$

∴ Area:

$$A = \frac{\pi}{4} d^2 = \frac{3.14}{4} (0.1)^2 = 7.85 \times 10^{-3} \, \text{m}^2$$

Stroke:

$$l = 120 \, \text{mm} = 0.12 \, \text{m}$$

Speed:

$$N = 1200 \, \text{rpm}$$

Brake torque:

$$T = 120 \, \text{Nm} = 0.12 \, \text{kNm}$$

Fuel consumption:

$$m_f = 5 \, \text{kg/h} = \frac{5}{3600} \, \text{kg/s}$$

Calorific value of fuel:

$$CV = 42000 \, \text{kJ/kg}$$

Pressure head drop across orifice:

$$\Delta h_w = 46 \text{ mm of water} = 0.046 \text{ m of water}$$

Pressure drop across orifice:

$$\Delta p = (\rho \, g \, \Delta h_w)_{\text{water}}$$
$$= 1000 \times 9.81 \times 0.046 \, \text{Pa} = 451.26 \, \text{Pa}$$

Ambient temperature:

$$T_a = 17\,^\circ\text{C} = 290 \, \text{K}$$

Ambient pressure:

$$p_a = 1 \, \text{bar} = 100 \, \text{kPa}$$

Brake power:

$$bp = \frac{2\pi NT}{60} \, \text{kW}$$

where

> N is in rpm
> T is in kNm

$\therefore$

$$bp = \frac{2 \times 3.14 \times 1200 \times 0.12}{60} = 15.072 \, \text{kW}$$

(i)   Brake thermal efficiency: η_{bth}

$$\eta_{bth} = \frac{bp}{m_f \times CV}$$
$$= \frac{15.072}{\frac{5}{3600} \times 42000} = \frac{15.072 \times 3600}{5 \times 42000} = 0.2583 = \mathbf{25.83\%}$$

(ii)   Brake mean effective pressure: p_{bm}
Brake power:

$$bp = \frac{p_{bm} \, Al \, nx}{60} \, \text{kW}$$

where

p_{bm} is in kPa

A is in m^2

l is in m

$$n = \frac{N}{2}\text{rpm} = \frac{1200}{2} = 600\,\text{rpm} \quad \text{for 4-s engine}$$

$$x = 4, \quad \text{number of cylinder}$$

$\therefore$

$$15.072 = \frac{p_{bm} \times 7.85 \times 10^{-3} \times 0.12 \times 600 \times 4}{60}$$

or

$$p_{bm} = 400\,\text{kPa} = \mathbf{4\,bar}$$

(iii) Volumetric efficiency: η_v

$$\eta_v = \frac{\text{actual volume flow rate sucked at ambient condition}: V_a}{\text{displacement volume}: V_s}$$

where actual volume flow rate sucked at ambient condition:

$$V_a = C_d\, a\sqrt{2g\,\Delta h_a}\,\text{m}^3/\text{s}$$

where

C_d = coefficient of discharge

a = cross-sectional area of orifice in m^2

$g = 9.81$ m/s^2, acceleration due to gravity

Δh_a = pressure head drop across orifice in m of air.

Applying the equation of state at ambient condition.

$$p_a v_a = RT_a$$
$$\frac{p_a}{\rho_a} = RT_a \quad \because v_a = \frac{1}{\rho_a}$$

or

$$\rho_a = \frac{p_a}{RT_a}\,\text{kg/m}^3$$

where

p_a is in kPa
$R = 0.287$ kJ/kgK
T_a is in K

$$\rho_a = \frac{100}{0.287 \times 290} = 1.20 \, \text{kg/m}^3$$

Pressure drop across orifice:

$$\Delta p = (\rho_a g \, \Delta h_a)_{\text{air}}$$
$$451.26 = 1.2 \times 9.81 \times \Delta h_a$$

or

$$\Delta h_a = 38.33 \, \text{m of air}$$

$\therefore$

$$V_a = 0.6 \times 1.96 \times 10^{-3}\sqrt{2 \times 9.81 \times 38.33}$$
$$= 0.03225 \, \text{m}^3/\text{s} = 1.935 \, \text{m}^3/\text{min}$$

Displacement volume:

$$V_s = Al\,n\,x \, \text{m}^3/\text{min}$$

where

A is in m^2
l is in m

$$n = \frac{N}{2}\text{rpm} = \frac{1200}{2} = 600 \, \text{rpm}$$
$$x = 4, \text{number of cylinder}$$

$\therefore$

$$V_s = 7.85 \times 10^{-3} \times 0.12 \times 600 \times 4 \, \text{m}^3/\text{min}$$
$$= 2.260 \, \text{m}^3/\text{min}$$

Volumetric efficiency:

$$\eta_v = \frac{1.935}{2.260} = 0.8562 = \mathbf{85.62\%}$$

Problem 3.10: A six cylinder, four–stroke SI engine of 10 cm $\times$ 12 cm with a compression ratio of 6 is tested at 4800 rpm on a dynamometer of arm 55 cm. During a 10 min test, dynamometer reads 45 kg and the engine consumed 5 kg of petrol of calorific value 45 MJ/kg. The carburetor receives the air at 29 °C and 1 bar at the rate of 10 kg/min. Calculate.

(i)   the brake power
(ii)   the brake mean effective pressure
(iii)   the brake specific fuel consumption
(iv)   the brake thermal efficiency
(v)   the air–fuel ratio

Solution: Given data:
Number of cylinder:

$$x = 6$$

Diameter of cylinder:

$$d = 10\,\text{cm} = 0.1\,\text{m}$$

Stroke length:

$$l = 12\,\text{cm} = 0.12\,\text{m}$$

Compression ratio:

$$r = 6$$

Speed:

$$N = 4800\,\text{rpm}$$

Length of dynamometer arm:

$$L = 55\,\text{cm} = 0.55\,\text{m}$$

Duration of test:

$$t = 10\,\text{min} = 10 \times 60 = 600\,\text{s}$$

Dynamometer reading:

$$W = 45 \, \text{kg} = 45 \times 9.81 \, \text{N} = 441.45 \, \text{N}$$

Mass of fuel:

$$m_f = 5 \, \text{kg}$$

Mass flow rate of fuel:

$$m_f = \frac{5}{600} \, \text{kg/s} = 0.008334 \, \text{kg/s}$$

Calorific value:

$$CV = 45 \, \text{MJ/kg} = 45 \times 10^6 \, \text{J/kg} = 45000 \, \text{kJ/kg}$$

Mass flow rate of air:

$$m_a = 10 \, \text{kg/min}$$
$$= \frac{10}{60} \, \text{kg/s} = 0.1667 \, \text{kg/s}$$

Torque:

$$T = WL = 441.45 \times 0.55 = 242.79 \, \text{Nm}$$

(i)   Brake Power:

$$bp = \frac{2\pi NT}{60} = \frac{2 \times 3.14 \times 4800 \times 242.79}{60}$$
$$121977.69 \, \text{W} = \mathbf{121.977 \, kW}$$

(ii)   Brake mean effective pressure: p_{bm}

$$p_{bm} = \frac{bp}{Al\,n\,x}$$

where

$$n = \frac{N}{2}$$

$\therefore$

$$p_{bm} = \frac{121.977}{0.12 \times \frac{\pi}{4}(0.1)^2 \times \frac{4800}{2\times60} \times 6} = \mathbf{539.53\,kPa}$$

(iii) Brake specific fuel consumption:

$$bsfc = \frac{m_f}{bp}$$

where

$$m_f = 0.008334\,kg/s = 30\,kg/h$$
$$bp = 121.977\,kW$$

$\therefore$

$$bsfc = \frac{30}{121.977}\ kg/kWh = \mathbf{0.2459\,kg/kWh}$$

(iv) Brake specific air consumption: $bsac$

$$bsac = \frac{m_a}{bp}$$

where

$$m_a = 10\,kg/min = 600\,kg/h$$

$\therefore$

$$bsac = \frac{600}{121.977} = 4.91\,kg/kWh$$

(v) Brake thermal efficiency: η_{bth}

$$\eta_{bth} = \frac{bp}{m_f \times CV} = \frac{121.977}{0.008334} \times 45000 = 0.3252 = \mathbf{32.52\%}$$

(vi) Air–fuel ratio: A/F

$$\frac{A}{F} = \frac{m_a}{m_f} = \frac{0.1667}{0.008334} = \mathbf{20}$$

Problem 3.11: A single cylinder four-stroke cycle oil engine was tested and the following particulars were obtained.

Bore = 240 mm, stroke = 360 mm, indicated mean effective pressure = 6.5 bar, speed = 250 rpm, fuel consumption = 4.5 kg/h, brake torque = 650 Nm, calorific value of fuel = 45,000 kJ/kg, flow rate of cooling water = 360 kg/h, rise of tempera-ture of jacket cooling water = 35 °C, mass of flow rate of air = 90 kg/h, temperature of exhaust gases = 330 °C and room air temperature = 20 °C, find the mechanical efficiency, indicated thermal efficiency, specific fuel consumption and draw up a heat balance sheet.

Solution: Given data:
   Diameter of bore:

$$d = 240\,\text{mm} = 0.24\,\text{m}$$

Stroke:

$$l = 360\,\text{mm} = 0.36\,\text{m}$$

Indicated mean effective pressure:

$$p_{im} = 6.5\,\text{bar} = 6.5 \times 10^5\,\text{N/m}^2$$

Speed:

$$N = 250\,\text{rpm}$$

Fuel consumption:

$$m_f = 4.5\,\text{kg/h} = \frac{4.5}{3600}\,\text{kg/s} = 1.25 \times 10^{-3}\,\text{kg/s}$$

Brake torque:

$$T = 650\,\text{Nm}$$
$$CV = 45000\,\text{kJ/kg} = 45000 \times 10^3\,\text{J/kg}$$

Flow of water:

$$m_w = 360\,\text{kg/h} = \frac{360}{3600} = 0.1\,\text{kg/s}$$

Rise in temperature of cooling water:

$$\Delta T_w = 35\,°\text{C}$$

Mas flow rate of air:

$$m_a = 90\,\text{kg/h} = \frac{90}{3600} = 0.025\,\text{kg/h}$$

Temperature of exhaust gases:

$$T_{exhaust} = 330\,°\text{C}$$

Room air temp:

$$T_0 = 20\,°\text{C}$$

Indicated power:

$$ip = \frac{p_{im}l\,An\,x}{60}$$

where

$$x = 1, \text{ number of cylinder}$$

and

$$n = \frac{N}{2} \text{ for four-stroke cycle}$$

$\therefore$

$$ip = \frac{p_{im}\,l}{60}\frac{\pi}{4}d^2 \times \frac{N}{2} \times 1$$
$$= \frac{6.5 \times 10^5 \times 0.36}{60} \times \frac{3.14 \times (0.24)^2}{4} \times \frac{250}{2}$$
$$= 22042.8\,\text{W}$$

Brake power:

$$bp = \frac{2\pi NT}{60} = \frac{2 \times 3.14 \times 250 \times 650}{60}$$
$$= 17008.33\,\text{W}$$

(i) Mechanical efficiency:

$$\eta_m = \frac{\text{brake power: bp}}{\text{indicated power: ip}}$$

$$= \frac{17008.33}{22042.8} = 0.7716 = \mathbf{77.16\%}$$

(ii)   Indicated thermal efficiency:

$$\eta_{ith} = \frac{\text{indicated power: ip}}{m_f \times CV}$$

$$= \frac{22042.8}{1.25 \times 10^{-3} \times 45000 \times 10^3} = 0.3918$$

$$= \mathbf{39.18\%}$$

(iii)   Specific fuel consumption:

$$sfc = \frac{\text{fuel consumption/h}}{bp \text{ in kW}} = \frac{4.5}{1.7008} \text{ kg/kWh}$$

$$= \mathbf{2.645} \text{ kg/kWh}$$

(iv)   Heat balance sheet:
        Heat supplied:

$$Q_s = m_f \times CV$$

$$= 4.5 \times 45000 \text{ kJ/h}$$

$$= \mathbf{202500 \, kJ/h}$$

Heat equivalent to brake power:

$$bp = 17008.33 \text{ W}$$

$$= \frac{17008.33 \times 3600}{1000} \text{ kJ/h} = \mathbf{61229.98 \, kJ/h}$$

Heat carried away by cooling water:

$$Q_w = c_{pw} \, m_w \, \Delta T_w$$

$$= 4.2 \times 360 \times 35 \text{ kJ/h} = \mathbf{52920 \, kJ/h}$$

Heat carried away by exhaust:

$$Q_e = c_{pg} \left(m_f + m_a \right) (T_{exhaust} - T_0)$$

$$= 1.005 \times (4.5 + 90)(330 - 20) = \mathbf{29441.47 \, kJ/h}$$

$$\text{Unaccounted loss} = Q_s - bp - Q_w - Q_e$$

$$= 202500 - 61229.98 - 52920 - 29441.47$$

$$= \mathbf{58908.55 \, kJ/h.}$$

Heat Balance Sheet

Heat input per hour	kJ	%	Heat expenditure per hour	kJ	%
Heat supplied by fuel: Q_s	202,500	100%	1. Heat equivalent to bp	61,229.98	30.23%
			2. Heat lost to cooling water	52,920	26.13%
			3. Heat carried away by exhaust gases	29,441.47	14.54%
			4. Unaccounted losses	58,908.55	29.10%
			Total	202,500	100%

Summary

1. **Heat Engine**. It is defined as a device which convert the heat energy into mechanical work. It can be broadly classified into two categories:

    (i) Internal combustion engines (IC engines)
    (ii) External combustion engines (EC engines)

2. **Internal Combustion Engine**. It is a heat engine which converts the chemical energy of fuel (liquid or gas) into heat energy due to the combustion of fuel that takes place inside the engine cylinder and further a part of this heat energy converted into useful work. Internal combustion engine more popularly known as IC engine.

3. **External Combustion Engine**. It is a heat engine which converts the heat energy into mechanical work. The conversion of chemical energy of fuel (solid, liquid or gas) into heat energy takes place outside the engine. External combustion engine more popularly known as EC engine.

4. **Classification of IC Engines**
    IC engines are classified based on:

    (a) Working cycles
         (i)   Otto cycle
         (ii) Diesel cycle
         (iii) Dual cycle
    (b) Types of fuel used
         (i)   Petrol (or gasoline) engine
         (ii) Diesel engine
         (iii) Gas engine
         (iv) Bi-fuel engine
    (c) Ignition system
         (i)   Spark ignition (SI) engines
         (ii) Compression ignition (CI) engines
    (d) Number of strokes per cycle
         (i)   Four-stroke engines
         (ii) Two-stroke engines

    (e)   Speed of engines
- (i)   Low speed engines
- (ii)   Medium speed engines
- (iii)   High speed engines

    (f)   Application of engines
- (i)   Automotive
- (ii)   Locomotive
- (iii)   Light aircraft
- (iv)   Marine
- (v)   Power generation
- (vi)   Agricultural
- (vii) Earthmoving

    (g)   Number of cylinder
- (i)   Single cylinder engine
- (ii)   Multi cylinder engine

    (h)   Cylinder arrangement
- (i)   Vertical engine
- (ii)   Horizontal engine
- (iii)   V-engine
- (iv)   In-line engine
- (v)   Opposed cylinder engine
- (vi)   Radial engine

    (i)   Cooling system
- (i)   Air-cooled engine
- (ii)   Water-cooled engine

5.   **Bore**. The inner diameter of the cylinder is called the bore.

6.   **Dead Centres**. The piston reciprocates in the cylinder between two fixed positions called dead centres. The name 'dead' means speed of piston is zero at these positions. At dead centre piston changes its direction.

    (a)   Top dead centre (TDC). The position of the piston towards the cylinder head when it forms the smallest volume in the cylinder is called top dead centre. For horizontal engine, this position is called inner dead centre (IDC).

    (b)   Bottom dead centre (BDC). The position of the piston towards the crank end side when it forms the largest volume tin the cylinder is called bottom dead centre (BDC). For horizontal engine, this position is called outer dead centre (ODC).

7.   **Stroke**: The distance travelled by the piston from top dead centre to bottom dead centre is called the stroke of the engine. It is equal to twice the radius of the crank. It is denoted by *l*.
Mathematically,
Stroke:

$$l = 2r$$

where

r = radius of the crank

8. **Clearance Volume**: V_c. The volume between the piston and cylinder head when the piston is at the top dead centre (TDC) is called the clearance volume. It is denoted by V_c.

9. **Swept or Displacement Volume**: V_s. The volume displaced by the piston as it moves between the top dead centre and the bottom dead centre is called the swept or displacement volume.
    Swept volume:

$$V_s = A\,l$$

10. **Cylinder Volume**: V. The volume between the BDC and cylinder head when the piston is at the bottom dead centre (BDC) is called the cylinder volume. Mathematically, it is equal to the sum of the clearance volume (V_c) and swept volume (V_s).
    Cylinder volume:

$$V = V_c + V_s$$

11. **Compression Ratio**: r. The compression ratio is a volume ratio. It is defined as the ratio of the volume in the cylinder when the piston at BDC to the volume in the cylinder when the piston is at TDC. It is denoted by r.
    Mathematically,
    Compression ratio:

$$r = \frac{V}{V_c}$$

where

$$V = V_c + V_s$$

∴

$$r = \frac{V_c + V_s}{V_c}$$
$$r = 1 + \frac{V_s}{V_c}$$

12. Working of Four-stroke Petrol (SI) Engine.
    The cycle of the four-stroke SI engine consists of the following four strokes.

    (i)   Suction stroke
    (ii)   Compression stroke

(iii)   Expansion or power stroke
(iv)   Exhaust stroke

13. **Indicated Power**: *ip*. The power produced inside the cylinder is called the indicated power.
Indicated power;

$$ip = \frac{p_m \, Al \, nx}{60} \text{ kW}$$

where

$$x = \text{number of cylinder}$$

$p_m =$ mean effective pressure in kPa,
$l =$ stroke length in m,
$A =$ cross-sectional of the cylinder in m^2,
$n =$ speed in rpm,

$\qquad = N$ for two-stroke engine,

$\qquad = N/2$ for four-stroke engine,

where

N is speed of the crank shaft in rpm,
$x =$ number of cylinder

14. **Brake Power**: *bp*. The power available at the crank shaft is called brake power.
Brake power:

$$bp = \frac{2\pi NT}{60} \text{ kW}$$

where

$N =$ speed of the crank shaft in rpm,
$T =$ torque applied on the crank shaft in kNm.

15. **Friction Power**: *fp*. The friction power is the difference between the *ip* and the *b*p.
That is friction power:

$$fp = ip - bp$$

16. **Mean Effective Pressure**: *mep*. The mean effective pressure is the constant hypothetical pressure that, if it acting on the piston during the entire expansion (or working) stroke, would have produced the same amount of net work as that produced during the actual cycle.

Net work produced per cycle $=$ means effective pressure $\times$ displacement volume
$$W_{net} = mep \times V_s$$

or

$$mep = \frac{W_{net}}{V_s}\,\text{kPa}$$

where

W_{net} is in kJ
V_s is in m^3

also

$$mep = \frac{w_{net}}{v_s}\,\text{kPa}$$

where

w_{net} is specific work; kJ/kg
v_s is specific displacement volume; m^3/kg

$\therefore$

$$\boldsymbol{mep = \frac{W_{net}}{V_s} = \frac{w_{net}}{v_s}}$$

The actual mean effective pressure is found by using the indicator diagram as follows:

$$mep = \frac{\text{spring value} \times \text{area}}{\text{base length}} = \frac{ka}{l}\,\text{N/m}^2$$

17. **Indicated Thermal Efficiency**: η_{ith}. It is defined as the ratio of the indicated power (ip) to the energy supplied by the fuel.

$$\eta_{ith} = \frac{ip}{m_f\,CV}$$

18. **Brake Thermal Efficiency**: η_{bth}. It is defined as the ratio of the brake power (bp) to the energy supplied by the fuel.

$$\eta_{bth} = \frac{bp}{m_f\,CV}$$

19. **Mechanical Efficiency**: η_m. It is defined as the ratio of the brake power to the indicated power.

$$\eta_m = \frac{bp}{ip} = \frac{\eta_{bth}}{\eta_{ith}}$$

20. **Relative Efficiency or Efficiency Ratio**: η_{ret}. It is defined as the ratio of actual thermal efficiency to the ideal thermal efficiency.

$$\eta_{ret} = \frac{\text{actual thermal efficiency}}{\text{air - standard efficiency}}$$

Air-standard efficiency of Otto cycle:

$$\eta_{Otto} = 1 - \frac{1}{r^{\gamma-1}}$$

where

$r =$ compression ratio,
$\gamma =$ adiabatic index,
$=1.4$ of air.

Air-standard efficiency of Diesel cycle:

$$\eta_{Diesel} = 1 - \frac{1}{r^{\gamma-1}} \frac{\left(r_c^{\gamma} - 1\right)}{\gamma(r_c - 1)}$$

where

$r =$ compression ratio,
$r_c =$ cut-off ratio,
$\gamma =$ adiabatic index,
$=1.4$ of air.

21. **Volumetric Efficiency**: η_v. It is defined as the ratio of the actual volume sucked at ambient condition to the displacement volume of the engine.

$$\eta_v = \frac{\text{actual volume sucked at ambient condition}}{\text{displacement volume}}$$
$$= \frac{V_a}{V_s}$$

22. **Specific Fuel Consumption**: *sfc*. It is the defined as the ratio of the mass of fuel consumed per hour to the power developed in kW. Thus Specific fuel consumption:

$$sf_c = \frac{\text{fuel consumed in kg/h}}{\text{power developed in kW}}$$

The units of sfc in kg/kWh.

It is reciprocal of the thermal efficiency.

Indicated specific fuel consumption: $isfc$

$$isfc = \frac{\text{fuel consumed in kg/h}}{\text{indicated power in kW}}$$

$$= \frac{m_f}{ip} \text{ kg/kWh}$$

Brake specific fuel consumption: $bsfc$.

$$bsfc = \frac{\text{fuel consumed in kg/h}}{\text{brake power in kW}}$$

$$= \frac{m_f}{ip} \text{ kg/kWh}$$

23. **Piston Speed**: The average distance travelled by the piston per minute in an engine is known as piston speed. Thus

$$\text{Piston speed} = 2\, lN \text{ m/min}$$

where

$l =$ stroke length in m,

$N =$ speed of the crankshaft in rpm.

24. **Heat Balance Sheet**: Heat balance sheet for IC engine given the information about a part of heat supplied is converted into work and the rest part of the heat supplied is carried away by the exhaust gases and the cooling medium.

The parameters are calculated for the heat balance sheet as

(a) Heat supplied; Q_s,

(b) Heat equivalent to brake power,

(c) Heat lost to cooling water, and

(d) Heat carried away by exhaust gases.

Assignment-1

1. Define the heat engine.
2. What is an IC engine?
3. Compare EC and IC engines. Give examples of EC and IC engines.
4. How are IC engines classified?
5. What are the main parts of an IC engine?
6. Define bore, stroke, swept volume, top dead centre, bottom dead centre, and compression ratio.
7. Differentiate between SI and CI engines.
8. What is a four-stroke engine?

9.   What is a two-stroke engine?
10. With the help of neat sketches explain the working of 4-s SI engine.
11. Describe with a neat sketch the working of 4-s diesel engine.
12. Explain the working of 2-s petrol engine.
13. Explain the working of 2-s diesel engine.
14. What is the basic difference between SI and CI engines?
15. What are the advantages and disadvantages of 2-s IC engine over 4-s IC engine?
16. Compare the working of 2-s internal combustion engines and 4-s internal combustion engines.
17. What are the advantages of 2-s engines over 4-s engines?
18. What are the advantages of 4-s engines over 2-s engines?
19. What is CI engine? Why it has more compression ratio compared to SI engine?
20. Define the following efficiencies:

   (i)   Indicated thermal efficiency,
   (ii)   Brake thermal efficiency,
   (iii) Mechanical efficiency,
   (iv)   Relative efficiency, and
   (v)   Volumetric efficiency.

21. Explain the following terms:

   (i)   Mean effective pressure,
   (ii)   Indicated power,
   (iii) Brake power,
   (iv)   Friction power, and
   (v)   Specific fuel consumption.

Assignment-2

1.   A certain engine produces 12 kW indicated power. Its mechanical efficiency is 80%. Determine the brake power and friction power. [**Ans.** 9.6 kW, 2.4 kW]
2.   A four-stroke petrol engine develops 40 kW brake power with an efficiency of 80%. The fuel consumption of the engine is 0.4 kg/kWh and the air–fuel ratio is 14.1. The heating value of the fuel is 43000 kJ/kg. Determine:

   (i)   the indicated power,
   (ii)   the friction power,
   (iii) the brake thermal efficiency,
   (iv)   the fuel consumption per hour, and.
   (v)   the air consumption per hour.
          [**Ans.** (i) 50 kW (ii) 10 kW (iii) 20.93% (iv) 26.16% (v) 16 kg/h (vi) 224 kg/h]

3.   A diesel engine has brake thermal efficiency of 26%. If the calorific value of fuel is 42000 kJ/kg, find its brake-specific fuel consumption. [**Ans.** 0.3296 kg/kWh]

4. A single-cylinder, four-stroke diesel engine having a swept volume of 1000 cc is tested at 350 rpm. When a braking torque of 60 Nm is applied, analysis of an indicator diagram gives a mean effective pressure of 10 bar.
    Determine:

    (i) Brake power,
    (ii) Mechanical efficiency.
        [**Ans.** (i) 2.198 kW (ii) 75.37%]

5. A gas engine has a cylinder 204 mm bore and 432 mm stroke. The clearance volume is $0.00348 \, m^3$. If the relative efficiency is 45%, find the ideal efficiency. Find also indicated specific fuel consumption if the calorific value of the gas is 17500 kJ/m^3.
    [**Ans.** 47.67%, 0.9589 m^3/kWh]

6. The relative efficiency of a 4-s petrol engine working on the ideal cycle is 60%. It has four-cylinder of 76 mm bore and 114 mm stroke, having a compression ratio of 5 and consuming 7 kg of petrol per hour when running at 2000 rpm. Determine the mean effective pressure.
    Take the calorific value of petrol as 42,200 kJ/kg.
    [**Ans.** $p_m = 6.78$ bar].

7. A 4-s petrol engine works on a mean effective pressure of 5 bar and engine speed of 1250 rpm. If the bore is 100 mm and stroke 150 mm, find the indicated power of the engine. [**Ans.** 6.13 kW]

8. A single-cylinder, 4-s cycle oil engine is filled with a rope brake. The diameter of the brake wheel is 600 mm and the rope diameter is 26 mm. Dead load on the brake is 200 N and spring balance reading as 30 N. If the engine runs at 450 rpm, find the brake power of the engine. [**Ans.** 2.50 kW]

9. The following data were noted in a test on a four-cylinder, 4-s engine:

    Diameter: 100 mm
    Stroke: 120 mm
    Speed: 1600 rpm
    Fuel consumption: 10 kg/h
    Calorific value of fuel: 44,000 kJ/kg
    Diameter of the brake wheel: 800 mm
    Net load on the wheel: 400 N
    Mechanical efficiency: 80%

    Determine:

    (i) Brake thermal efficiency,
    (ii) Indicated thermal efficiency,
    (iii) Indicated mean effective pressure, and
    (iv) Brake-specific fuel consumption.
        [**Ans.** (i) 21.92% (ii) 27.40% (iii) 6.66 bar (iv) 0.3732 kg/kWh]

10. The following data were recorded during a trial of a 4-s, single-cylinder oil engine.

Duration of trial: 30 min
Specific gravity of oil: 0.8
Oil consumption: 4 L
Calorific volume of oil: 43,000 kJ/kg
Average area of the indicator diagram: 85 cm^2
Length of the indicator: 8.5 cm
Spring constant: 5.5 bar/cm
Brake load: 150 kg
Spring balance reading: 20 kg
Effective brake wheel diameter: 1.5 m
Speed: 200 rpm
Cylinder diameter: 300 mm
Stroke: 450 mm
Jacket cooling water: 10 kg/min
Temperature rise of cooling water: 36 °C

Determine:

(i)   Indicated power,
(ii)   Brake power,
(iii)   Mechanical efficiency,
(vi)   Brake-specific fuel consumption, and.
(v)   Indicated thermal efficiency.
        [**Ans.** (i) 29.14 kW (ii) 20.02 kW (iii) 68.70%, (iv) 0.3196 kg/kWh (v)
        38.20%]

11. The airflow to a four-cylinder, 4-stroke gasoline engine was measured by means
     of a 80 mm diameter of orifice with the coefficient of discharge is 0.65.
     During a test the following data were recorded:

     Bore: 100 mm
     Stroke: 150 mm
     Speed: 2500 rpm
     Brake power: 36 kW
     Fuel consumption: 10 kg/h
     Calorific value of fuel: 42,000 kJ/kg
     Pressure drop across the orifice: 40 mm of water
     Ambient temperature: 17 °C
     Ambient pressure: 100 kPa

     Determine:

(i)   Brake thermal efficiency,
(ii)   Brake mean effective pressure,
(iii)   Volumetric efficiency based on free air condition.
        [**Ans.** (i) 30.85% (ii) 3.66 bar (iii) 84.98%]

12. During the trial of a single-cylinder, 4-s oil engine, the following results were obtained:

Bore: 200 mm
Stroke: 400 mm
Mean effective pressure: 6 bar
Torque: 407 Nm
Speed: 250 rpm
Oil consumption: 4 kg/h
Calorific value of fuel: 43,000 kJ/kg
Cooling water flow rate: 4.5 kg/min
Air used per kg of fuel: 30 kg
Rise is cooling water temperature: 45 °C
Temperature of exhaust gases: 420 °C
Room temperature: 20 °C
Mean specific heat of exhaust gas: 1 kJ/kgK
Specific heat of water: 4.18 kJ/kgK

Determine:

(i)   Indicated power,
(ii)   Brake power,
(iii) Mechanical efficiency, and
(iv)   Draw up a heat balance sheet for the test on a basis of one hour and express each term as a percentage of the heat supplied.
       [**Ans.** (i) 15.7 kW (ii) 10.64% (iii) 67.77% (iv) ....]

Chapter 4
Air-Standard Cycles

Nomenclature

The following is the nomenclature introduced in this chapter:

p	kPa	Pressure
V	m^3	Volume
v	m^3/kg	Specific volume
T	K	Temperature
s	kJ/kgK	Specific entropy
Q	kW	Rate of heat transfer
q	kJ/kg	Specific heat transfer
W	kW	Rate of work
w	kJ/kg	Specific work
w_{net}	kJ/kg	Specific work output
m	kg/s	Mass flow rate
c_v	kJ/kgK	Specific heat at constant volume
c_p	kJ/kgK	Specific heat at constant pressure
r	—	Compression ratio
γ	—	Adiabatic index
η_{Otto}	—	Thermal efficiency of the Otto cycle
mep or p_m	kPa	Mean effective pressure
V_s	m^3	Swept or displacement volume
v_s	m^3/kg	Specific displacement volume
r_c	—	Cut-off ratio
η_{Diesel}	—	Thermal efficiency of the Diesel cycle
η_{Dual}	—	Thermal efficiency of the dual cycle

(continued)

© The Author(s) 2022
S. Kumar, *Thermal Engineering Volume 2*,
https://doi.org/10.1007/978-3-030-89216-6_4

(continued)

r_p	—	Pressure ratio
R	kJ/kgK	Gas constant
$\eta_{Stirling}$	—	Thermal efficiency of the Stirling cycle
η_{Carnot}	—	Thermal efficiency of the Carnot cycle
$\eta_{Ericsson}$	—	Thermal efficiency of the Ericsson cycle
η_{th}	—	Thermal efficiency
BWR	—	Back work ratio
W_c	kW	Rate of compressor work
W_T	kW	Rate of turbine work
w_c	kJ/kg	Specific compressor work
w_T	kJ/kg	Specific turbine work
T_{min}	K	Minimum temperature
T_{max}	K	Maximum temperature
η_c	—	Compressor efficiency or isentropic efficiency of the compressor
η_T	—	Turbine efficiency or isentropic efficiency of the turbine
A/F	—	Air-fuel ratio
η_o	—	Overall efficiency
η_m	—	Mechanical efficiency

4.1   Introduction

In the previous chapter, we discussed the IC engines (i.e., Otto and Diesel cycles). The word 'cycle' used in context with an IC engine is technically not much accurate because the working fluid does not undergo a thermodynamic cycle. That is, in IC engine air enters the engines, gets mixed with a fuel, undergoes combustion, and exits the engine as exhaust gas. This cycle is called an open cycle, a thermodynamic cycle does not really take place.

The operation of an IC engine can be analysed by assuming that the working fluid is air, no inlet and exhaust are there, heat transfer process is used instead of the combustion process, and air is recirculated in the cycle. Such a cycle is called air-standard cycle. The air-standard cycles are based on certain assumptions:

- Air is the working fluid throughout the entire cycle. There is no change in the mass of the air.
- All processes in the cycle are assumed to be reversible.
- The air is assumed to be an ideal gas with constant specific heat throughout the cycle.
- There is no inlet and exhaust process.

- A heat transfer process is used instead of the combustion process with heat energy transfer from an external source.

The air-standard cycles are used as standard (i.e., ideal) cycles. By using these cycles, we can compare the performance of the actual cycles.

4.2 The Otto Cycle

The Otto cycle is used for petrol engines. It consists of two isentropic processes and two isochoric (i.e., constant volume) processes. The four processes that form the Otto cycle are shown in the p–v and T–s diagrams of Fig. 4.1. The four processes are

1. Process 1–2   Isentropic (or reversible adiabatic) compression,
2. Process 2–3   Heat supplied at constant volume,
3. Process 3–4   Isentropic (or reversible adiabatic) expansion, and
4. Process 4–1   Heat rejected at constant volume.

For process 2–3: Heat supplied at

$$V = C$$

Heat supplied:

$$Q_{2-3} = m\, c_v(T_3 - T_2)$$

For unit mass
Heat supplied:

$$q_{2-3} = c_v(T_3 - T_2)$$

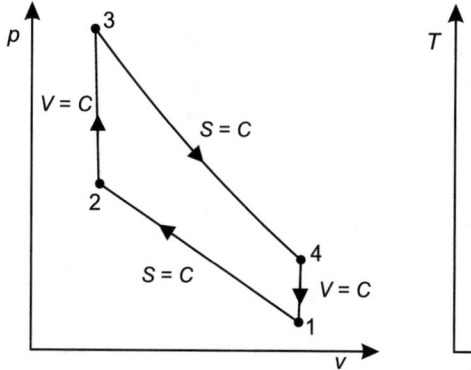 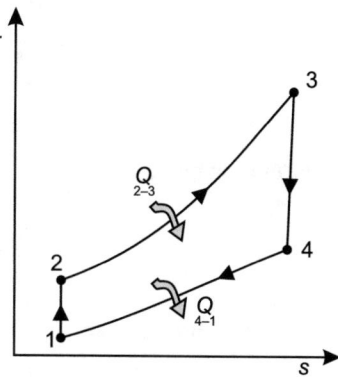

Fig. 4.1 The Otto cycle

For process 4–1: Heat rejected at

$$V = C$$

Heat rejected:

$$Q_{4-1} = m\ c_v(T_4 - T_1)$$

For unit mass
Heat rejected:

$$q_{4-1} = c_v(T_4 - T_1)$$

Net work output:

$$w_{net} = \text{heat supplied} - \text{heat rejected}$$

$$w_{net} = q_{2-3} - q_{4-1}$$

The thermal efficiency of the Otto cycle is defined as the ratio of net work output to heat supplied.
Mathematically,
Thermal efficiency:

$$\eta_{Otto} = \frac{\text{net work output} : w_{net}}{\text{heat supplied} : q_{2-3}}$$

$$\eta_{Otto} = \frac{w_{net}}{q_{2-3}} = \frac{q_{2-3} - q_{4-1}}{q_{2-3}}$$

$$= 1 - \frac{q_{4-1}}{q_{2-3}} = 1 - \frac{c_v(T_4 - T_1)}{c_v(T_3 - T_2)}$$

$$= 1 - \frac{(T_4 - T_1)}{(T_3 - T_2)} = 1 - \frac{T_1\left(\frac{T_4}{T_1} - 1\right)}{T_2\left(\frac{T_3}{T_2} - 1\right)} \qquad (4.1)$$

For isentropic process 1–2:

$$\frac{T_2}{T_1} = \left(\frac{V_1}{V_2}\right)^{\gamma-1} \qquad (4.2)$$

For isentropic process 3–4:

$$\frac{T_3}{T_4} = \left(\frac{V_4}{V_3}\right)^{\gamma-1}$$

As $V_4 = V_1$ and $V_3 = V_2$

$\therefore$

$$\frac{T_3}{T_4} = \left(\frac{V_1}{V_2}\right)^{\gamma - 1}$$
(4.3)

From Eqs. (4.2) and (4.3), we get

$$\frac{T_2}{T_1} = \frac{T_3}{T_4}$$

or

$$\frac{T_4}{T_1} = \frac{T_3}{T_2}$$

Substituting the value of $\frac{T_4}{T_1} = \frac{T_3}{T_2}$ in Eq. (4.1), we get

$$\eta_{\text{Otto}} = 1 - \frac{T_1\left(\frac{T_3}{T_2} - 1\right)}{T_2\left(\frac{T_3}{T_2} - 1\right)} = 1 - \frac{T_1}{T_2}$$

$$\eta_{\text{Otto}} = 1 - \frac{1}{T_2/T_1} = 1 - \frac{1}{\left(\frac{V_1}{V_2}\right)^{\gamma - 1}}$$

$$\boldsymbol{\eta_{\text{Otto}} = 1 - \frac{1}{r^{\gamma - 1}}}$$
(4.4)

where

$r = \frac{V_1}{V_2}$, compression ratio which is defined as the ratio of the maximum volume in the cylinder to the minimum (or clearance) volume,
$\gamma = \frac{c_p}{c_v}$, adiabatic index which is defined as the ratio of the specific heat at constant pressure to the specific heat at constant volume.

The thermal efficiency of the Otto cycle is called the air-standard efficiency. Equation (4.4), clearly indicates that the thermal efficiency of the Otto cycle depends only on the compression ratio r; higher the compression ratio, higher the thermal efficiency (Fig. 4.2).

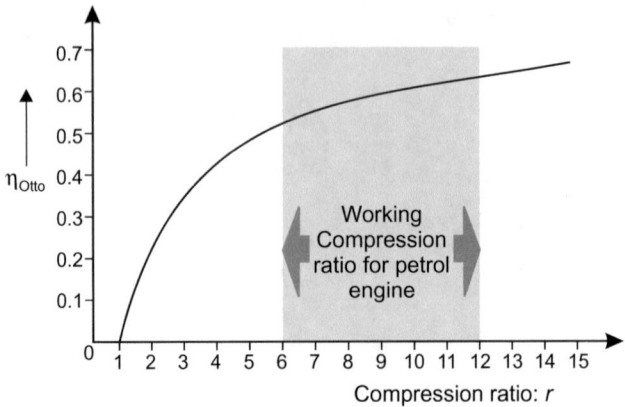

Fig. 4.2 Thermal efficiency of the Otto cycle as a function of compression ratio r at $\gamma = 1.4$

Mean effective pressure: p_m.

The mean effective pressure is the constant hypothetical pressure that, if it acting on the piston during the entire expansion (or working) stroke, would have produced the same amount of net work that produced during the actual cycle.

Net work produced per cycle = mean effective pressure × displacement volume.

$$W_{\text{net}} = p_m \times V_s$$

or

$$p_m = \frac{W_{\text{net}}}{V_s} \ \text{kPa}$$

where

W_{net} = net work output; kJ,
V_s = swept or displacement volume; m³,
$= V_1 - V_2,$

also

$$p_m = \frac{w_{\text{net}}}{v_s} \ \text{kPa}$$

where

w_{net} = net specific work output; kJ/kg,

v_s = specific displacement volume; m³/kg.
= $v_1 - v_2$,

∴

$$p_m = \frac{W_{net}}{V_s} = \frac{w_{net}}{v_s}.$$

4.3 The Diesel Cycle

The diesel cycle is used for diesel engines. It consists of two isentropic processes, one constant pressure process, and one constant volume process. The difference between this cycle and the Otto cycle is that, in diesel cycle, the heat is supplied during a constant pressure process. The four processes that form the diesel cycle are shown in the p–v and T–s diagrams of Fig. 4.3. The four processes are

1. Process 1–2   Isentropic (or reversible adiabatic) compression,
2. Process 2–3   Heat supplied at constant pressure,
3. Process 3–4   Isentropic (or reversible adiabatic) expansion, and
4. Process 4–1   Heat rejected at constant volume.

Process 2–3: Heat supplied at $p = C$.
Heat supplied:

$$Q_{2-3} = m \, c_p (T_3 - T_2)$$

For unit mass
Heat supplied:

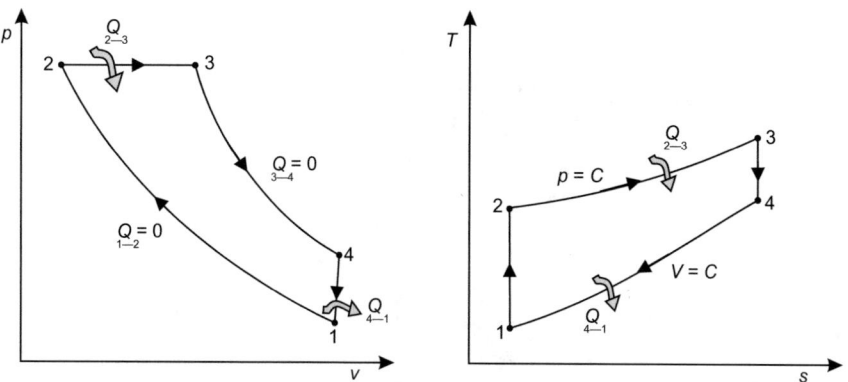

Fig. 4.3   The diesel cycle

$$q_{2-3} = c_p(T_3 - T_2)$$

Cut-off ratio: r_c. It is defined as the ratio of volumes at the end and starting of heat supplied at constant pressure process.

Cut-off ratio:

$$r_c = \frac{V_3}{V_2}$$

Process 4–1: Heat rejected at

$$V = C$$

Heat rejected:

$$Q_{4-1} = m\, c_v(T_4 - T_1)$$

For unit mass
Heat rejected:

$$q_{4-1} = c_v(T_4 - T_1)$$

Net work output:

$$w_{\text{net}} = \text{heat supplied} - \text{heat rejected}$$

$$w_{\text{net}} = q_{2-3} - q_{4-1}$$

The thermal efficiency of the diesel cycle is defined as the ratio of net work output to heat supplied.

Thermal efficiency:

$$\eta_{\text{Diesel}} = \frac{\text{net work output: } w_{net}}{\text{heat supplied: } q_{2-3}}$$

$$= \frac{w_{net}}{q_{2-3}} = \frac{q_{2-3} - q_{4-1}}{q_{2-3}}$$

$$= 1 - \frac{q_{4-1}}{q_{2-3}} = 1 - \frac{c_v(T_4 - T_1)}{c_p(T_3 - T_2)}$$

$$= 1 - \frac{c_v\, T_1 \left(\frac{T_4}{T_1} - 1\right)}{c_p\, T_2 \left(\frac{T_3}{T_2} - 1\right)} = 1 - \frac{1}{\frac{T_2}{T_1} \frac{c_p}{c_v}} \frac{\left(\frac{T_4}{T_1} - 1\right)}{\left(\frac{T_3}{T_2} - 1\right)}$$

$$= 1 - \frac{\left(\frac{T_4}{T_1} - 1\right)}{\left(\frac{T_2}{T_2}\right)\gamma\left(\frac{T_3}{T_2} - 1\right)}$$

Process 1–2:

$$\frac{T_2}{T_1} = \left(\frac{V_1}{V_2}\right)^{\gamma-1} \tag{4.5}$$

$$\frac{T_2}{T_1} = r^{\gamma-1}$$

$\therefore$

$$\eta_{\text{diesel}} = 1 - \frac{1}{r^{\gamma-1}} \frac{\left(\frac{T_4}{T_1} - 1\right)}{\gamma\left(\frac{T_3}{T_2} - 1\right)} \tag{4.6}$$

Process 2–3:

$$\frac{T_3}{V_3} = \frac{T_2}{V_2}$$

or

$$\frac{T_3}{T_2} = \frac{V_3}{V_2}$$

$$\frac{T_3}{T_2} = \mathbf{r_c}$$

Process 3–4:

$$\frac{T_3}{T_4} = \left(\frac{V_4}{V_3}\right)^{\gamma-1}$$

As

$$V_4 = V_1$$

$\therefore$

$$\frac{T_3}{T_4} = \left(\frac{V_1}{V_3}\right)^{\gamma-1} \tag{4.7}$$

Dividing Eq. (4.5) by Eq. (4.7), we get

$$\frac{T_2/T_1}{T_3/T_4} = \frac{(V_1/V_2)^{\gamma-1}}{(V_1/V_3)^{\gamma-1}}$$

$$\frac{T_2}{T_3} \times \frac{T_4}{T_1} = \left(\frac{V_3}{V_2}\right)^{\gamma-1}$$

or

$$\frac{T_4}{T_1} = \frac{T_3}{T_2}\left(\frac{V_3}{V_2}\right)^{\gamma-1}$$

$$\frac{T_4}{T_1} = \frac{V_3}{V_2}\left(\frac{V_3}{V_2}\right)^{\gamma-1} \qquad \because \frac{T_3}{T_2} = \frac{V_3}{V_2}$$

$$\frac{T_4}{T_1} = \left(\frac{V_3}{V_2}\right)^{\gamma}$$

$$\frac{T_4}{T_1} = r_c^{\gamma}$$

Substituting the values of $\frac{T_3}{T_2} = r_c$ and $\frac{T_4}{T_1} = r_c^{\gamma}$ in Eq. (4.6), we get

$$\eta_{\text{Diesel}} = 1 - \frac{1}{r^{\gamma-1}}\left[\frac{(r_c^{\gamma} - 1)}{\gamma(r_c - 1)}\right] \tag{4.8}$$

Equation (4.8) clearly indicates that the thermal efficiency of a diesel cycle differs from the thermal efficiency of an Otto cycle by the quantity in the brackets. The quantity in the brackets is always greater than one. For a given compression ratio, the thermal efficiency of the diesel cycle is less than that of an Otto cycle. For example:

If $r = 12$ and $r_c = 2$,

$$\eta_{\text{Diesel}} = 1 - \frac{1}{12^{1.4-1}}\left[\frac{(2^{1.4} - 1)}{1.4(2 - 1)}\right] = 1 - 0.37[1.1707]$$

$$= 0.5668 = 56.68\%$$

If $r = 12$ and $r_c = 4$

$$\eta_{\text{Diesel}} = 1 - \frac{1}{12^{1.4-1}}\left[\frac{(4^{1.4} - 1)}{1.4(4 - 1)}\right] = 1 - 0.37[1.42]$$

$$= 0.4746 = 47.46\%$$

As r_c increases, the thermal efficiency of the diesel cycle decreases. In practice, diesel engines can be designed to operate at a much higher compression ratio between 12 to 24. Thus, because of the higher compression ratio, a diesel engine operates at a significantly higher thermal efficiency than a petrol engine.

4.4 The Dual Cycle

The dual cycle is used for diesel engines. In this cycle, heat is supplied partly at constant volume and partly at constant pressure. The net heat is supplied in dual process in this cycle. That is why it is called the dual cycle. Heat is rejected at constant volume. The five processes that form the dual cycle are shown in the p–v and T–s diagrams of Fig. 4.4. The five processes are

1. Process 1–2   Isentropic (or reversible adiabatic) compression,
2. Process 2–3   Heat supplied at constant volume,
3. Process 3–4   Heat supplied at constant pressure,
4. Process 4–5   Isentropic (or reversible adiabatic) expansion, and
5. Process 5–1   Heat rejected at constant volume.

Process 2–3: Heat supplied at

$$V = C$$

Heat supplied:

$$Q_{2-3} = m \, c_v (T_3 - T_2)$$

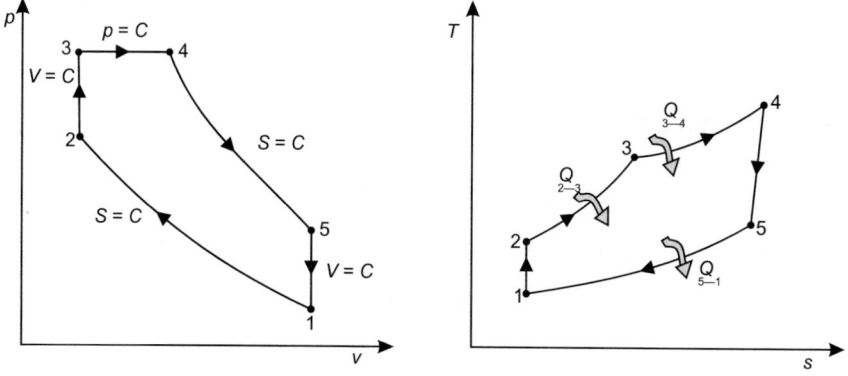

Fig. 4.4 The dual cycle

For unit mass
Heat supplied:

$$q_{2-3} = c_v(T_3 - T_2)$$

Pressure ratio: r_p. It is defined as the ratio of pressure at the end and starting of heat supplied at constant volume process.
Pressure ratio:

$$r_p = \frac{p_3}{p_2}$$

Process 3–4: Heat supplied at

$$p = C$$

Heat supplied:

$$Q_{3-4} = m\,c_p(T_4 - T_3)$$

For unit mass
Heat supplied:

$$q_{3-4} = c_p(T_4 - T_3)$$

Cut-off ratio: r_c. It is defined as the ratio of volumes at the end and starting of heat supplied at constant pressure process.
Cut-off ratio:

$$r_c = \frac{V_4}{V_3}$$

Net heat supplied:

$$q_{net} = q_{2-3} + q_{3-4}$$

For process 5–1: Heat rejected at

$$V = C$$

Heat rejected:

$$Q_{5-1} = m\,c_v(T_5 - T_1)$$

For unit mass

Heat rejected:

$$q_{5-1} = c_v(T_5 - T_1)$$

Net work output:

$$w_{\text{net}} = q_{\text{net}} - q_{5-1}$$

Thermal efficiency:

$$\eta_{\text{Dual}} = \frac{\text{net work output} : w_{\text{net}}}{\text{heat supplied} : q_{\text{net}}}$$

$$= \frac{w_{net}}{q_{net}} = \frac{q_{net} - q_{5-1}}{q_{net}}$$

$$= 1 - \frac{q_{5-1}}{q_{net}} = 1 - \frac{q_{5-1}}{q_{2-3} + q_{3-4}}$$

$$= 1 - \frac{c_v(T_5 - T_1)}{c_v(T_3 - T_2) + c_p(T_4 - T_3)} = 1 - \frac{(T_5 - T_1)}{(T_3 - T_2) + \gamma(T_4 - T_3)}$$

$$= 1 - \frac{T_1\left[\frac{T_5}{T_1} - 1\right]}{T_2\left[\left(\frac{T_3}{T_2} - 1\right) + \frac{\gamma}{T_2}(T_4 - T_3)\right]}$$

$$= 1 - \frac{\left[\frac{T_5}{T_1} - 1\right]}{\frac{T_2}{T_1}\left[\left(\frac{T_3}{T_2} - 1\right) + \frac{\gamma T_3}{T_2}\left(\frac{T_4}{T_3} - 1\right)\right]}$$

For process 1–2:

$$\frac{T_2}{T_1} = \left(\frac{V_1}{V_2}\right)^{\gamma-1} \tag{4.9}$$

or

$$\frac{T_2}{T_1} = r^{\gamma-1}$$

$$\therefore$$

$$\eta_{\text{Dual}} = 1 - \frac{1}{r^{\gamma-1}} \frac{\left[\frac{T_5}{T_1} - 1\right]}{\left[\left(\frac{T_3}{T_2} - 1\right) + \frac{\gamma T_3}{T_2}\left(\frac{T_4}{T_3} - 1\right)\right]} \tag{4.10}$$

For process 4–5

$$\frac{T_4}{T_5} = \left(\frac{V_5}{V_4}\right)^{\gamma-1}$$

As

$$V_5 = V_1$$

∴

$$\frac{T_4}{T_5} = \left(\frac{V_1}{V_4}\right)^{\gamma-1} \tag{4.11}$$

Dividing Eq. (4.9) by Eq. (4.11), we get

$$\frac{T_2/T_1}{T_4/T_5} = \frac{\left(\frac{V_1}{V_2}\right)^{\gamma-1}}{\left(\frac{V_1}{V_4}\right)^{\gamma-1}}$$

$$\frac{T_2}{T_1} \times \frac{T_5}{T_4} = \left(\frac{V_4}{V_2}\right)^{\gamma-1}$$

As

$$V_2 = V_3$$

$$\frac{T_2}{T_4} \times \frac{T_5}{T_1} = \left(\frac{V_4}{V_3}\right)^{\gamma-1}$$

$$\frac{T_5}{T_1} = \frac{T_4}{T_2}\left(\frac{V_4}{V_3}\right)^{\gamma-1}$$

$$\frac{T_5}{T_1} = \frac{T_4}{T_3} \times \frac{T_3}{T_2}\left(\frac{V_4}{V_3}\right)^{\gamma-1} = r_c r_p r_c^{\gamma-1}$$

where

$$\frac{T_4}{T_3} = \frac{V_4}{V_3} = r_c, \text{ cut-off ratio}$$

and

$$\frac{T_3}{T_2} = \frac{p_3}{p_2} = r_p, \text{ pressure ratio}$$

$\therefore$

$$\frac{T_5}{T_1} = r_p r_c^{\gamma}$$

Process 2–3: Heat supplied at

$$V = C$$

$$\frac{p_3}{p_2} = \frac{T_3}{T_2} = r_p, \text{ pressure ratio}$$

or

$$\frac{T_3}{T_2} = r_p$$

Process 3–4: Heat supplied at $p = C$

$$\frac{V_4}{V_3} = \frac{T_4}{T_3} = r_c, \text{ cut-off ratio}$$

or

$$\frac{T_4}{T_3} = r_c$$

Substituting the values of $\frac{T_5}{T_1} = r_p r_c^{\gamma}$, $\frac{T_2}{T_1} = r_p$ and $\frac{T_4}{T_3} = r_c$ in Eq. (4.10), we get

$$\eta_{\text{Dual}} = 1 - \frac{1}{r^{\gamma-1}} \frac{\left(r_p r_c^{\gamma} - 1\right)}{\left[(r_p - 1) + \gamma r_p (r_c - 1)\right]}$$

or

$$\eta_{\text{Dual}} = 1 - \frac{1}{r^{\gamma-1}} \frac{\left(r_p r_c^{\gamma} - 1\right)}{\left[\gamma r_p (r_c - 1) + (r_p - 1)\right]}$$

where

$r = \frac{V_1}{V_2}$, compression ration.

$\gamma = \frac{c_p}{c_v}$, ratio of specific heats. It is called adiabatic index,

$r_c = \frac{V_4}{V_1}$, cut-off ratio,

$r_p = \frac{p_3}{p_2}$, pressure ratio,

Mean effective pressure:

$$p_m = \frac{\text{net work output per cycle}}{\text{displacement volume}}$$
$$= \frac{w_{net}}{v_s} \text{ kPa}$$

where

$$w_{net} = q_{net} - q_{5-1} \text{ kJ/kg}$$
$$= q_{2-3} + q_{3-4} - q_{5-1}$$

and

$$v_s = v_1 - v_2 \text{ m}^3/\text{kg}$$

4.5   Comparison of the Otto, Diesel, and Dual Cycles

The comparison of the Otto, diesel and dual cycle is made on the basic of the following parameters:

(i)   Compression ratio,
(ii)   Maximum pressure,
(iii)   Maximum temperature,
(iv)   Heat supplied, and
(v)   Heat rejection.

The cycles can be analysed for these parameters.

4.5.1   Same Compression Ratio and Heat Supplied

The Otto, dual, and diesel cycles are shown in the p–v and T–s diagrams of Fig. 4.5 according to the given condition of the same compression ratio and heat supplied. These cycles are shown in Fig. 4.5 as

Otto cycle: 1–2–3–4–1,
Dual cycle: 1–2–2′–3′–4′–1, and
Diesel cycle: 1–2–3″–4″–1.

We know that the thermal efficiency of the cycle:

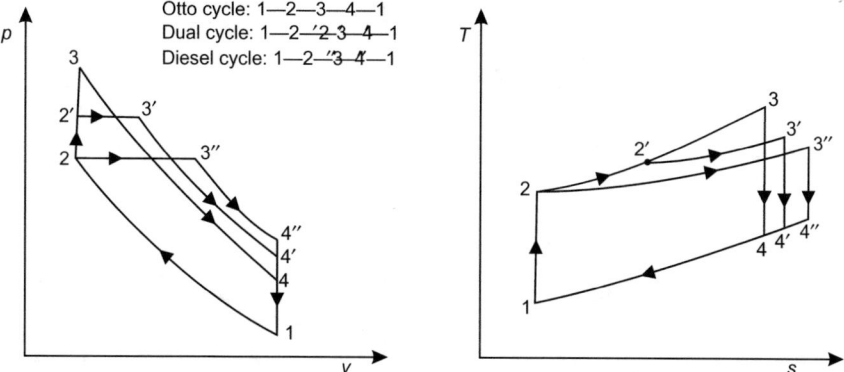

Fig. 4.5 Same compression ratio and heat supplied

$$\eta = 1 - \frac{\text{heat rejection}}{\text{heat supplied}}$$

As the heat supplied is constant. The thermal efficiency of the cycle is dependent only on the heat rejection, higher the heat rejection, lower the thermal efficiency.

It is cleared from T–s diagram of Fig. 4.5, heat rejection is higher in the diesel cycle, lower in the Otto cycle and heat rejection in the dual cycle lies between the Otto cycle and the diesel cycle.

i.e.,

$$(Q_{4''-1})_{\text{Diesel}} > (Q_{4'-1})_{\text{Dual}} > (Q_{4-1})_{\text{Otto}}$$

Thus the thermal efficiency of the Otto cycle is maximum, the thermal efficiency of the diesel cycle is minimum and the dual cycle's thermal efficiency lies between the Otto and the diesel cycles, at given condition of the same compression ratio $\left(\frac{V_1}{V_2}\right)$ and heat supplied.

That is,

$$\boxed{\eta_{\text{Otto}} > \eta_{\text{Dual}} > \eta_{\text{Diesel}}}$$

4.5.2 Same Compression Ratio and Heat Rejection

The Otto, diesel, and dual cycles are shown in the p–v and T–s diagrams of Fig. 4.6 according to given condition of the same compression ratio and heat rejection. These cycles are shown in Fig. 4.6 as

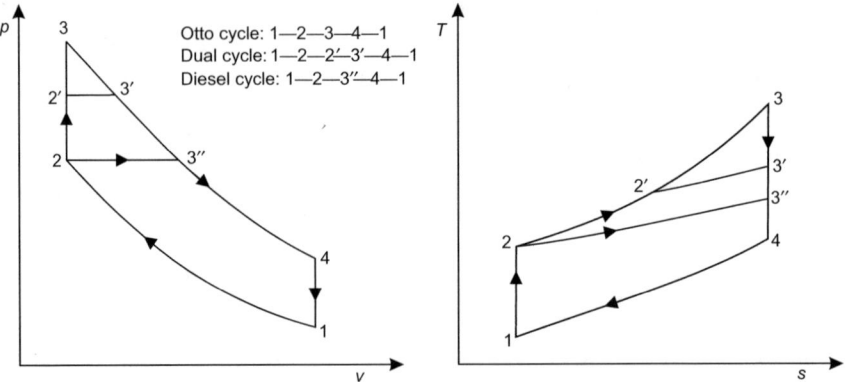

Fig. 4.6 Same compression ratio and heat rejection

Otto cycle: 1–2–3–4–1,
Dual cycle: 1–2–2′–3′–4–1, and
Diesel cycle: 1–2–3″–4–1.

The thermal efficiency:

$$\eta = 1 - \frac{\text{heat rejection}}{\text{heat supplied}}$$

As the heat rejection is constant. The thermal efficiency of the cycle is dependent only on the heat supplied, higher the heat supplied, higher the thermal efficiency.

It is cleared from T–s diagram of Fig. 4.6, heat supplied is lower in the diesel cycle, higher in the Otto cycle. The amount of heat supplied is more in Otto cycle as compared to diesel cycle and is equal to area under 2–3–3–2. The heat supplied to the dual cycle lies between the Otto cycle and the diesel cycle. That is,

$$(Q_{2-3})_{\text{Otto}} > (Q_{2-2'-3'})_{\text{Dual}} > (Q_{2-3''})_{\text{Diesel}}$$

Thus, the thermal efficiency of the Otto cycle is maximum, the thermal efficiency of the diesel cycle is minimum and the dual cycle's thermal efficiency lies between the Otto and the diesel cycles, at given condition of the same compression ratio and heat rejection

$$\boxed{\eta_{\text{Otto}} > \eta_{\text{Dual}} > \eta_{\text{Diesel}}}$$

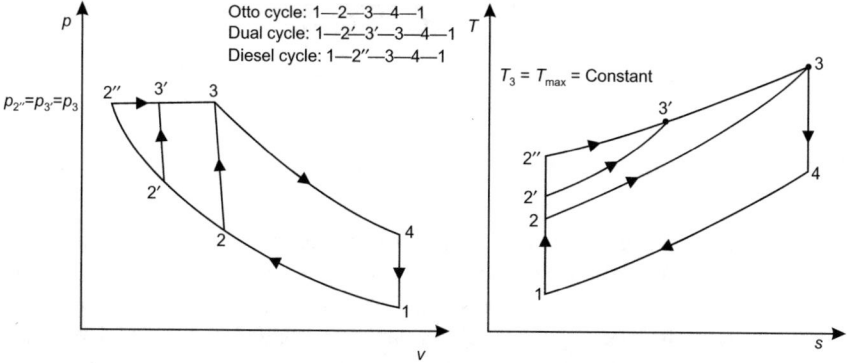

Fig. 4.7 Same maximum pressure, maximum temperature, and heat rejection

4.5.3 Same Maximum Pressure, Maximum Temperature, and Heat Rejection

The Otto, dual and diesel cycles are shown in p–v and T–s diagrams of Fig. 4.7 according to given condition of the same maximum pressure, maximum temperature, and heat rejection. These cycles are shown in Fig. 4.7 as

Otto cycle: 1–2–3–4–1,
Dual cycle: 1–2′–3′–3–4–1, and
Diesel cycle: 1–2″–3–4–1.

The thermal efficiency:

$$\eta = 1 - \frac{\text{heat rejection}}{\text{heat supplied}}$$

As the heat rejection is constant. The thermal efficiency of the cycle is dependent only on the heat supplied, higher the heat supplied, higher the thermal efficiency.

It is cleared from T–s diagram of Fig. 4.7, heat supplied is lower in the Otto cycle, higher in the diesel cycle. The amount of heat supplied in diesel cycle is more than heat supplied in Otto cycle and is equal to area under 2″–3–2–2″. The heat supplied to the dual cycle lies between the Otto and the diesel cycle.

That is,

$$(Q_{2''-3})_{\text{Diesel}} > (Q_{2'-3'-3})_{\text{Dual}} > (Q_{2-3})_{\text{Otto}}$$

Thus the thermal efficiency of the diesel cycle is maximum, the thermal efficiency of the Otto cycle is minimum and the dual cycle's thermal efficiency lies between the Otto and the diesel cycles, at given condition of the same maximum pressure, maximum temperature, and heat rejection.

$$\boxed{\eta_{\text{Diesel}} > \eta_{\text{Dual}} > \eta_{\text{Otto}}}$$

4.6 The Stirling Cycle

The Stirling cycle consists of two isothermal processes and two constant volume processes. The four processes that form the Stirling cycle are shown in the p–v and T–s diagrams of Fig. 4.8. This cycle is made reversible cycle by using regeneration. In regeneration, the working fluid rejects heat to a thermal storage device (called regenerator) during a constant volume heat rejection process 4–1. This heat is stored in the regenerator and supplied back to the working fluid during a constant volume heat supplied process 2–3.

A cycle heat utilizes this type of heating is called regenerative cycle, and the process is referred to as regeneration.

∴

$$Q_{4-1} = Q_{2-3} \text{ i.e., the area under process } 4-1 \text{ is equal}$$
$$\text{to the area under process } 2-3$$

Heat supplied during process 3–4 at $T = C$,

$$Q_{3-4} = m R T_3 \log_e \frac{V_4}{V_3}$$

As

$$V_4 = V_1 \text{ and } V_3 = V_2$$

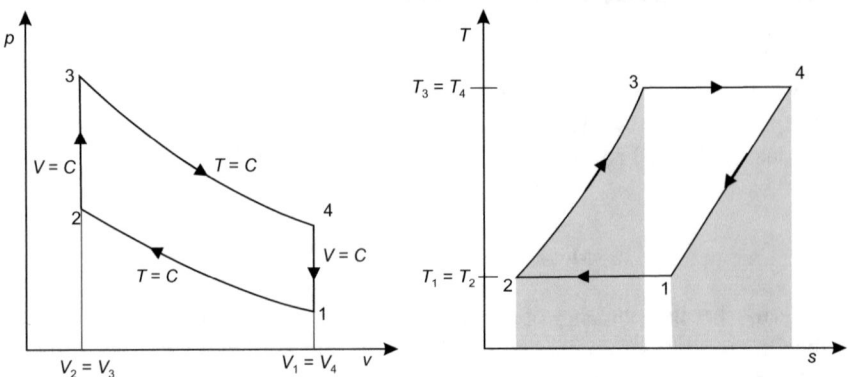

Fig. 4.8 The Stirling cycle

$\therefore$

$$Q_{3-4} = mRT_3 \log_e \frac{V_1}{V_2}$$

Heat rejected during process 1–2 at $T = C$,

$$Q_{1-2} = mRT_1 \log_e \frac{V_1}{V_2}$$

Net work output:

$$W_{net} = \text{heat supplied} - \text{heat rejected}$$
$$= Q_{3-4} - Q_{1-2}$$

Thermal efficiency:

$$\eta_{Striling} = \frac{\text{net work output}}{\text{heat supplied}}$$
$$= \frac{W_{net}}{Q_{3-4}} = \frac{Q_{3-4} - Q_{1-2}}{Q_{3-4}}$$
$$= 1 - \frac{Q_{1-2}}{Q_{3-4}} = 1 - \frac{mRT_1 \log_e \frac{V_1}{V_2}}{mRT_3 \log_e \frac{V_1}{V_2}}$$

$$\eta_{Stirling} = 1 - \frac{T_1}{T_3}$$
$$= \eta_{Carnot}$$

Thus, the thermal efficiency of the regeneration Stirling cycle is equal to the Carnot cycle when operating between same temperature limits.

4.7 The Ericsson Cycle

The Ericsson cycle consists of two isothermal processes and two constant pressure processes. The four processes that form the Ericsson cycle are shown in Fig. 4.9. Like Stirling cycle, it is made reversible cycle by using regeneration. In regeneration, the working fluid rejects heat to a thermal storage device (called regenerator) during a constant pressure heat rejection process 4–1. This heat is stored in the regenerator and supplied back to the working fluid during a constant pressure heat supplied process 2–3.

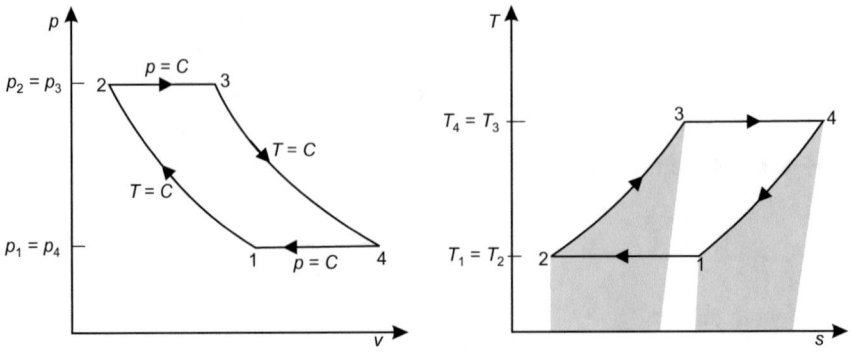

Fig. 4.9 The Ericsson cycle

$$\therefore$$

$$Q_{4-1} = Q_{2-3} \text{ i.e., the area under process } 4 - 1 \text{ is equal}$$
$$\text{to the area under process } 2 - 3$$

Heat supplied during process 3–4 at $T = C$:

$$Q_{3-4} = mRT_3 \log_e \frac{V_4}{V_3}$$

$$= mRT_3 \log_e \frac{p_3}{p_4} \qquad \left| \therefore \frac{V_4}{V_3} = \frac{p_3}{p_4} \right.$$

Heat rejected during process 1–2 at $T = C$

$$Q_{1-2} = mRT_1 \log_e \frac{V_1}{V_2}$$

$$= mRT_1 \log_e \frac{p_2}{p_1} \qquad \left| \therefore \frac{V_1}{V_2} = \frac{p_2}{p_1} \right.$$

As

$$p_2 = p_3, \quad p_1 = p_4$$

$$\therefore$$

$$Q_{1-2} = RT_1 \log_e \frac{p_3}{p_4}$$

Net work output:

$$W_{net} = \text{heat supplied} - \text{heat rejected}$$
$$= Q_{3-4} - Q_{1-2}$$

Thermal efficiency:

$$\eta_{\text{Ericsson}} = \frac{\text{net work output}}{\text{heat supplied}}$$
$$= \frac{W_{net}}{Q_{3-4}} = \frac{Q_{3-4} - Q_{1-2}}{Q_{3-4}}$$
$$= 1 - \frac{Q_{1-2}}{Q_{3-4}} = 1 - \frac{m\,RT_1 \log_e \frac{p_3}{p_4}}{m\,RT_3 \log_e \frac{p_3}{p_4}}$$

$$\eta_{\text{Ericsson}} = 1 - \frac{T_1}{T_3}$$
$$= \eta_{\text{Carnot}}$$

Thus, the thermal efficiency of the regeneration Ericsson cycle is equal to the Carnot cycle when operating between same temperature limits.

Problem 4.1: An ideal petrol engine has a diameter of 15 cm and a stroke of 45 cm. If the clearance volume is 2000 cm^3, determine the air-standard efficiency.

Solution: Given data:
Diameter:

$$d = 15\,\text{cm}$$

Stroke:

$$l = 45\,\text{cm}$$

Clearance volume:

$$V_c = 2000\,\text{cm}^2$$

∴ Swept volume:

$$V_S = \frac{\pi}{4}d^2 l = \frac{3.14}{4} \times (15)^2 \times 45 = 7948.125\,\text{cm}^3$$

Total volume:

$$V_1 = V_C + V_S = 2000 + 7948.125 = 9948.125\,\text{cm}^3$$

Compression ratio:

$$r = \frac{V_1}{V_C} = \frac{9948.125}{2000} = 4.97$$

Air standard efficiency of an ideal petrol engine:

$$\eta = 1 - \frac{1}{r^{\gamma-1}} = 1 - \frac{1}{(4.97)^{1.4-1}} = 0.4734 = \mathbf{47.34\%}$$

Problem 4.2: In an air-standard Otto cycle, the compression ratio is 7 and compression begins at 1 bar and 313 K. The heat added is 251 kJ/kg. Find (i) Maximum temperature and pressure of the cycle (ii) Work done per kg of air (iii) Cycle efficiency (iv) Mean effective pressure.

Solution: Given data for Otto cycle: (refer Fig. 4.10)
Compression ratio:

$$r = \frac{v_1}{v_2} = 7$$

At the beginning of compression:

$$p_1 = 1 \text{ bar} = 100\text{kPa}$$
$$T_1 = 313 \text{ K}$$

Heat added:

$$q_{2-3} = 251 \text{ kJ/kg}$$

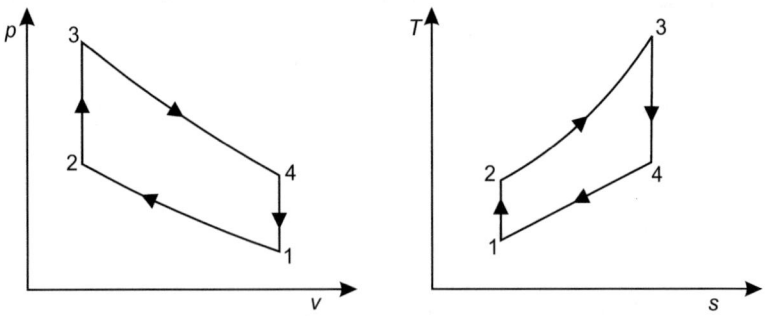

Fig. 4.10 p-v and T-s diagrams for Problem 4.2

Remember

$$\left.\begin{array}{l} c_v = 0.718 \text{ kJ/kgK} \\ c_p = 1.005 \text{ kJ/kgK} \\ R = 0.287 \text{ kJ/kgK} \end{array}\right\} \text{ for air}$$

At state 1:

$$p_1 v_1 = RT_1$$

$$100 \times v_1 = 0.287 \times 313$$

or

$$v_1 = 0.898 \text{ m}^3/\text{kg}$$

also

$$v_1 = 7v_2$$

$\therefore$

$$0.898 = 7 \times v_2$$

or

$$v_2 = 0.1282 \text{ m}^3/\text{kg}$$

Process 1–2 (Isentropic compression):

$$\frac{T_2}{T_1} = \left(\frac{v_1}{v_2}\right)^{\gamma-1}$$

$$T_2 = T_1 \left(\frac{v_1}{v_2}\right)^{\gamma-1} = 313(7)^{1.4-1} = 681.68 \text{ K}$$

and

$$\frac{T_2}{T_1} = \left(\frac{p_2}{p_1}\right)^{\frac{\gamma-1}{\gamma}}$$

$$\frac{681.68}{313} = \left(\frac{p_2}{1}\right)^{\frac{1.4-1}{1.4}}$$

or

$$p_2 = 15.24 \text{ bar}$$

Process 2–3 (Constant volume heat addition):

$$q_{2-3} = c_v(T_3 - T_2)$$

$$251 = 0.718(T_3 - 681.68)$$

or

$$T_3 = 1031.26 \text{ K}$$

Process 3–4 (Isentropic expansion)

$$\frac{T_3}{T_4} = \left(\frac{v_4}{v_3}\right)^{\gamma-1}$$

$$\frac{1031.26}{T_4} = \left(\frac{v_1}{v_2}\right)^{1.4-1} = (7)^{0.4} \qquad \because v_1 = v_4, v_2 = v_3$$

or

$$T_4 = 473.50 \text{ K}$$

$$\frac{p_2 v_2}{T_2} = \frac{p_3 v_3}{T_3}$$

$$\frac{p_2}{T_2} = \frac{p_3}{T_3} \qquad \because v_2 = v_3$$

$$\frac{15.24}{681.68} = \frac{p_3}{1031.26}$$

or

$$p_3 = 23.05 \text{ bar}$$

(i)   Maximum temperature: $T_{\max} = T_3 = \mathbf{1031.26\ K}$
      Maximum pressure: $p_{\max} = p_3 = \mathbf{23.05\ bar}$.

(ii)   Work done per kg of air:

$$w = q_{2-3} - q_{4-1}$$
$$= 251 - c_v(T_4 - T_1)$$
$$= 251 - 0.718(473.5 - 313) = \mathbf{135.76\,kJ/kg}$$

(iii)   Cycle efficiency:

$$\eta_{th} = \frac{w}{q_{2-3}} = \frac{135.76}{251} = 0.5408 = \mathbf{54.08\%}$$

Also determine by the following expression

$$\eta_{th} = 1 - \frac{1}{r^{\gamma-1}}$$

(iv)   Mean effective pressure:

$$mep = \frac{\text{work done}}{\text{displacement volume}}$$
$$= \frac{w}{v_1 - v_2} = \frac{135.76}{0.898 - 0.1282} = \mathbf{176.35\,kPa}$$

Problem 4.3: Calculate the percentage loss in the ideal efficiency of a diesel engine with compression ratio 14, if the fuel cut-off is delayed from 5 to 8%.

Solution: Given data:
   Compression ratio:

$$r = 14$$

Case-I: Cut-off ratio:

$$r_c = 5\% = 0.05$$

We know that the ideal efficiency of a diesel engine:

$$\eta' = 1 - \frac{1}{r^{\gamma-1}}\left[\frac{r_c^{\gamma} - 1}{\gamma(r_c - 1)}\right] = 1 - \frac{1}{(14)^{1.4-1}}\left[\frac{0.05^{1.4} - 1}{1.4(0.05 - 1)}\right]$$
$$= 1 - 0.3479 \times [0.7405] = 1 - 0.2576 = 0.7424$$

Case-II: Cut-off ratio:

$$r_c = 8\% = 0.08$$

$$\therefore$$

$$\eta' = 1 - \frac{1}{r^{\gamma-1}}\left[\frac{r_c^{\gamma} - 1}{\gamma(r_c - 1)}\right] = 1 - \frac{1}{(14)^{1.4-1}}\left[\frac{0.08^{1.4} - 1}{1.4(0.08 - 1)}\right]$$

$$= 1 - 0.3479[0.7537] = 0.7377.$$

Loss in efficiency

$$= \frac{\eta - \eta'}{\eta} = \frac{0.7428 - 0.7377}{0.7428} = 0.0068 = \mathbf{0.68\%}$$

Problem 4.4: In an air-standard diesel cycle, the compression ratio is 16 and at the beginning of isentropic compression, the temperature is 15 °C and the pressure is 0.1 MPa. Heat is added until the temperature at the end of the constant pressure process is 1480 °C. Calculate (i) the cut-off ratio, (ii) the heat supplied per kg of air, (iii) the cycle efficiency, and (iv) the mean effective pressure.

Solution: Given data for a Diesel cycle: (refer Fig. 4.11)
Compression ratio:

$$r = \frac{v_1}{v_2} = 16$$

$$T_1 = 15\,°C = 288\ K$$

$$p_1 = 0.1\,MPa = 100\,kPa$$

$$T_3 = 1480\,°C = 1753\ K$$

Now applying the equation of state at state 1, we get

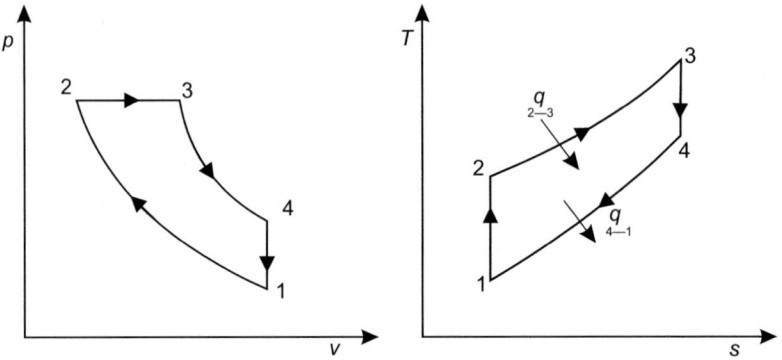

Fig. 4.11 p-v and T-s diagrams for Problem 4.4

$$p_1 v_1 = RT_1$$

$$100 \times v_1 = 0.287 \times 288$$

or

$$v_1 = 0.8265 \,\text{m}^3/\text{kg}$$

also

$$v_1 = 16\, v_2$$

$$\therefore$$

$$0.8265 = 16\, v_2$$

or

$$v_2 = 0.0516 \,\text{m}^3/\text{kg}$$

For process 1–2, adiabatic compression:

$$\frac{T_2}{T_1} = \left(\frac{v_1}{v_2}\right)^{\gamma-1}$$

$$\frac{T_2}{288} = (16)^{1.4-1}$$

or

$$T_2 = 873.05 \text{ K}$$

also

$$\frac{p_2}{p_1} = \left(\frac{v_1}{v_2}\right)^{\gamma}$$

$$\frac{p_2}{100} = (16)^{1.4}$$

or

$$p_2 = 4850.29 \, \text{kPa}$$

For process 2–3, heat supplied at $p = C$

$$\frac{T_3}{T_2} = \frac{v_3}{v_2}$$

$$\frac{1753}{873.05} = \frac{v_3}{0.0516}$$

or

$$v_3 = 0.1036 \, \text{m}^3/\text{kg}$$

Heat supply per kg of air: q_{2-3}

$$q_{2-3} = c_p(T_3 - T_2)$$
$$= 1.005(1753 - 873.05) = \mathbf{884.35 \, kJ/kg}$$

For process 3–4, adiabatic expansion:

$$\frac{T_3}{T_4} = \left(\frac{v_4}{v_3}\right)^{\gamma-1}$$

$$\frac{1753}{T_4} = \left(\frac{0.8265}{0.1036}\right)^{1.4-1} \qquad \because v_4 = v_1$$

or

$$T_4 = 763.88 \, \text{K}$$

For process 4–1, heat rejection at

$$V = C$$

Heat rejection per unit mass of air:

$$q_{4-1} = c_v(T_4 - T_1)$$
$$= 0.718(763.88 - 288) = 341.68 \, \text{kJ/kg}$$

(i)   The cut-off ratio: r_c

$$r_c = \frac{v_3}{v_2} = \frac{0.1036}{0.0516} = \mathbf{2}$$

(ii)   The heat supplied per kg of air:

$$q_{2-3} = \textbf{884.35 kJ/kg}$$

(iii)   The cycle efficiency: η_{Diesel}

$$\eta_{\text{Diesel}} = \frac{\text{work output per kg of air}}{\text{heat supplied per kg of air}}$$

$$= \frac{w}{q_{2-3}} = \frac{q_{2-3} - q_{4-1}}{q_{2-3}} = 1 - \frac{q_{4-1}}{q_{2-3}}$$

$$= 1 - \frac{341.68}{884.35} = 0.6136 = \textbf{61.36\%}$$

(iv)   The mean effective pressure: *mep*

$$mep = \frac{\text{work done per cycle}}{\text{swept volume}} = \frac{w}{v_1 - v_2}$$

$$= \frac{q_{2-3} - q_{4-1}}{v_1 - v_2} = \frac{884.34 - 341.68}{0.8265 - 0.0516} = \textbf{700.29kPa}$$

Problem 4.5: An ideal diesel engine has compression ratio of 20 and use air as the working fluid. The state of air at the beginning of the compression process is 95 kPa and 20 °C. If the maximum temperature in the cycle is not to exceed 2200 K, determine (i) the heat addition, (ii) the heat rejection, (iii) the thermal efficiency, and (iv) the mean effective pressure.

Solution: Given data: (refer Fig. 4.12)
  Compression ratio:

$$r = \frac{v_1}{v_2} = 20$$

$$p_1 = 95 \, \text{kPa}$$

$$T_1 = 20\,^\circ\text{C} = 20 + 273 = 293 \, \text{K}$$

$$T_3 = 2200 \, \text{K}$$

Process 1–2 (Isentropic compression):

$$\frac{T_2}{T_1} = \left(\frac{v_1}{v_2}\right)^{\gamma-1}$$

$$T_2 = T_1 r^{\gamma-1} = 293 \times (20)^{1.4-1} = 971.13 \, \text{K}$$

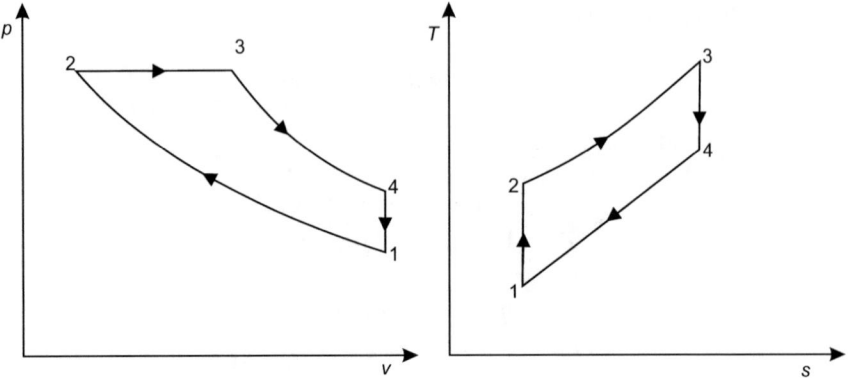

Fig. 4.12 p-v and T-s diagrams for Problem 4.5

Process 2–3 (Constant pressure heat addition):

$$\frac{p_2 v_2}{T_2} = \frac{p_3 v_3}{T_3}$$

or

$$\frac{v_2}{T_2} = \frac{v_3}{T_3} \qquad \because p_1 = p_3$$

$$\frac{v_3}{v_2} = \frac{T_3}{T_2} = \frac{2200}{971.13}$$

Cut-off ratio:

$$r_c = \frac{v_3}{v_2} = 2.26$$

At state 1,

$$p_1 v_1 = R\,T_1$$

$$v_1 = \frac{RT_1}{p_1} = \frac{0.287 \times 293}{95} = 0.8851\,\text{m}^3/\text{kg}$$

and

$$\frac{v_1}{v_2} = 20$$

or

$$v_2 = \frac{v_1}{20} = \frac{0.8851}{20} = 0.0442 \, \text{m}^3/\text{kg}.$$

Cut-off ratio:

$$\frac{v_3}{v_2} = 2.26$$

or

$$v_3 = 2.26 v_2 = 2.26 \times 0.0442 = 0.09989 \, \text{m}^3/\text{kg}.$$

Process 3–4 (Isentropic expansion):

$$\frac{T_3}{T_4} = \left(\frac{v_4}{v_3}\right)^{\gamma-1} \qquad \because v_1 = v_4$$

$$\frac{T_3}{T_4} = \left(\frac{v_1}{v_3}\right)^{\gamma-1}$$

$$\frac{2200}{T_4} = \left(\frac{0.8851}{0.09989}\right)^{1.4-1}$$

or

$$T_4 = 919.25 \, \text{K}.$$

(a) Heat addition (Process 2–3 is a constant pressure heat addition process):

$$Q_A = m \, c_p (T_3 - T_2)$$

or

$$q_A = \frac{Q_A}{m} = c_p (T_3 - T_2)$$
$$= 1.005(2200 - 971.13) = 1235.01 \, \text{kJ/kg}$$

(b) Heat rejection (Process 4–1 is a constant volume heat rejection):

$$Q_R = m \, c_v (T_4 - T_1)$$

$$q_R = \frac{Q_A}{m} = c_v (T_4 - T_1)$$
$$= 0.718(919.25 - 293) = 449.64 \, \text{kJ/kg}$$

(c)   Thermal efficiency:

$$\eta_{th} = \frac{w_{net}}{q_A} = \frac{q_A - q_R}{q_A} = 1 - \frac{q_R}{q_A}$$

$$= 1 - \frac{449.64}{1235.01} = 1 - 0.3640 = 0.636 = \textbf{63.60\%}$$

The thermal efficiency of ideal diesel cycle could also be determined by following expression:

$$\eta_{th} = 1 - \frac{1}{r^{\gamma-1}}\left[\frac{r_c^{\gamma} - 1}{\gamma(r_c - 1)}\right]$$

(d)   The mean effective pressure: *mep*

$$mep = \frac{\text{work output/cycle}}{\text{swept volume}}$$

$$= \frac{w_{net}}{v_1 - v_2} = \frac{q_A - q_R}{v_1 - v_2} = \frac{1235.01 - 449.64}{0.8851 - 0.0442} = \textbf{933.96 kPa}$$

Problem 4.6: In an oil engine, working on dual cycle, the temperature and pressure at the beginning of compression are 90 °C and 100 kPa. The compression ratio of cycle is 13: 1. Heat supplied per kg of air is 1000 kJ, half of which is supplied at constant volume and half at constant pressure. Calculate:

(i)   maximum pressure in the cycle,
(ii)   percentage of stroke at which cut-off occurs.

Assume $\gamma = 1.4$, $c_v = 0.7106$ kJ/kgK and $R = 0.287$ kJ/kgK.

Solution: Given data:(refer Fig. 4.13)

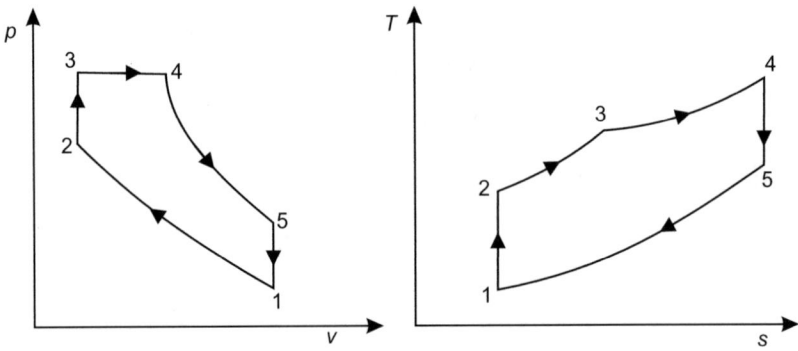

Fig. 4.13 p-v and T-s diagrams for Problem 4.6

$$T_1 = 90\,°C = 363 \ K$$

$$p_1 = 100\,kPa$$

Compression ratio:

$$r = \frac{v_1}{v_2} = 13$$

Heat supplied:

$$q_{2-3-4} = 1000 \ kJ/kg.$$

Heat supplied:

$$q_{2-3} = \frac{q_{2-3-4}}{2} = \frac{1000}{2} = 500\,kJ/kg \qquad \text{at } v = C.$$

Heat supplied:

$$q_{3-4} = \frac{q_{2-3-4}}{2} = \frac{1000}{2} \qquad \text{at } p = C$$

$$q_{3-4} = 500\,kJ/kg$$

$$\gamma = 1.4$$

$$c_v = 0.7106\,kJ/kgK$$

$$R = 0.287\,kJ/kgK$$

also

$$R = c_p - c_v.$$

$$\therefore$$

$$0.287 = c_p - 0.7106$$

or

$$c_p = 0.9976\,kJ/kgK$$

Now applying the equation of state at state 1, we get

$$p_1 v_1 = RT_1$$

$$100 \times v_1 = 0.287 \times 363$$

or

$$v_1 = 1.0418\,\text{m}^3/\text{kg}$$

also

$$v_1 = 13 v_2$$

$\therefore$

$$1.0418 = 13 \times v_2.$$

or

$$v_2 = 0.0801\,\text{m}^3/\text{kg}$$

For process 1–2, adiabatic compression:

$$\frac{T_2}{T_1} = \left(\frac{v_1}{v_2}\right)^{\gamma-1}$$

$$\frac{T_2}{363} = (13)^{1.4-1}$$

or

$$T_2 = 1012.70\,\text{K}$$

and

$$\frac{p_2}{p_1} = \left(\frac{v_1}{v_2}\right)^{\gamma}$$

$$\frac{p_2}{100} = (13)^{1.4}$$

or

$$p_2 = 3626.77\,\text{kPa}.$$

For process 2–3, heat supplied at $v = C$

$$q_{2-3} = c_v(T_3 - T_2)$$

$$500 = 0.7106(T_3 - 1012.70)$$

or

$$T_3 = 1716.33\,\text{K}$$

and

$$\frac{T_3}{T_2} = \frac{p_3}{p_2}$$

$$\frac{1316.33}{1012.70} = \frac{p_3}{3626.77}$$

or

$$p_3 = \mathbf{4714.15\,kPa}$$
$$= p_4.$$

(i)   The maximum pressure in the cycle is **4714.15 kPa**.
(ii)   The percentage of stroke at which cut-off occurs:

For process 3–4, heat supplied at $p = C$

$$q_{3-4} = c_p(T_4 - T_3)$$

$$500 = 0.9976(T_4 - 1316.33)$$

or

$$T_4 = 1817.53\ \text{K}.$$

and

$$\frac{T_4}{T_3} = \frac{v_4}{v_3}$$

$$\frac{1817.53}{1316.33} = \frac{v_4}{0.0801} \qquad | \because v_3 = v_2$$

or

$$v_4 = 0.11059 \, \text{m}^3/\text{kg}.$$

and swept volume:

$$v_s = v_1 - v_2 = 1.0418 - 0.0801 = 0.9617 \, \text{m}^3/\text{kg}.$$

The percentage of stroke at which cut-off occurs:

$$= \frac{v_4}{v_s} \times 100 = \frac{0.11059}{0.9617} \times 100 = 11.499\% \approx \mathbf{11.50\%}$$

Problem 4.7: A diesel engine running on dual cycle takes in air at 100 kPa and 15 °C and has a compression ratio of 16. The maximum pressure in the cycle is 5 MPa. It takes two-thirds of the total heat supply at constant volume and one-third at constant pressure. Calculate (i) the temperatures and pressures at all cardinal points of the cycle, (ii) the ideal thermal efficiency of the cycle, and (iii) the mean effective pressure of the cycle.

Solution: Given data:(refer Fig. 4.14)
   At state 1,

$$p_1 = 100 \, \text{kPa}$$
$$T_1 = 15 \, ^\circ \text{C} = 288 \, \text{K}.$$

Compression ratio:

$$r = \frac{v_1}{v_2} = 16.$$

Maximum in the cycle:

$$p_3 = p_4 = 5 \, \text{MPa} = 5000 \, \text{kPa}.$$

Let q_A is total heat addition per unit mass at constant volume and constant pressure, Heat addition at constant volume:

$$q_{2-3} = \frac{2}{3} q_A.$$

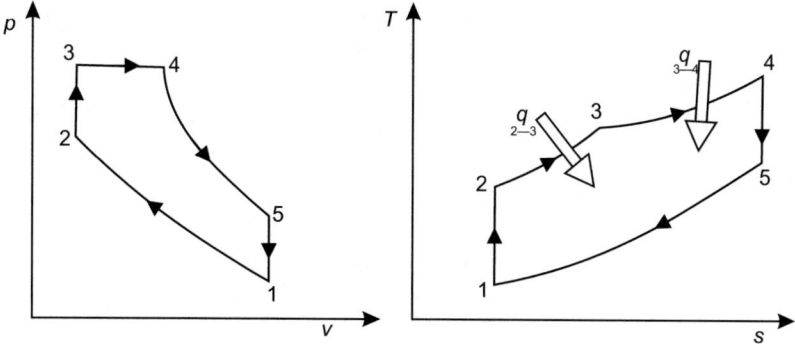

Fig. 4.14 p-v and T-s diagrams for Problem 4.7

Heat addition at constant pressure:

$$q_{3-4} = \frac{1}{3}q_A.$$

Applying the equation of state at state 1,

$$p_1 v_1 = RT_1$$

$$100 \times v_1 = 0.287 \times 288$$

$$v_1 = 0.8265 \,\mathrm{m^3/kg}$$

also

$$v_1 = 16\, v_2.$$

$$\therefore$$

$$0.8265 = 16 \times v_2.$$

or

$$v_2 = 0.0516\,\mathrm{m^3/kg}.$$

(i)   The temperatures and pressures at all cardinal points of the cycle.

For process 1–2, adiabatic compression.

$$\frac{T_2}{T_1} = \left(\frac{v_1}{v_2}\right)^{\gamma-1}$$

$$\frac{T_2}{288} = (16)^{1.4-1}$$

or

$$T_2 = \mathbf{873.05K}$$

also

$$\frac{p_2}{P_1} = \left(\frac{v_1}{v_2}\right)^{\gamma}$$

$$\frac{p_2}{100} = (16)^{1.4}$$

or

$$p_2 = \mathbf{4850.29\,kPa}$$

For process 2–3, heat addition at $v = C$

$$\frac{T_3}{T_2} = \frac{p_3}{p_2}$$

$$\frac{T_3}{873.05} = \frac{5000}{4850.29}$$

$$T_3 = \mathbf{899.99\,K}$$

$$q_{2-3} = c_v(T_3 - T_2) = 0.718(899.99 - 873.05) = 19.34\,kJ/kg$$

also

$$q_{2-3} = \frac{2}{3}q_A.$$

∴

$$19.34 = \frac{2}{3}q_A.$$

or

$$q_A = 29.01 \text{ kJ/kg}.$$

For process 3–4, heat addition at $p = c$

$$q_{3-4} = \frac{1}{3} q_A = \frac{1}{3} \times 29.01 = 9.67 \text{ kJ/kg}$$

also

$$q_{3-4} = c_p (T_4 - T_3)$$

$$9.67 = 1.005(T_4 - 899.99)$$

or

$$T_4 = \mathbf{909.61 \text{ K}}.$$

and

$$\frac{T_4}{T_3} = \frac{v_4}{v_3}$$

$$\frac{909.61}{899.99} = \frac{v_4}{0.0516} \qquad \because v_3 = v_2$$

or

$$v_4 = 0.05215 \text{ m}^3/\text{kg}.$$

For process 4–5, adiabatic expansion:

$$\frac{T_4}{T_5} = \left(\frac{v_5}{v_4} \right)^{\gamma - 1}$$

$$\frac{909.61}{T_5} = \left(\frac{0.8265}{0.05215} \right)^{1.4-1} \qquad \because v_5 = v_1$$

$$\frac{909.61}{T_5} = 3.02$$

or

$$T_5 = \mathbf{301.19 \text{ K}}.$$

and

$$\frac{p_4}{p_5} = \left(\frac{v_5}{v_4}\right)^{\gamma}$$

$$\frac{50,000}{p_5} = \left(\frac{v_1}{v_4}\right)^{\gamma} \qquad \because v_5 = v_1$$

$$\frac{5000}{p_5} = \left(\frac{0.8265}{0.05215}\right)^{1.4}$$

$$\frac{5000}{p_5} = 47.86$$

or

$$p_5 = \mathbf{104.47\,kPa}.$$

(ii)   The ideal thermal efficiency of the cycle: η_{Dual}

$$\eta_{\mathrm{Dual}} = \frac{\text{work output per unit mass of air}}{\text{heat addition per unit mass of air}} = \frac{w}{q_A}$$

where

$$w = \text{heat addition} - \text{heat rejection}$$
$$= q_A - q_R = 29.01 - c_v(T_5 - T_1)$$
$$= 29.01 - 0.718(301.19 - 288)$$
$$= 29.01 - 9.47 = 19.54\,\mathrm{kJ/kg}$$

$$\eta_{\mathrm{Dual}} = \frac{19.54}{29.01} = 0.6735 = \mathbf{67.35\%}$$

(iii)   The mean effective pressure of the cycle: *mep*

$$mep = \frac{\text{work done per cycle}}{\text{swept volume}}$$

$$= \frac{w}{v_1 - v_2} = \frac{19.54}{0.8265 - 0.0516} = \mathbf{25.216\,kPa}$$

Problem 4.8: The compression ratio of an ideal dual cycle is 16. Air is at 1 bar and 17 °C at beginning of the compression process, and the mean effective pressure is 10 bar. The thermal efficiency of the cycle is 60%. If maximum pressure in the cycle is 60 bar, determine the maximum temperature in the cycle.

Solution: Given data: (refer Fig. 4.15)

$$mep = 10\,\text{bar} = 10 \times 10^2\,\text{kPa}$$
$$p_1 = 1\,\text{bar} = 1 \times 10^2\,\text{kPa}$$
$$T_1 = 17\,^\circ\text{C} = 290\,\text{K}$$

Compression ratio:

$$r = \frac{v_1}{v_2} = 16$$

$\eta_{\text{th}} = 60\%$
$p_3 = p_4 = 60\,\text{bar} = 60 \times 10^2\,\text{kPa}$
$T_3 = ?$

$$p_1 v_1 = R\,T_1$$

$$1 \times 10^2 \times v_1 = 0.287 \times 290$$

$$v_1 = 0.8323\,\text{m}^3/\text{kg}$$

$$v_2 = \frac{v_1}{16} = \frac{0.8323}{16} = 0.0520\,\text{m}^3/\text{kg}$$

$$mep = \frac{\text{work done/cycle}}{v_1 - v_2}$$

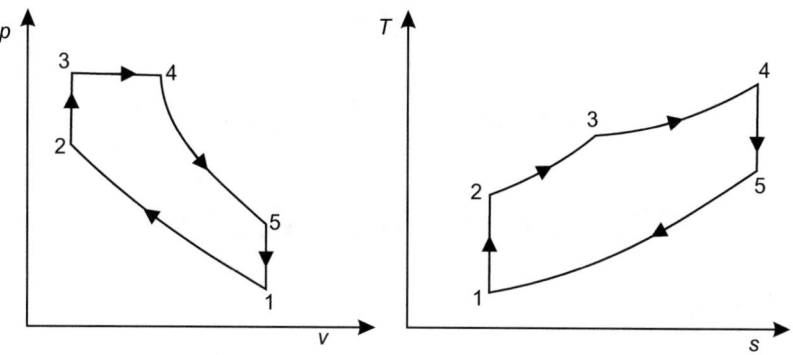

Fig. 4.15 p-v and T-s diagrams for Problem 4.8

or

$$\text{work done/cycle} = mep \times v_1 - v_2$$
$$= 10 \times 10^2 \times (0.8323 - 0.0520) = 780.3 \,\text{kJ/kg}.$$

For process 1–2,

$$\frac{T_2}{T_1} = \left(\frac{v_1}{v_2}\right)^{\gamma-1}$$

$$T_2 = T_1\left(\frac{v_1}{v_2}\right)^{\gamma-1} = 290 \times (16)^{1.4-1} = 879.11 \,\text{K}$$

and

$$\frac{p_2}{p_1} = \left(\frac{v_1}{v_2}\right)^{\gamma}$$

$$p_2 = p_1\left(\frac{v_1}{v_2}\right)^{\gamma} = 1 \times (16)^{1.4} = 48.5 \,\text{bar}$$

For process 2–3,

$$\frac{T_3}{T_2} = \frac{p_3}{p_2}$$

or

$$T_3 = p_1\left(\frac{v_1}{v_2}\right)^{\gamma} = 879.11 \times \frac{60}{48.5} = 1087.55 \,\text{K}$$

$$\eta_{th} = \frac{\text{work done}}{q_{2-3} + q_{3-4}}$$

$$0.6 = \frac{780.3}{c_v(T_3 - T_2) + c_p(T_4 - T_3)}$$

$$0.6 = \frac{780.3}{0.718(1087.55 - 879.11) + 1.005(T_4 - 1087.55)}$$

or

$$T_4 = 2232.66 \, \text{K} = \mathbf{1959.66} \, °\mathbf{C}$$

Summary

1. **A Thermodynamic Cycle** using air as the working fluid with constant mass flow rate (m), specific heats (c_p and c_v), and gas constant (R) is known as **air-standard cycle**. The efficiency of air-standard cycle is called air-standard efficiency.

2. **The Otto Cycle:**
   Thermal efficiency:

$$\eta_{\text{Otto}} = 1 - \frac{1}{r^{\gamma-1}}.$$

   where

   $r = $ compression ratio,
   $\gamma = \frac{c_p}{c_v} = $ adiabatic index.

   The thermal efficiency is also called air-standard efficiency.

3. **Mean Effective Pressure: p_m**

$$p_m = \frac{W_{\text{net}}}{V_s} = \frac{w_{\text{net}}}{v_s} \, \text{kPa}$$

   where

   $W_{\text{net}} = $ net work output; kJ,
   $V_s = $ swept or displacement volume; m^3,
   $w_{\text{net}} = $ net specific work output; kJ/kg,
   $v_s = $ specific displacement volume; m^3/kg.

4. **The Diesel Cycle:**
   Thermal efficiency:

$$\eta_{\text{Diesel}} = 1 - \frac{1}{r^{\gamma-1}} \left[\frac{\left(r_c^{\gamma} - 1\right)}{\gamma \left(r_c - 1\right)} \right].$$

   where

   $r = $ compression ratio,
   $r_c = $ cut-off ratio,
   $\gamma = $ adiabatic index.

5. **The Dual Cycle**:
   Thermal efficiency:

$$\eta_{Dual} = 1 - \frac{1}{r^{\gamma-1}} \frac{(r_p r_c^{\gamma} - 1)}{[\gamma r_p(r_c - 1) + (r_p - 1)]}.$$

6. **Compression of the Otto, Diesel, and Dual cycles**:

S. No.	Condition	Compression of efficiencies
1.	Same compression ratio and heat supplied	$\eta_{Otto} > \eta_{Dual} > \eta_{Diesel}$
2.	Same compression ratio and heat rejection	$\eta_{Otto} > \eta_{Dual} > \eta_{Diesel}$
3.	Same maximum pressure, maximum temperature, and heat rejection	$\eta_{Diesel} > \eta_{Dual} > \eta_{Otto}$

7. **The Stirling Cycle**:
   The thermal efficiency of the regeneration Stirling cycle is equal to the Carnot cycle when operating between the same temperature limits.

$$\eta_{Striling} = \eta_{Camot} = 1 - \frac{T_1}{T_3}$$

8. **The Ericsson Cycle**:
   Similarly, the thermal efficiency of the regeneration Ericsson cycle is equal to the Carnot cycle when operating between the same temperature limits.

$$\eta_{Ericsson} = \eta_{Carnot} = 1 - \frac{T_1}{T_3}$$

Assignment–1

1. What is an air-standard cycle? What is the use of air-standard cycle?
2. What assumptions are made in air-standard cycle analysis?
3. Explain the Otto cycle. Derive an expression for its thermal efficiency. What is the effect of compression ratio on thermal efficiency?
4. Draw the Diesel cycle on p–v and T–s diagrams. Derive an expression for its thermal efficiency.
5. Explain the dual cycle. Derive an expression for its thermal efficiency.
6. Define the following terms associated with dual cycle:
   (i) Compression ratio, (ii) Pressure ratio, and (iii) Cut-off ratio.
7. Sketch 'dual cycle' on T–s and p–v planes.
8. Shows that the air-standard efficiency of the Otto cycle depends on the compression ratio only.
9. Compare the Otto, diesel, and dual cycles for the following conditions:

   (i)   Same compression ratio and heat supplied,

(ii)   Same compression ratio and heat rejection, and
(iii) Same maximum pressure, maximum temperature, and heat rejection.

10. With the help of p–v and T–s diagrams, show that for the same maximum pressure and temperature of the cycle and the same heat rejection

$$\eta_{Diesel} > \eta_{Dual} > \eta_{Otto}$$

11. Draw the Stirling cycle on p–v and T–s diagrams. Show how it made a reversible cycle.
12. Draw the Ericsson cycle on p–v and T–s diagrams. Show how it made a reversible cycle.
13. State the four processes of the Stirling cycle. Show that the regenerative Stirling cycle has the same efficiency as the Carnot cycle?
14. State the four processes of the Ericsson cycle. Show that the regenerative Ericsson cycle has the same efficiency as the Carnot cycle?

Assignment–2

1.   The efficiency of the Otto cycle is 50% and the adiabatic index is 1.4. What is the compression ratio? [**Ans.** 5.65]
2.   In an ideal Otto cycle the temperature at the beginning and end of the compression process are 25 °C and 350 °C. Find the compression ratio and the thermal efficiency for the cycle. [**Ans.** 6.31, 52.14%]
3.   In an Otto cycle air at 17 °C and 1 bar is compressed adiabatically until the pressure is 15 bar. Heat is added at constant volume until the pressure rises to 40 bar. Determine (i) the compression ratio (ii) the air-standard efficiency, and (iii) the mean effective pressure for the cycle. Assume $\gamma = 1.4$, $c_v = 0.718$ kJ/kgK, $R = 0.287$ kJ/kgK.
     [**Ans.** (i) 6.91 (ii) 53.84% (iii) 5.68 bar)].
4.   A gas engine working on the Otto cycle has 120 mm stroke and 80 mm diameter of the cylinder. The clearance volume is 80 cm³. Determine the swept volume and the air-standard efficiency. [Take $\pi = 3.14$] [**Ans.** 602.88 cm³, 57.57%]
5.   An ideal Otto cycle has a compression ratio of 8. At the beginning of the compression process, air is at 100 kPa and 17 °C, and 500 kJ/kg of heat is supplied to air during the constant volume process. Determine (i) the maximum temperature and pressure in the cycle (ii) the net work output (iii) the thermal efficiency, and (iv) the mean effective pressure for the cycle.
     Take $c_v = 0.718$ kJ/kgK, $\gamma = 1.4$, $R = 0.287$ kJ/kgK.
     [**Ans.** (i) 1362.61 K, 18.379 bar (ii) 282.37 kJ/kg (iii) 56.47% (iv) 3.87 bar].
6.   A diesel engine has a compression ratio of 18 and cut-off takes place at 5% of the stroke. Determine the air-standard efficiency. Take $\gamma = 1.4$ [**Ans.** 63.63%]
7.   Determine the thermal efficiency of the diesel engine having a cylinder with bore 250 mm, stroke 350 mm, a clearance volume of 1000 cm³, and cut-off takes place at 5% of the stroke. Take $\gamma = 1.4$. [**Ans.** 63.96%]

8. The compression ratio of an ideal Diesel cycle is 15. The heat transfer is 1465 kJ/kg of air. The temperature and pressure at the beginning of the compression process are 27 °C and 100 kPa. Determine (i) the air-standard efficiency, and (ii) the mean effective pressure of the cycle. Take $c_p = 1.005$ kJ/kgK, $c_v = 0.718$ kJ/kgK
    [**Ans.** (i) 57.26% (ii) 10.439 bar].

9. An ideal Diesel cycle with air as the working fluid has a compression ratio of 20 and a cut-off ratio of 2. At the beginning of the compression process, the working fluid is at 100 kPa, 25 °C, and 2000 cc. Determine (i) the temperature and pressure of air at the end of each process, (ii) the thermal efficiency, and (iii) the mean effective pressure.
    [**Ans.** (i) $T_2 = 987.70$ K, $p_2 = 6628.90$ kPa, $T_3 = 1975.40$ K, $p_3 = p_2 = 6628.90$ kPa, $T_4 = 786.42$ K, $p_4 = 264.90$ kPa (ii) 64.66% (iii) 786.84 kPa].

10. An oil engine working on the dual combustion cycle has a swept volume of 9900 cc and a compression ratio of 14. The fuel has a calorific value of 44350 kJ/kg and the maximum pressure in the cycle is 63.27 bar. At the beginning of the compression process, the working fluid is at 102 kPa and 366.4 K. Determine (i) the work done per cycle at an air–fuel ratio of 21:1. (ii) the thermal efficiency, and (iii) the mean effective pressure.
    [**Ans.** (i) 13.13 kJ (ii) 60.28% (iii) 13.26 bar].

11. A high-speed diesel engine working on the dual combustion cycle has a compression ratio of 15. The engine draws in air at 100 kPa, 27 °C and the maximum pressure in the cycle is limited to 55 bar. If the heat transfer at constant volume is twice that at constant pressure, determine (i) the pressure ratio (ii) the cut-off ratio, and (iii) the thermal efficiency of the cycle. Assume $\gamma = 1.4$, $c_p = 1.005$ kJ/kgK, and $c_v = 0.718$ kJ/kgK.
    [**Ans.** (i) 1.24 (ii) 1.07 (iii) 65.99%].

Chapter 5
Gas Turbines and Jet Propulsion

Nomenclature

The nomenclature introduced in this chapter is as follows:

T	°C or K	Temperature
p	kPa	Pressure
V	m^3	Volume
v	m^3/kg	Specific volume
T	K	Temperature
Q	kW	Rate of heat transfer
W	kW	Rate of work
q	kJ/kg	Specific heat transfer
w	kJ/kg	Specific work
h	kJ/kg	Specific enthalpy
c_p	kJ/kgK	Specific heat at constant pressure
m	kg/s	Mass flow rate
s	kJ/kgK	Specific entropy
η_{Joule}	—	Thermal efficiency of Joule's cycle
r_p	—	Pressure ratio
η_{th}	—	Thermal efficiency
γ	—	Adiabatic index
C_v	kJ/kgK	Specific heat at constant volume
BWR	—	Back work ratio
CC	—	Combustion chamber
$\in$	—	Effectiveness
t	—	$\frac{T_{max}}{T_{min}}$, temperature ratio
η_{Carnot}	—	Efficiency of Carnot cycle

(continued)

© The Author(s) 2022
S. Kumar, *Thermal Engineering Volume 2*,
https://doi.org/10.1007/978-3-030-89216-6_5

(continued)

HP	—	High pressure
LP	—	Low pressure
IC	—	Internal combustion
Ca	m/s	Absolute velocity of aircraft OR Velocity of approach of air
CJ	m/s	Relative velocity of jet
C	m/s	Absolute velocity of jet
F	N	Propulsive force
η_p	—	Propulsive efficiency
σ	—	$\frac{C_a}{C_J}$, Ratio of aircraft velocity to jet velocity
CV	kJ/kg	Calorific value of fuel
η_o	—	Overall efficiency

5.1   Introduction

A gas turbine is a type of rotodynamic machine that converts the thermal energy of gas into mechanical work. In an open-cycle gas turbine, the atmospheric air is compressed in a rotary compressor and the compressed air is supplied to a combustion chamber where fuel is burnt. The high-temperature gas is then directed into the gas turbine which may be either of impulse or reaction type. The high-velocity gas, produced by expansion in the turbine nozzle, strikes on the turbine moving blades, and the kinetic energy of the gas is converted into mechanical work of turbine rotor.

5.2   Classification of Gas Turbines

Gas turbines are classified into two categories as follows:

1. Closed-cycle gas turbine,
2. Open-cycle gas turbine.

5.2.1   Closed-Cycle Gas Turbine

In 1851 Dr. Joule proposed a closed-cycle gas turbine. It consists of four main parts (i) Compressor (ii) Heat exchanger (iii) Turbine (iv) Intercooler as shown in Fig. 5.1. In closed-cycle gas turbine, the same working fluid (air, hydrogen, helium, or any other gas) circulates through its various components. Heat cannot be supplied to the working fluid by internal combustion, instead, it is supplied externally by using a

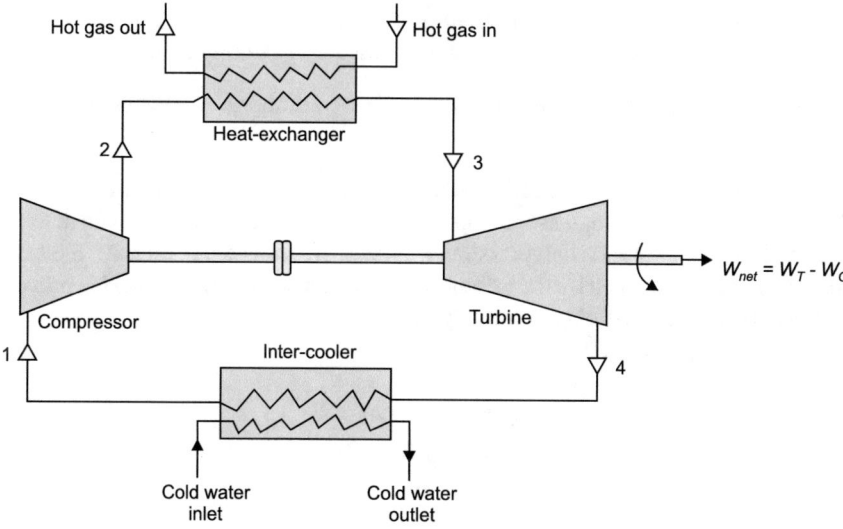

Fig. 5.1 Closed-cycle gas turbine

heat exchanger. An inter-cooler is used between the turbine exit and the compressor inlet. This decreases the specific volume of the working fluid (air or gas) entering the compressor. The lower value of the specific volume decreases the compressor work $(-\int v dp)$ and its size.

A closed-cycle gas turbine works on the principle of the Joule cycle. It consists of two isentropic processes and two constant pressure processes. The four processes that form the Joule cycle as shown in Fig. 5.2 on p–v and T-s diagrams.

The four processes are the following:

1. **Process 1–2**: Isentropic (or reversible) adiabatic compression in a compressor.
2. **Process 2–3**: Heat supplied at constant pressure in a heat exchanger.

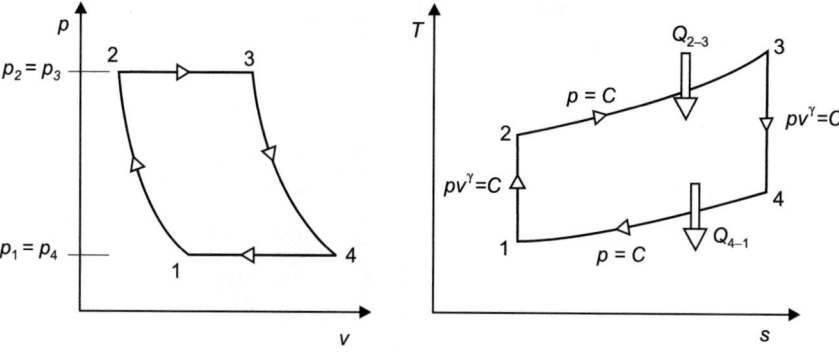

Fig. 5.2 The Joule cycle

3. **Process 3–4**: Isentropic (or reversible) adiabatic expansion in a turbine.
4. **Process 4–1**: Heat rejected at constant pressure in an intercooler.

1. **Process 1–2: Isentropic (or reversible) adiabatic compression in a compressor**

The low-pressure air is entered in the compressor at state 1 from the inter-cooler (Fig. 5.3). The compressor compresses the air adiabatic isentropic and delivers it to the heat exchanger at high pressure at state 2. To achieve this process, the large amount of work produced by the turbine is supplied to the compressor. This process is shown on p–v and T-s diagram in Fig. 5.2.

Work input per second:

$$W_{1-2} = m(h_2 - h_1)\text{kW}$$
$$= mc_p(T_2 - T_1)\text{kW}$$

where

$$m = \text{mass flow rate of air in kg/s}$$
$$h_1,\ h_2 = \text{specific enthalpies of air at inlet and outlet, respectively, in kJ/kg}$$
$$c_p = \text{specific heat at constant pressure of air in kJ/kgK}$$
$$= 1.005 \text{ kJ/kgK for air}$$

For unit mass flow rate

$$w_{1-2} = c_p(T_2 - T_1)\text{kJ/kg}.$$

2. **Process 2–3: Heat supplied at constant pressure in a heat exchanger**

Inlet of the heat exchanger at state 2, air at high temperature and pressure (Fig. 5.4). The air is further heated at constant pressure by supplying the heat of hot gas. This process is shown on *p–v* and *T-s* diagram in Fig. 5.2.

Heat supplied per second:

Fig. 5.3 Compressor

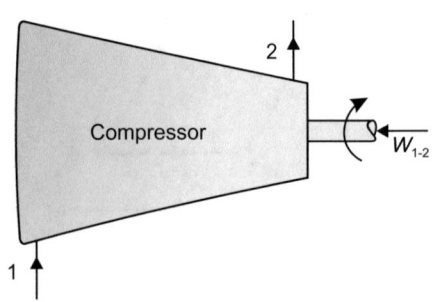

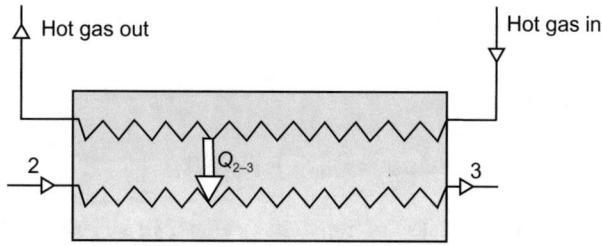

Fig. 5.4 Heat exchanger

$$Q_{2-3} = m(h_3 - h_2)\text{kW}$$
$$= mc_p(T_3 - T_2)\text{kW}$$

where h_2, h_3 are in kJ/kg

and c_p is in kJ/kgK.
  For unit mass flow rate

$$q_{2-3} = c_p(T_3 - T_2)\text{kJ/kg}$$

3. **Process 3–4: Isentropic (or reversible) adiabatic expansion in a turbine**

Inlet of the turbine at state 3, the air is at high temperature and pressure (Fig. 5.5). The high pressure and temperature air expands adiabatic isentropic in the turbine at low temperature and pressure. During this process, work is produced which large amount supply to the compressor and remaining work is used to drive the generator or used other purpose. This process is shown on p–v and T-s diagram in Fig. 5.2.
  Work output per second:

$$W_{3-4} = m(h_3 - h_4)$$
$$= mc_p(T_3 - T_4)$$

Fig. 5.5 Turbine

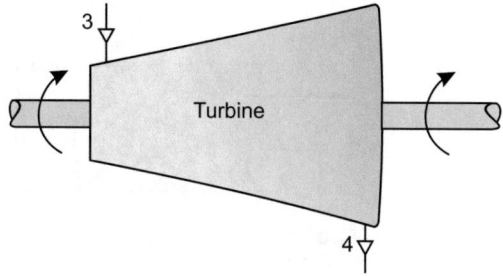

where h_2, h_3 are in kJ/kg

and c_p is in kJ/kgK.

For unit mass flow rate

$$w_{3-4} = c_p(T_3 - T_2)\text{kJ/kg}$$

4.   **Process 4–1**: **Heat rejected at constant pressure in an intercooler**

The temperature at the outlet of turbine air is still high, it is required to decrease
the temperature of air through an intercooler before supply to the compressor, i.e.,
air comes again at the inlet of the compressor with the same temperature and pressure (Fig. 5.6). That is why this cycle is called closed-cycle gas turbine. In the intercooler, a considerable amount of heat is absorbed by cold water and decreases temperature and increases the density of air, the low temperature and dense air is required
small work input for the same compression ratio as compared to high temperature
and low dense air. The process 4–1 is shown on p–v and T-s diagram in Fig. 5.2.

Heat rejected per second:

$$Q_{4-1} = m(h_4 - h_1)$$
$$= mc_p(T_4 - T_1)$$

Net work output per second:

$$W_{net} = \text{Turbine output work} - \text{Compressor input work.}$$

$$W_{net} = W_{3-4} - W_{2-1}$$

also

$$W_{net} = \text{Heat added} - \text{Heat rejected}$$
$$= Q_{2-3} - Q_{4-1}$$

Thermal efficiency:

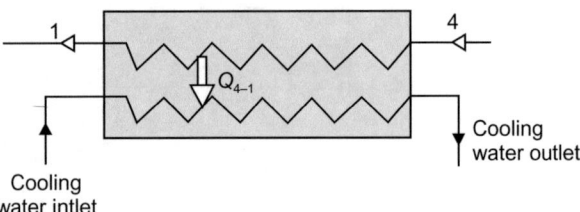

Fig. 5.6   Intercooler

$$\eta_{\text{Joule}} = \frac{\text{Net work output per second}}{\text{Heat added per second}}$$

$$= \frac{W_{\text{net}}}{Q_{2-3}}$$

$$= 1 - \frac{Q_{4-1}}{Q_{2-3}} = 1 - \frac{mc_p(T_4 - T_1)}{mc_p(T_3 - T_2)}$$

$$= 1 - \frac{(T_4 - T_1)}{(T_3 - T_2)}$$

$$= 1 - \frac{T_1\left(\frac{T_4}{T_1} - 1\right)}{T_2\left(\frac{T_3}{T_2} - 1\right)} \tag{5.1}$$

For powers 1–2,

$$\frac{p_2}{p_1} = \left(\frac{T_2}{T_1}\right)^{\frac{\gamma-1}{\gamma}}$$

For process 4–1,

$$\frac{p_2}{p_1} = \left(\frac{T_3}{T_4}\right)^{\frac{\gamma-1}{\gamma}}$$

$$\therefore \qquad \left(\frac{T_2}{T_1}\right)^{\frac{\gamma-1}{\gamma}} = \left(\frac{T_3}{T_4}\right)^{\frac{\gamma-1}{\gamma}}$$

or

$$\frac{T_2}{T_1} = \frac{T_3}{T_4}$$

or

$$\frac{T_4}{T_1} = \frac{T_3}{T_2}$$

Substituting $\frac{T_4}{T_1} = \frac{T_3}{T_2}$ in Eq. (5.1), we get

$$\eta_{\text{Joule}} = 1 - \frac{T_1\left(\frac{T_4}{T_1} - 1\right)}{T_2\left(\frac{T_4}{T_1} - 1\right)}$$

$$= 1 - \frac{T_1}{T_2} = 1 - \frac{1}{\frac{T_2}{T_1}} = 1 - \frac{1}{\left(\frac{p_2}{p_1}\right)^{(\gamma-1)/\gamma}}$$

$$\eta_{\text{Joule}} = 1 - \frac{1}{r_p^{(\gamma-1)/\gamma}}$$

where $r_p = \frac{p_2}{p_1}$, pressure ratio.

The thermal efficiency of the Joule cycle increases with (i) increasing adiabatic index, and (ii) increasing pressure ratio.

5.2.2   Open-Cycle Gas Turbine

The closed-cycle gas turbine had a serious drawback that air had to be heated through a metal wall. In order to obtain high efficiency, a high temperature had to be provided. But in earlier times the thermal properties of the material were not good, i.e., heat transfer between the hot gases and air was very low. As a resulting rise in temperature of the air was low. This resulted in a very low efficiency of the cycle. So to remove the drawback, Brayton in 1872 invented open-cycle gas turbine. He used closed-cycle gas turbine but instead of circulating a given quantity of air in a closed cycle, he admitted air into the compressor directly from the atmosphere and compressed it and delivered it to the combustion chamber. In the combustion chamber, fuel was injected burned with air. The production of combustion was fed to a turbine where it expands to the atmospheric pressure. The exhaust from turbine delivered to atmosphere. The intercooler that was in between the turbine and compressor in the closed-cycle gas turbine was eliminated. And fresh air from the atmosphere was drawn by compressor and cycle repeat.

An open-cycle gas turbine consists of three main parts: (i) compressor, (ii) combustion, chamber, and (iii) turbine, as shown in Fig. 5.7. The open-cycle gas turbine works on the principle of the Brayton cycle.

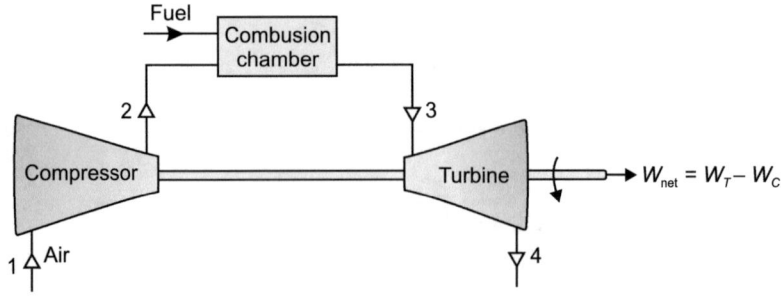

Fig. 5.7   The Brayton cycle components

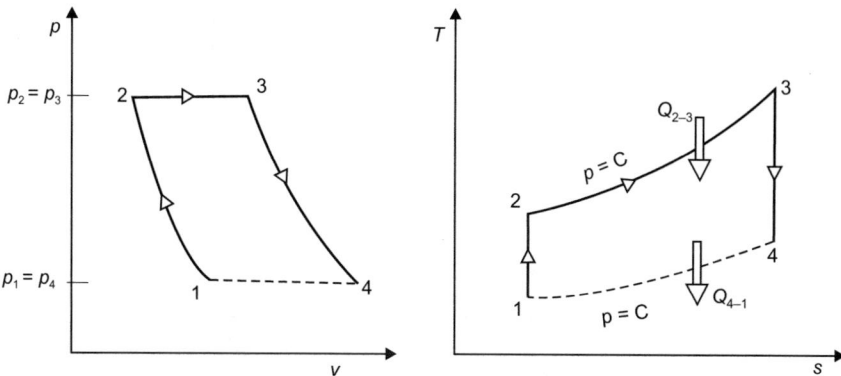

Fig. 5.8 The Brayton cycle

It consists of two isentropic processes and two constant pressure processes. The four processes that form the Brayton cycle are shown in the p–v and T-s diagrams of Fig. 5.8. The four processes are the following.

1. Process 1–2 Isentropic (or reversible) adiabatic compression in a compressor.
2. Process 2–3 Heat supplied at constant pressure
3. Process 3–4 Isentropic (or reversible) adiabatic expansion in a turbine.
4. Process 4–1 Heat rejected at constant pressure.

Process 2–3: Heat supplied at $p = C$
  Heat supplied:

$$Q_{2-3} = mc_p(T_3 - T_2)$$

  For unit mass
  Heat supplied:

$$q_{2-3} = c_p(T_3 - T_2)$$

For process 4–1: Heat rejected at $p = C$
  Heat rejected:

$$Q_{4-1} = mc_p(T_4 - T_1)$$

  For unit mass
  Heat rejected:

$$q_{4-1} = c_p(T_4 - T_1)$$

  Net work output:

$$w_{\text{net}} = \text{heat supplied} - \text{heat rejected}$$

$$w_{\text{net}} = q_{2-3} - q_{4-1}$$

Thermal efficiency:

$$\eta_{\text{th}} = \frac{\text{Net work output} : w_{\text{net}}}{\text{Heat supplied} : q_{2-3}}$$

$$\eta_{\text{th}} = \frac{w_{\text{net}}}{q_{2-3}} = \frac{q_{2-3} - q_{4-1}}{q_{2-3}} = 1 - \frac{q_{4-1}}{q_{2-3}}$$

$$= 1 - \frac{c_p(T_4 - T_1)}{c_p(T_3 - T_2)} = 1 - \frac{(T_4 - T_1)}{(T_3 - T_2)}$$

$$\eta_{\text{th}} = 1 - \frac{T_1\left[\frac{T_4}{T_1} - 1\right]}{T_2\left[\frac{T_3}{T_2} - 1\right]} \tag{5.2}$$

Process 1–2: Reversible adiabatic compression

$$\frac{T_2}{T_1} = \left(\frac{p_2}{p_1}\right)^{\frac{\gamma-1}{\gamma}} \tag{5.3}$$

Process 3–4: Reversible adiabatic expansion

$$\frac{T_3}{T_4} = \left(\frac{p_3}{p_4}\right)^{\frac{\gamma-1}{\gamma}}$$

As

$$p_3 = p_2 \text{ and } p_4 = p_1$$

$$\therefore \qquad \frac{T_3}{T_4} = \left(\frac{p_2}{p_1}\right)^{\frac{\gamma-1}{\gamma}} \tag{5.4}$$

From Eqs. (5.3) and (5.4), we get

$$\frac{T_2}{T_1} = \frac{T_3}{T_4}$$

or

$$\frac{T_4}{T_1} = \frac{T_3}{T_2}$$

Substituting $\frac{T_4}{T_1} = \frac{T_3}{T_2}$ in Eq. (5.2), we get

$$\eta_{th} = 1 - \frac{T_1\left[\frac{T_3}{T_2} - 1\right]}{T_2\left[\frac{T_3}{T_2} - 1\right]} = 1 - \frac{T_1}{T_2} = 1 - \frac{1}{T_2/T_1}$$

$$= 1 - \frac{1}{\left(\frac{p_2}{p_1}\right)^{\frac{\gamma-1}{\gamma}}}$$

$$\boldsymbol{\eta_{th} = 1 - \frac{1}{r_p^{(\gamma-1)/\gamma}}} \tag{5.5}$$

where

$r_p = \frac{p_2}{p_1}$, pressure ratio,

$\gamma = \frac{c_p}{c_v}$, specific heat ratio. It is called the adiabatic index.

Equation (5.5) clearly indicates that the thermal efficiency of the Brayton cycle depends on the pressure ratio (r_p) and the adiabatic index (γ) of the working fluid. The thermal efficiency increases with an increase of pressure ratio (r_p) and the adiabatic index (γ).

Back Work Ratio: BWR

It is defined as the ratio of the compressor work to the turbine work.

Back work ratio:

$$\text{BWR} = \frac{\text{Compressor work} : W_C}{\text{Turbine work} : W_T}$$

$$\text{BWR} = \frac{W_C}{W_T} \text{ or } \frac{w_{1-2}}{w_{3-4}}$$

where compressor work:

$$W_C = W_{1-2} = mc_p(T_2 - T_1)$$

and turbine work:

$$W_T = W_{3-4} = mc_p(T_3 - T_4)$$

For unit mass

$$\text{BWR} = \frac{w_C}{w_T} \text{ or } \frac{w_{1-2}}{w_{3-4}}$$

where

$$w_C = w_{1-2} = c_p(T_2 - T_1)$$

and

$$w_T = w_{3-4} = c_p(T_3 - T_4)$$

The main demerit of the gas turbine is low thermal efficiency which is due to the large amount of work required to drive the compressor. The compressor may require up to 80% of the turbine's output work (back work ratio is 0.8), leaving only 20% net work output.

Work Ratio: WR

It is defined as the ratio of the net work output to the turbine work.

$$\text{Work ratio}: WR = \frac{\text{Net work output}: W_{\text{net}}}{\text{Turbine work}: W_T}$$

$$WR = \frac{W_{\text{net}}}{W_T}$$

where

$$W_{\text{net}} = W_T - W_C$$

5.3   Condition for Maximum Work Output in the Brayton Cycle for Given Minimum and Maximum Temperatures Limit

Let

T_1 = minimum temperature at the inlet of a compressor in the Brayton cycle

$= T_{\text{min}}$

and

T_3 = maximum temperature at the inlet of a turbine in the Brayton cycle.

$T_3 = T_{\text{max}}$

Process 2–3: Heat supplied at $p = C$

$$Q_{2-3} = mc_p(T_3 - T_2)$$

For unit mass flow rate

$$q_{2-3} = c_p(T_3 - T_2)$$

Process 4–1: Heat rejected at $p = C$

$$Q_{4-1} = mc_p(T_4 - T_1)$$

For unit mass flow rate

$$q_{4-1} = c_p(T_4 - T_1)$$

Net work output (Fig. 5.9):

$$w_{net} = \text{heat supplied} - \text{heat rejected}$$
$$= q_{2-3} - q_{4-1} = c_p(T_3 - T_2) - c_p(T_4 - T_1)$$
$$w_{net} = c_p[T_3 - T_2 - T_4 + T_1] \tag{5.6}$$

Process 1–2

$$\frac{T_2}{T_1} = \left(\frac{p_2}{p_1}\right)^{\frac{\gamma-1}{\gamma}}$$

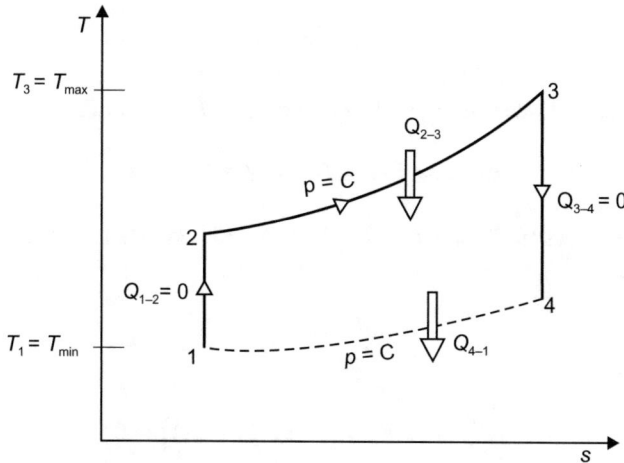

Fig. 5.9   The Brayton cycle

$$\frac{T_2}{T_1} = r_p^{\frac{\gamma-1}{\gamma}}$$

where $r_p = \frac{p_2}{p_1}$, pressure ratio
   Let

$$z = \frac{\gamma - 1}{\gamma}$$

$$\therefore \qquad \frac{T_2}{T_1} = r_p^z$$

or

$$T_2 = T_1 r_p^z$$

Process 3–4

$$\frac{T_3}{T_4} = \left(\frac{p_2}{p_1}\right)^{\frac{\gamma-1}{\gamma}}$$

$$\frac{T_3}{T_4} = r_p^z$$

or

$$T_4 = \frac{T_3}{r_p^z}$$

$$T_4 = T_3 r_p^{-z}$$

Substituting the values of $T_2 = T_1 r_p^z$ and $T_4 = T_3 r_p^{-z}$ in Eq. (5.6), we get

$$w_{net} = c_p\left(T_3 - T_1 r_p^z - T_3 r_p^{-z} + T_1\right) \qquad (5.7)$$

For given values of T_1 and T_3, Eq. (5.7) can be optimized. Thus for maximum work output.

$$\frac{dw_{net}}{dr_p} = 0$$

$$\frac{d}{dr_p}\left[c_p\left(T_3 - T_1 r_p^z - T_3 r_p^{-z} + T_1\right)\right] = 0$$

$$c_p\left[0 - T_1 z r_p^{z-1} - T_3(-z) r_p^{-z-1} + 0\right] = 0$$

or

$$-T_1 z r_p^{z-1} + z T_3 r_p^{-z-1} = 0$$

or

$$T_3 r_p^{-z-1} = T_1 r_p^{z-1}$$

$$\frac{T_3}{T_1} = \frac{r_p^{z-1}}{r_p^{-z-1}}$$

$$\frac{T_3}{T_1} = r_p^{z-1+z+1} = r_p^{2z}$$

$$\frac{T_3}{T_1} = r_p^{\frac{2(\gamma-1)}{\gamma}} = r_p^{2(\gamma-1)/\gamma} \tag{5.8}$$

The Maximum work output condition for ideal cycle.

$$\frac{T_{max}}{T_{min}} = \frac{T_3}{T_1} = r_p^{2(\gamma-1)/\gamma}$$

Process 1–2: Isentropic compression.

$$\frac{T_2}{T_1} = \left(\frac{p_2}{p_1}\right)^{\frac{\gamma-1}{\gamma}}$$

$$\frac{T_2}{T_1} = r_p^{\frac{\gamma-1}{\gamma}}$$

Squaring both sides, we get

$$\left(\frac{T_2}{T_1}\right)^2 = (r_p)^{2\left(\frac{\gamma-1}{\gamma}\right)} \tag{5.9}$$

Right-hand side terms of Eqs. (5.8) and (5.9) are the same.

$$\therefore \qquad \frac{T_3}{T_1} = \left(\frac{T_2}{T_1}\right)^2$$

or

$$T_2^2 = T_1^2 \times \frac{T_3}{T_1} = T_1 T_3$$

$$T_2 = \sqrt{T_1 T_3}$$

Similarly for process 3–4: Isentropic expansion

$$\frac{T_3}{T_4} = \left(\frac{p_2}{p_1}\right)^{\frac{\gamma-1}{\gamma}}$$

$$\frac{T_3}{T_4} = r_p^{\frac{\gamma-1}{\gamma}}$$

Squaring both sides, we get

$$\left(\frac{T_3}{T_4}\right)^2 = r_p^{2\left(\frac{\gamma-1}{\gamma}\right)} \tag{5.10}$$

Right-hand side terms of Eqs. (5.8) and (5.10) are the same.

$$\therefore \qquad \frac{T_3}{T_1} = \left(\frac{T_3}{T_4}\right)^2$$

or

$$T_4^2 = T_1 T_3$$

$$T_4 = \sqrt{T_1 T_3}$$

We can get maximum net work output by substituting the values of $T_1 = \sqrt{T_1 T_3}$ = and $T_4 = \sqrt{T_1 T_3}$ in Eq. (5.6)

$$w_{\max} = c_p\left(T_3 - \sqrt{T_1 T_3} - \sqrt{T_1 T_3} + T_1\right) = c_p\left(T_1 + T_3 - 2\sqrt{T_1 T_3}\right)$$

$$w_{\max} = c_p\left(\sqrt{T_1} - \sqrt{T_3}\right)^2$$

5.4   Actual Gas Turbine Cycle

In an actual gas turbine, the compressor and the turbine are not working as adiabatic isentropic, some losses do occur due to internal friction.

For actual compression process 1–2 in the compressor:

Compressor efficiency: η_C. It is defined as the ratio of isentropic increase in temperature to the actual increase in temperature in the compressor (Fig. 5.10).

Mathematically,

Compressor efficiency:

$$\eta_C = \frac{(dT)_{\text{isentropic}}}{(dT)_{\text{actual}}}$$

$$\eta_C = \frac{T_{2s} - T_1}{T_2 - T_1} \tag{5.11}$$

The compressor efficiency is also called the isentropic efficiency of the compressor.

For isentropic process 1–2 s.

$$\frac{T_{2s}}{T_2} = \left(\frac{p_2}{p_1}\right)^{\frac{\gamma-1}{\gamma}}$$

or

$$T_{2s} = T_1 \left(\frac{p_2}{p_1}\right)^{\frac{\gamma-1}{\gamma}} \tag{5.12}$$

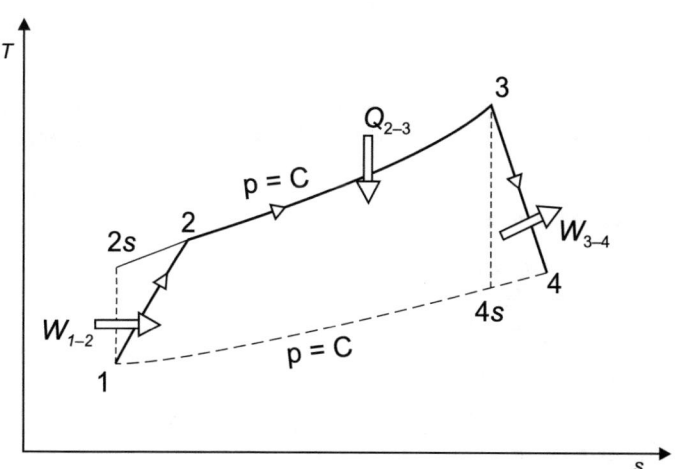

Fig. 5.10 Actual cycle without pressure losses

For given values of $T_1, p_1,$ and p_2, we can determine the value of T_{2s}. By substituting the values of T_{2s} in Eq. (5.11), we can find out the value of T_2 at a given value of η_C.

For actual expansion process 3–4 in the turbine:

Turbine efficiency: η_T. It is defined as the ratio of the actual decrease in temperature to the isentropic decrease in temperature in the turbine.

Mathematically,

Turbine efficiency:

$$\eta_T = \frac{(dT)_{\text{actual}}}{(dT)_{\text{isentropic}}}$$

$$\eta_T = \frac{T_3 - T_4}{T_3 - T_{4s}} \qquad (5.13)$$

The turbine efficiency is also called the isentropic efficiency of the turbine.

For isentropic process 3–4 s

$$\frac{T_3}{T_{4s}} = \left(\frac{p_2}{p_1}\right)^{\frac{\gamma-1}{\gamma}}$$

or

$$T_{4s} = \frac{T_3}{\left(\frac{p_2}{p_1}\right)^{\frac{\gamma-1}{\gamma}}} \qquad (5.14)$$

For given values of $T_3, p_1,$ and p_2, we can determine the value of T_{4s}. By substituting the values of T_{4s} in Eq. (5.13), we can find out the value of T_4 at a given value of η_T.

Process 2–3: Heat supplied at $p = C$

Heat supplied: $Q_{2-3} = mc_p\,(T_3 - T_2)$
    For unit mass flow rate
    Heat supplied: $q_{2-3} = c_p\,(T_3 - T_2)$

Process 3–4: Actual turbine work output

Actual turbine work output: $W_{3-4} = W_T = mc_p\,(T_3 - T_4)$
    For unit mass flow rate

$$w_{3-4} = w_T = c_p(T_3 - T_4).$$

Actual compressor work required: $W_{1-2} = W_C = mc_p\,(T_2 - T_1)$.
    For unit mass flow rate

$$w_{1-2} = w_C = c_p(T_2 - T_1).$$

Net work output:

$$w_{net} = \text{turbine work} - \text{compressor work},$$
$$= w_T - w_C.$$

The thermal efficiency of the actual cycle:

$$\eta_{th} = \frac{\text{net work output}}{\text{heat supplied}}$$

$$\eta_{th} = \frac{w_{net}}{q_{2-3}}$$

$$\eta_{th} = \frac{\boldsymbol{w_T - w_C}}{\boldsymbol{q_{2-3}}}$$

Maximum work output condition for the actual cycle

$$w_{net} = w_T - w_C,$$
$$= c_p(T_3 - T_4) - c_p(T_2 - T_1).$$

$$\eta_T = \frac{T_3 - T_4}{T_3 - T_{4s}} \qquad\qquad \eta_C = \frac{T_{2s} - T_1}{T_2 - T_1}$$
$$\text{or} \quad T_3 - T_4 = \eta_T(T_3 - T_{4s}) \;\Big|\; \text{or} \quad T_2 - T_1 = \frac{T_{2s} - T_1}{\eta_C}$$

$$\therefore \qquad w_{net} = c_p\eta_T(T_3 - T_{4s}) - \frac{c_p}{\eta_C}(T_{2s} - T_1)$$

For ideal process 3–4 s,

$$\frac{T_3}{T_{4s}} = r_p^{\frac{\gamma-1}{\gamma}} = r_p^z \qquad\qquad \because \frac{\gamma-1}{\gamma} = z$$

or

$$T_{4s} = \frac{T_3}{r_p^z}$$

$$\boldsymbol{T_{4s} = T_3 r_p^{-z}}$$

For ideal process 1–2 s,

$$\frac{T_{2s}}{T_1} = r_p^{\frac{\gamma-1}{\gamma}} = r_p^z$$

or

$$T_{2s} = T_1 r_p^z$$

$$\therefore \qquad w_{\text{net}} = c_p \eta_T \left(T_3 - T_3 r_p^{-z}\right) - \frac{c_p}{\eta_C}\left(T_1 r_p^z - T_1\right)$$

For maximum work output condition for the actual cycle,

$$\frac{dw_{\text{net}}}{dr_p} = 0$$

$$c_p \eta_T \left[0 - T_3(-z)r_p^{-z-1}\right] - \frac{c_p}{\eta_C}\left(T_1 z r_p^{z-1} - 0\right) = 0$$

or

$$\eta_T T_3 z r_p^{-z-1} - \frac{T_1 z r_p^{z-1}}{\eta_C} = 0$$

or

$$\eta_T T_3 z r_p^{-z-1} = \frac{T_1 r_p^{z-1}}{\eta_C}$$

$$\eta_T \eta_C \frac{T_3}{T_1} = \frac{r_p^{z-1}}{r_p^{-z-1}} = r_p^{z-1+z+1} = r_p^{2z}$$

or

$$\eta_T \eta_C \frac{T_3}{T_1} = r_p^{2z}$$

or

$$\sqrt{\eta_T \eta_C \frac{T_3}{T_1}} = z_p^z$$

or

$$\sqrt{\eta_T \eta_C \frac{T_3}{T_1}} = r_p^{(\gamma-1)/\gamma} \qquad\qquad \because z = \frac{\gamma-1}{\gamma}$$

$$\sqrt{\eta_T \eta_C \frac{T_{\max}}{T_{\min}}} = r_p^{(\gamma-1)/\gamma}$$

5.5   Refinement of Open-Cycle Gas Turbine

Performance of the open-cycle gas turbine can be improved by decreasing input like compressor work, heat supplied, and increasing turbine output work. The processes are used in open cycle gas turbine to improve the performance of the cycle is called refinement of the open cycle gas turbine. The following processes are used for refinement of the cycle:

(i)   Regeneration,
(ii) Reheat, and
(iii) Intercooling.

5.5.1   Cycle with Regeneration: ⇩Heat Supplied, Wnet = C, ⇧η_{th}

The temperature of gases leaving the turbine is very high and usually, the heat energy in the exhaust go waste in a simple cycle. A counter-flow heat exchanger, called a regenerator, is used to transfer some of this heat energy to the air leaving the compressor, as shown in Fig. 5.11, results in a decrease of heat supplied in the combustion chamber thereby increasing of the thermal efficiency (Fig. 5.12).

The regeneration is only used when the exhaust temperature of gases from the turbine is greater than the exit temperature of air from the compressor, i.e., $T_5 > T_2$.

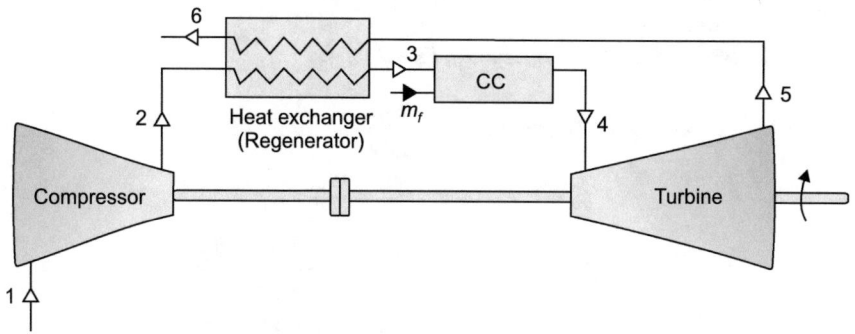

Fig. 5.11   Cycle with regeneration

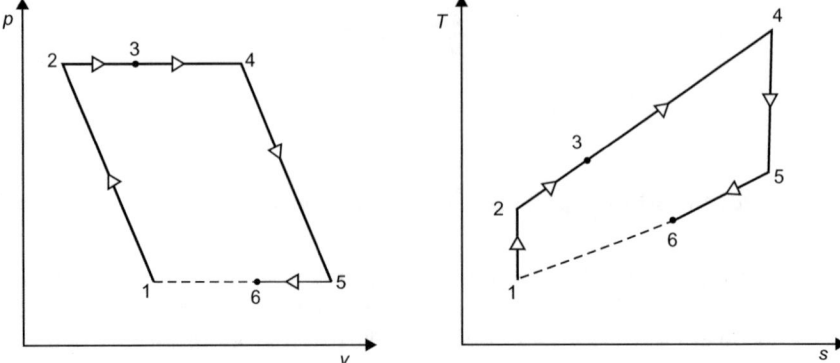

Fig. 5.12   Brayton cycle with regeneration

Effectiveness:$\in$

The effectiveness of a heat exchanger is defined as the heat transfer to air to the maximum heat transfer capacity of the heat exchanger.

Mathematically,

$$\text{Effectiveness} : \in = \frac{\text{Heat transfer to air}}{\text{Maximum heat transfer capacity}}$$

$$= \frac{mc_p(T_3 - T_2)}{mc_p(T_5 - T_2)}$$

$$= \frac{T_3 - T_2}{T_5 - T_2}$$

$$= 1 \text{ for ideal regeneration when } T_2 = T_6 \text{ and } T_3 = T_5$$

The effectiveness is also called efficiency or thermal ratio of heat exchanger.

Specific turbine work: $w_T = c_p(T_4 - T_5),$
Specific compressor work: $w_C = c_p(T_2 - T_1).$
Specific work output: $w_{net} = w_T - w_C,$
Specific heat supplied: $q_{3-4} = c_p(T_4 - T_3),$

$$\text{Thermal efficiency} : \eta_{th} = \frac{w_{net}}{q_{3-4}}$$

$$= \frac{w_T - w_C}{q_{3-4}}$$

$$= \frac{c_p(T_4 - T_5) - c_p(T_2 - T_1)}{c_p(T_4 - T_3)}$$

$$= \frac{(T_4 - T_5) - (T_2 - T_1)}{T_4 - T_3}$$

For ideal regeneration,

$$T_3 = T_5$$

$$\therefore \quad \eta_{th} = \frac{(T_4 - T_5) - (T_2 - T_1)}{T_4 - T_5}$$

$$\eta_{th} = 1 - \frac{T_2 - T_1}{T_4 - T_5}$$

$$= 1 - \frac{T_1\left[\frac{T_2}{T_1} - 1\right]}{T_4\left[1 - \frac{T_3}{T_4}\right]} = 1 - \frac{1}{\frac{T_4}{T_1}}\frac{\left[\frac{T_2}{T_1} - 1\right]}{\left[1 - \frac{T_3}{T_4}\right]}$$

where

$$\frac{T_2}{T_1} = (r_p)^{(\gamma-1)/\gamma}, \ \frac{T_4}{T_3} = (r_p)^{(\gamma-1)/\gamma}$$

and $\frac{T_4}{T_1} = t$, temperature ratio. It is the ratio of the maximum temperature to the minimum temperature in the cycle.

$$\therefore \quad \eta_{th} = 1 - \frac{1}{t} \frac{\left[r_p^{(\gamma-1)/\gamma} - 1\right]}{\left[1 - \frac{1}{r_p^{(\gamma-1)/\gamma}}\right]}$$

$$= 1 - \frac{1}{t} \frac{\left[r_p^{(\gamma-1)/\gamma} - 1\right]}{\left[\frac{r_p^{(\gamma-1)/\gamma} - 1}{r_p^{(\gamma-1)/\gamma}}\right]}$$

$$\eta_{th} = 1 - \frac{r_p^{(\gamma-1)/\gamma}}{t} \tag{5.15}$$

where

r_p = Pressure ratio,
$t = \frac{T_{max}}{T_{min}}$, the ratio of maximum temperature to minimum temperature.

Equation (5.15) for thermal efficiency is quite different from that for the Brayton cycle. For a given temperature ratio, as the pressure ratio increases the efficiency decreases, an effect opposite to that of the Brayton cycle. At a particular pressure ratio, the efficiency of the cycle is equal to the efficiency of the Brayton cycle. When $r_p = 1$, the cycle efficiency is equal to the Carnot efficiency.

i.e., At $r_p = 1$,

$$\eta_{th} = 1 - \frac{1}{t}$$

$$\eta_{th} = 1 - \frac{1}{\frac{T_{max}}{T_{min}}}$$

$$\eta_{th} = 1 - \frac{T_{min}}{T_{max}}$$

$$\eta_{th} = \eta_{Carnot}$$

For given pressure ratio, the efficiency increases as the temperature ratio increases. At points C, $r_p = 1$ with different values of t,

$$\eta_{Regeneration} = \eta_{Carnot}$$

At points B, different particular values of r_p and t (Fig. 5.13),

$$\eta_{Regeneration} = \eta_{Brayton}$$

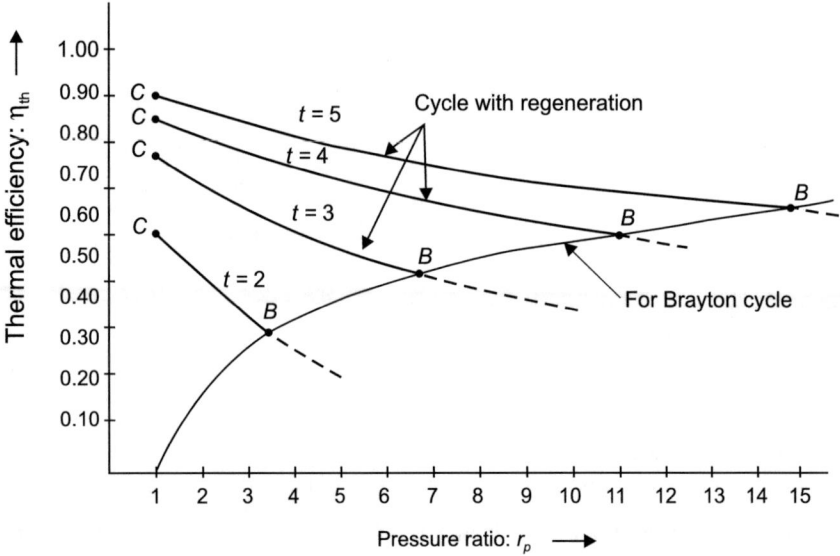

Fig. 5.13   Efficiency versus Pressure ratio for the Brayton cycle and the cycle with regeneration

5.5.2 *Cycle with Reheat:* ⇧*W*~net~, ⇧*Heat Supplied,* ⇩*η*~th~

In the cycle with reheat, the gas is reheated after expanding partially in the high-pressure turbine. The gas is then sent to the reheater where it is reheated at constant pressure, usually to the inlet temperature of the high-pressure turbine. The gas then supplied to the low-pressure turbine in which it expands isentropically to the atmospheric pressure, as shown in Fig. 5.14, resulting in the increase of net work output and heat supplied thereby decreasing thermal efficiency. This cycle is shown on p-v and T-s diagrams in Fig. 5.15.

Specific compressor work: $w_{1-2} = c_p(T_2 - T_1)$,
Specific HP turbine work: $w_{3-4} = c_p(T_3 - T_4)$,
Specific LP turbine work: $w_{5-6} = c_p(T_5 - T_6)$,
Net specific work output: $w_{net} = w_{3-4} + w_{5-6} - w_{1-2}$,

Net specific heat supplied:

$$q_{net} = \text{Specific heat addition in combustion chamber}$$

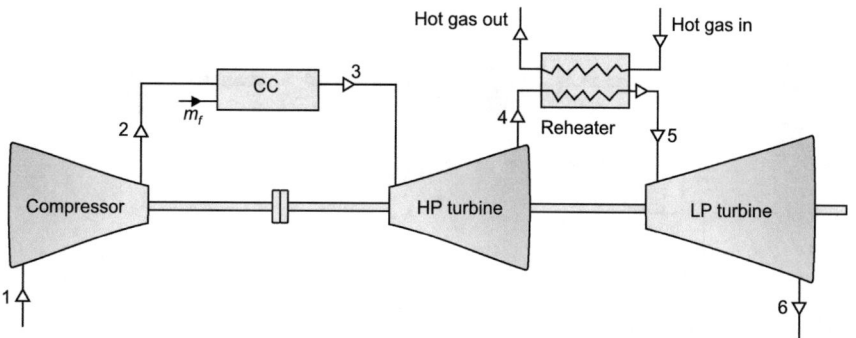

Fig. 5.14 Cycle with reheat

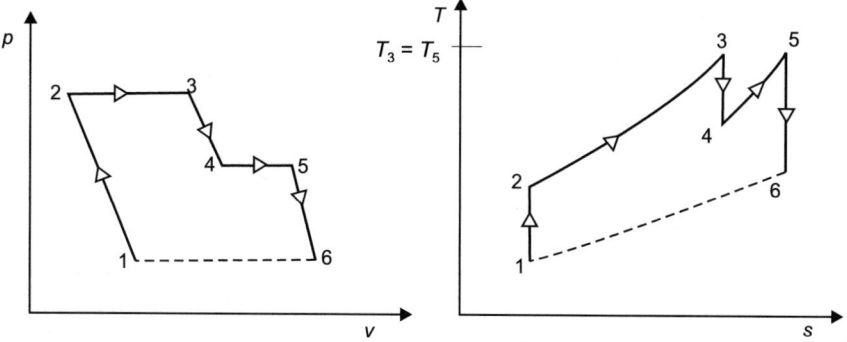

Fig. 5.15 Brayton cycle with reheat

+ specific heat supplied in reheater

$$= q_{2-3} + q_{4-5} = c_p(T_3 - T_2) + c_p(T_5 - T_4)$$

Thermal efficiency:

$$\eta_{th} = \frac{w_{net}}{q_{net}} = \frac{w_{3-4} + w_{5-6} - w_{1-2}}{q_{2-3} + q_{4-5}}$$

5.5.3   Cycle with Intercooling: ⇧W_{net}, ⇧Heat Supplied, ⇩η_{th}

The cooling of air between two compressors is called intercooling. The air compressed in low-pressure compressor to intermediate pressure is now passed to the intercooler where it is cooled at constant pressure (Figs. 5.16 and 5.17).

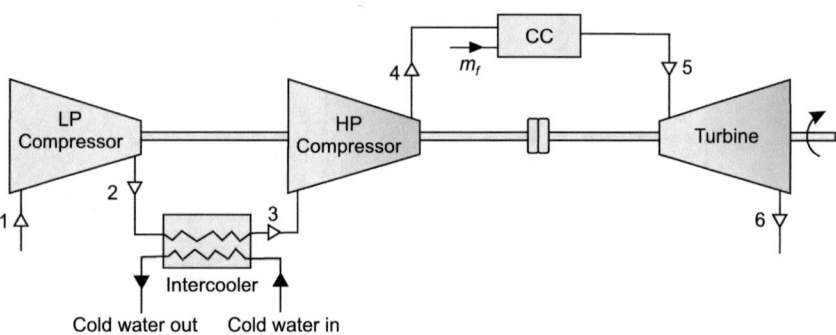

Fig. 5.16   Cycle with intercooling

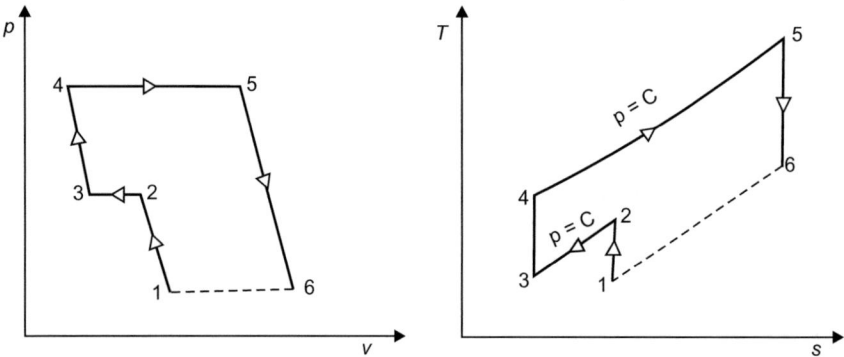

Fig. 5.17   Brayton cycle with intercooling

The cooled compressed air from the intercooler is passed to the high-pressure cylinder. The cooled air reduces the work and size of HP compressor thereby increasing the net output work and decreasing the thermal efficiency due to increasing the heat supplied.

Specific work of LP compressor: $w_{1-2} = c_p(T_2 - T_1)$,
Specific work of HP compressor: $w_{3-4} = c_p(T_4 - T_3)$,
Heat supplied: $q_{4-5} = c_p(T_5 - T_4)$,
Specific work of turbine: $w_{5-6} = c_p(T_5 - T_6)$,
Net specific work output: $w_{net} = w_{5-6} + w_{1-2} - w_{3-4}$,
Thermal efficiency: $\eta_{th} = \frac{w_{net}}{q_{4-5}}$

5.6 Cycle with Regeneration, Reheat, and Intercooling ⇧ W_{net}, ⇧Heat supplied, ⇧η_{th}

The cycle with regeneration, reheat, and intercooling increases both net work output and thermal efficiency. This cycle is shown in Fig. 5.18. The air after compression in the LP compressor is cooled at a constant pressure in the intercooler. This reduces the input work to the HP compressor. The counter-flow heat exchanger is used to transfer some of the heat energy in the exhaust of the LP turbine to the air leaving the HP compressor, resulting decrease in heat supplied in the combustion chamber. In the counter-flow heat exchanger, heat transfer at constant pressure, Further heat addition at constant pressure when air burns with fuel in the combustion chamber. The gas is reheated at constant pressure in the reheater after expanding partially in the HP turbine. The gas is then supplied to the LP turbine in which it expands isentropically to the atmospheric pressure. The exhaust of the LP turbine is passing through the counter-flow heat exchanger where heats the air coming from HP compressor and goes to the surrounding (Fig. 5.19).

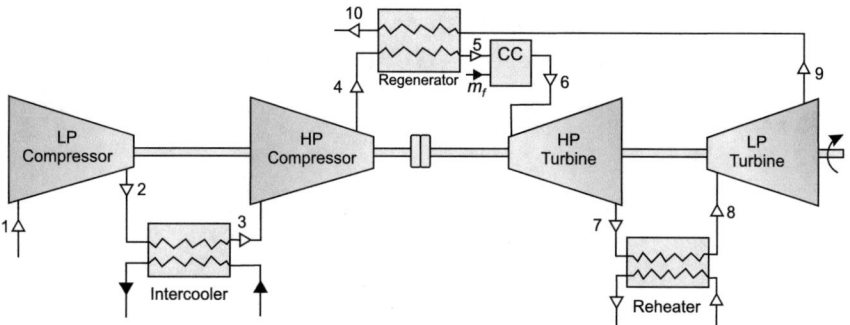

Fig. 5.18   Cycle with regeneration, reheat, and intercooling

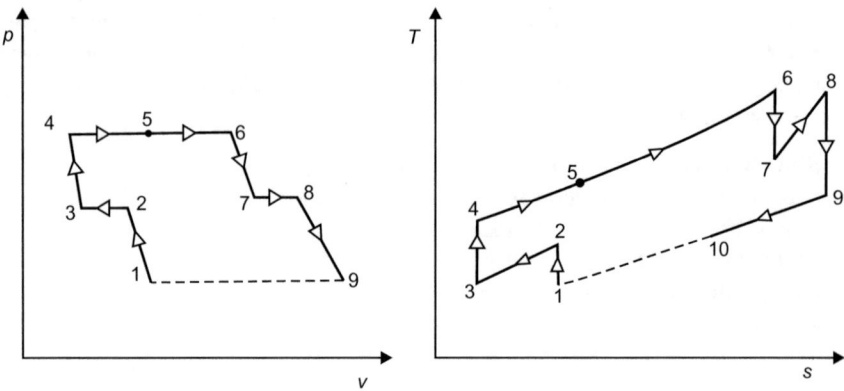

Fig. 5.19 Brayton cycle with regeneration, reheat, and intercooling

Net specific compressors work:

$$w_C = w_{1-2} + w_{3-4}$$
$$= c_p(T_2 - T_1) + c_p(T_4 - T_3)$$

Net specific turbine work:

$$w_T = c_p(T_6 - T_7) + c_p(T_8 - T_9)$$

Net specific work output:

$$w_{\text{net}} = w_T - w_C$$

Net specific heat supplied:

$$q_{\text{net}} = q_{5-6} - q_{7-8}$$
$$= c_p(T_6 - T_5) + c_p(T_8 - T_7)$$

Thermal efficiency:

$$\eta_{\text{th}} = \frac{w_{\text{net}}}{q_{\text{net}}}$$

5.7 Gas Turbines Versus IC Engines

Advantages of Gas Turbines over IC Engines:

1. **Mechanical efficiency**: The mechanical efficiency of the gas turbine is more than the IC engine. For gas turbines, the mechanical efficiency ranging from 95 to 97% and for IC engines, the mechanical efficiency ranging from 85 to 95%.
2. **Weight per unit power**: The weight per unit power of the gas turbine is lower than the diesel engine. The gas turbine has a weight of 15 kg/HP whereas 85 kg/HP for a diesel engine.
3. **Speed**: Due to the absence of a reciprocating part, the gas turbine can be driven at a very high speed (40,000 rpm) where as this is not possible in an IC engine.
4. **Fuels**: Cheap and different types of fuel such as benzene, powdered coal, heavy hydrocarbons, natural gas, etc. can be used in gas turbines whereas special grade fuels, such as petrol, diesel, and CNG, are used in IC engines.
5. **Pollution**: The exhaust from the gas turbine is less polluting than the IC engine because of excess air–fuel ratio is used.
6. **Operation at high altitudes**: The gas turbines are used in aircraft because of its low specific weight.
7. **Low operating pressure**: The gas turbines operate at much lower pressure up to 10 bar as compared with IC engines pressure up to 40 bar.
8. **Maintenance and installation costs**: The maintenance cost of the gas turbine is low because of its purely rotational parts. The installation cost of the gas turbine is also low due to its compact size which needs smaller floor foundations and the building to house the plant.

Disadvantages of Gas Turbines over IC engines

1. **Low thermal efficiency**: The thermal efficiency of a simple gas turbine cycle is low about 17% to 20% as compared with IC engines about 25%–35% because a large part of the power produced by the gas turbine is used to drive the compressor.
2. **Difficult to blade cooling**: The gas turbine blades need a special cooling system.
3. **Poor part-load efficiency**: The part-load efficiency of the gas turbine is very poor, as the quantity of air remains the same irrespective of load, and output is reduced by decreasing the quantity of fuel supplied.
4. **Difficult and costly of blades' manufacture**: The manufacture of nickel–chromium alloy blades is difficult and costly.
5. **Difficult to start**: It is difficult to start a gas turbine as compared to an IC engine.

5.8 Jet Propulsion

The word 'propulsion' means moving forward, the propulsion produced due to action up jet is called jet propulsion. The direction of propulsion is always in the opposite direction of the jet. The principle of jet propulsion is used in propelling the ships, aircraft, and missiles.

5.9 Propulsive Force, Propulsive Efficiency, and Thermal Efficiency

Let

$$C_a = \text{Absolute velocity of aircraft.}$$
$$\text{OR}$$
$$\text{Velocity approach of air}$$
$$C_j = \text{Relative velocity of the jet with respect to aircraft.}$$
$$C = \text{Absolute velocity of the jet}$$
$$= C_j - C_a$$
$$m_a = \text{Mass of air enter to the air - craft, kg/s}$$
$$m_f = \text{Mass of fuel enter to combustion chamber, kg/s}$$

See Fig. 5.20.

Propulsive force or Thrust

Propulsive force exerted on the aircraft is equal to the change of momentum.
    Propulsive force:

$$F = \text{Momentum at exit} - \text{Momentum at inlet}$$
$$= (m_a + m_f)C_j - m_a C_a$$

where m_f is neglected because $m_a \gg m_f$.

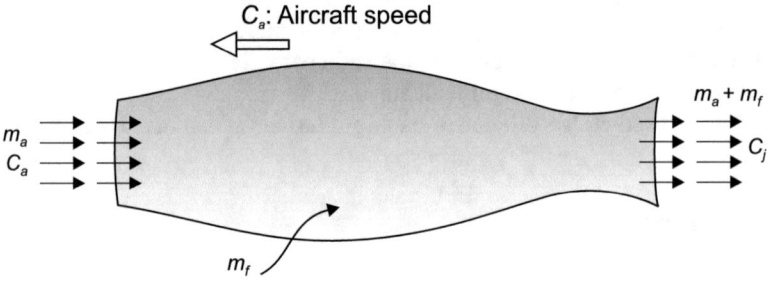

Fig. 5.20 Propulsive engine

$$\therefore \qquad F = m_a(C_j - C_a)$$

Propulsive power:

$$P = \text{Propulsive force} \times \text{speed of aircraft}$$
$$= F \cdot C_a$$
$$= m_a(C_j - C_a)C_a$$

Power produced by the jet engine = Kinetic energy of gases per second at exit − Kinetic energy of air per second at inlet

$$= \frac{m_a C_j^2}{2} - \frac{m_a C_a^2}{2}$$
$$= \frac{m_a}{2}\left(C_j^2 - C_a^2\right)$$

Propulsive efficiency: η_P. It is defined as the ratio of propulsive power to power produced by the jet engine.

Mathematically,

$$\text{Propulsive efficiency} : \eta_P = \frac{\text{Propulsive power}}{\text{Power produced by the jet engine}}$$

$$\eta_P = \frac{m_a(C_j - C_a)C_a}{\frac{m_a}{2}\left(C_j^2 - C_a^2\right)}$$

$$= \frac{2(C_j - C_a)C_a}{(C_j - C_a)(C_j + C_a)} = \frac{2C_a}{C_j + C_a}$$

$$= \frac{2C_a}{C_a\left(\frac{C_j}{C_a} + 1\right)} = \frac{2}{\frac{C_j}{C_a} + 1} \qquad (5.16)$$

Equation (5.16) indicates that the propulsive efficiency increases with an increase in the aircraft velocity C_a. When $C_a = C_J$, the propulsive efficiency becomes 100%, but the propulsive power becomes zero.

$$\eta_P = \frac{2C_a}{C_j + C_a}$$

$$\eta_P = \frac{2\frac{C_a}{C_j}}{1 + \frac{C_a}{C_j}} = \frac{2\sigma}{1 + \sigma}$$

$$\eta_P = \frac{2\sigma}{1 + \sigma}$$

where $\sigma = \frac{C_a}{C_j}$, the ratio of aircraft velocity to jet velocity.

Thermal efficiency: η_{th}. It is defined as the ratio of power produced by the jet engine to heat energy released by the combustion of fuel.

Mathematically,

$$\text{Thermal efficiency} : \eta_{th} = \frac{\text{Power produce by the jet engine}}{\text{Heat energy released per second by the cobustion of fuel}}$$

$$= \frac{m_a\left(C_j^2 - C_a^2\right)}{2m_f C.V}$$

where $C.V. = $ Calorific value of fuel.

Overall Efficiency: η_o. It is defined as the ratio of propulsive power to heat energy released per second by the combustion of fuel.

Methematically,

Overall efficiency:

$$\eta_o = \frac{\text{Propulsive power}}{\begin{array}{c}\text{Heat energy released per second}\\ \text{by the combustion of fuel}\end{array}}$$

$$\eta_o = \frac{m_a(C_j - C_a)C_a}{m_f C.V.}$$

$$\eta_o = \frac{m_a(C_j - C_a)C_a}{\frac{m_a}{2}\left(C_j^2 - C_a^2\right)} \times \frac{\frac{m_a}{2}\left(C_j^2 - C_a^2\right)}{m_f C.V.}$$

$$\eta_o = \eta_P \times \eta_{th}$$

The overall efficiency is also defined as the product of propulsive efficiency and thermal efficiency.

For maximum overall efficiency,

$$\frac{d\eta_o}{dC_a} = 0$$

$$\frac{d}{dC_a}(C_j - C_a)C_a = 0$$

$$\frac{d}{dC_a}(C_j C_a - C_a^2) = 0$$

$$C_j - 2C_a = 0$$

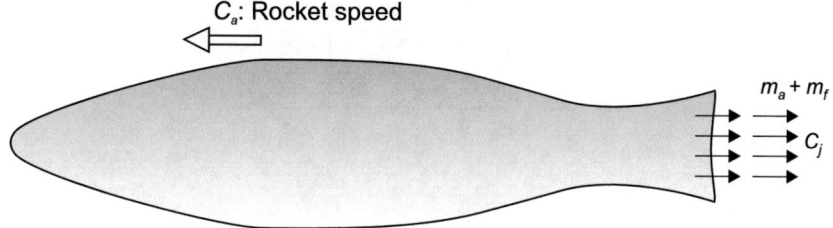

Fig. 5.21 Rocket propulsion

or

$$C_a = \frac{C_j}{2}$$

The overall efficiency of the propulsive engine is maximum when the velocity of the aircraft (C_a) is 1/2 times the of the jet at exit.

For Rocket

Rocket engines are non-breathing engines that do not use atmospheric air i.e., no atmospheric air enter the rocket.

Propulsive force (Fig. 5.21):

$$F = \text{Momentum at exit} - \text{Momentum of air in the rocket}$$
$$= (m_a + m_j)C_j - m_a C_a$$

where m_f is neglected because $m_a \gg m_f$.

$$\therefore \quad F = m_a C_j - m_a C_a$$
$$= m_a (C_j - C_a)$$

Propulsive power:

$$P = \text{Propulsive force} \times \text{speed of aircraft}$$
$$= F C_a$$
$$= m_a (C_j - C_a) C_a$$

Power produced by the rocket engine = Kinetic energy of gases per second at exit.

$$= \frac{m_a C_j^2}{2}$$

Propulsive efficiency:

$$\eta_P = \frac{\text{Propulsive power}}{\text{Power produced by the rocket engine}}$$

$$= \frac{m_a(C_j - C_a)C_a}{\frac{1}{2}m_a C_j^2} = \frac{2(C_j - C_a)C_a}{C_j^2}$$

where $C_j = C + C_a$,

C = Absolute velocity of jet at exit

$$\therefore \eta_p = \frac{2(C + C_a - C_a)C_a}{(C + C_a)^2}$$

$$\eta_p = \frac{2CC_a}{(C + C_a)^2}$$

5.10 Classification of Jet Propulsion Engines

Jet propulsion engines may be classified into two categories:

1.   Air-Breathing Engine,
2.   Non-Air-Breathing Engine.

1.   **Air-Breathing Engine**: In the air-breathing engine, the atmospheric air enters in the engine for the combustion of fuel. The air-breathing engines can further be classified as follows:

    (i)   Turbo Jet,
    (ii)   Turbo Prop, and
    (iii) Pulse Jet.

2.   **Non-Air-Breathing Engine**: In non-air-breathing engine, no atmospheric air is used for the combustion of fuel. The oxidizer is used for combustion which it carries in its own tanks. The non-air-breathing engine is

    (i)   Rocket.

5.11 Turbo Jet

The turbo jet consists of the following main components:

1.   Diffuser,
2.   Compressor,
3.   Combustion chamber,

4. Turbine, and
5. Nozzle.

The diffuser, compressor, combustion chamber, turbine, nozzle are steady flow devices, and all the processes in this system can be analysed as steady-flow processes. The arrangement of components used in turbo jet is shown in Fig. 5.22, and the ideal cycle is shown on *T-s* diagram in Fig. 5.24.

1. Diffuser: Process 1–2, adiabatic isentropic compression in a diffuser

The high velocity of air equal to the speed of aircraft is entered in the diffuser from the ambient. A diffuser increases the pressure of the air and simultaneously it decreases the velocity of the air i.e., it is used to transfer the kinetic energy into pressure energy (Fig. 5.23). As the air flows through the diffuser it compressor to a higher pressure and in this process velocity falls and the pressure increases from the entrance to the exit of the diffuser. In other words, the pressure increases in a diffuser due to the ramming effect. The air of exit of the diffuser at state 2 is called ram air.

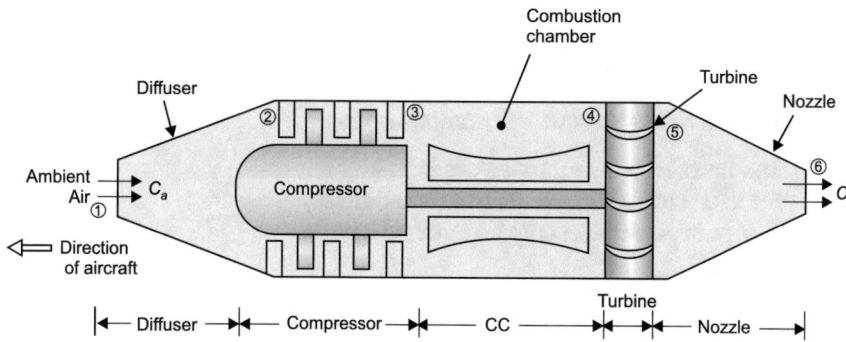

Fig. 5.22 Turbo Jet

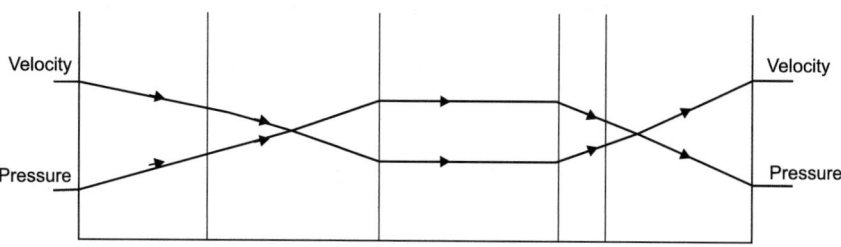

Fig. 5.23 Pressure and velocity variation in various components

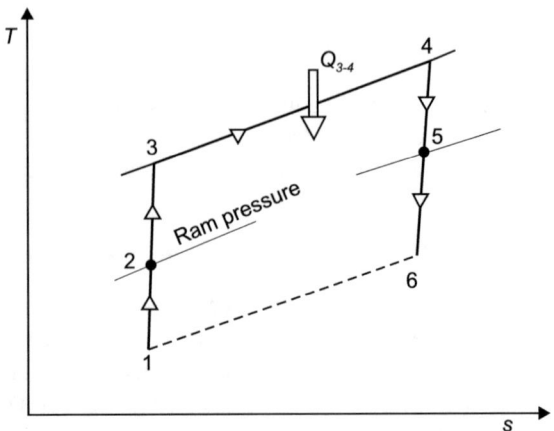

Fig. 5.24 Ideal cycle

The adiabatic isentropic compression 1–2 is shown on T-s diagram in Fig. 5.24. Applying steady flow energy equation,

$$h_1 + \frac{C_a^2}{2} + gz_1 + q_{1-2} = h_2 + \frac{C_2^2}{2} + g_2 + w_{1-2} \qquad (5.17)$$

Diffuser can be analysed with the following assumptions:

(i)   Heat transfer: $q_{1-2} = 0$ for adiabatic flow,
(ii) Work interaction: $w_{1-2} = 0$ always for diffuser,
(iii) Change in potential energy is neglected.

Equation (5.17) reduced to

$$h_1 + \frac{C_a^2}{2} = h_2 + \frac{C_2^2}{2}$$

If C_2 is very small as compared to the velocity C_a and C_2 is neglected

$$\therefore h_1 + \frac{C_a^2}{2} = h_2$$

$$c_p T_1 + \frac{C_a^2}{2} = c_p T_2$$

or

$$T_2 = T_1 + \frac{C_a^2}{2c_p}$$

or

$$\frac{T_2}{T_1} = 1 + \frac{C_a^2}{2c_p T_1}$$

$$\frac{T_2}{T_1} = 1 + \frac{C_a^2}{\frac{2\gamma RT_1}{\gamma - 1}} \qquad \left| \because c_p = \frac{\gamma R}{\gamma - 1} \right.$$

$$= 1 + \frac{\gamma - 1}{2} \frac{C_a^2}{\gamma RT_1}$$

$$= 1 + \frac{\gamma - 1}{2} \frac{C_a^2}{a^2}$$

where $a = \sqrt{\gamma RT_1}$, velocity of sound

or

$$a^2 = \gamma RT_1$$

$$\therefore \frac{T_2}{T_1} = 1 + \frac{(\gamma - 1)}{2} M^2$$

where $M = \frac{C_a}{a}$, Mach number

Also for process 1–2,

$$\frac{T_2}{T_1} = \left(\frac{p_2}{p_1} \right)^{\frac{\gamma - 1}{\gamma}}$$

where $\gamma =$ adiabatic index

2.   Compressor: Process 2–3, adiabatic isentropic compression in a compressor

The ram air at state 2 is entered in the compressor where the pressure of air increases further adiabatic isentropically. During this process, the power output of the gas turbine is used to drive the compressor. The adibatic isentropic compression 2–3 is shown on T-s diagram in Fig. 5.24. Applying steady flow energy equation,

$$h_2 + \frac{C_2^2}{2} + gz_2 + q_{2-3} = h_3 + \frac{C_3^2}{2} + gz_3 + w_{2-3} \qquad (5.18)$$

Compressor can be analysed with the following assumptions:

(i)   Heat transfer: $q_{2-3} = 0$ for adiabatic flow,
(ii) Change in kinetic and potential energies are neglected.

Equation (5.18) is reduced to

$$h_2 = h_3 + w_{2-3}$$

or

$$w_{2-3} = h_2 - h_3$$
$$= c_p(T_2 - T_3)\text{kJ/kg}$$

where c_p = specific heat at constant pressure in kJ/kgK.
   T_2 and T_3 are temperature in °C or K.

3.   **Combustion Chamber: Process 3–4, heat addition at constant pressure in a combustion chamber**

The high-pressure air from the compressor enters the combustion chamber where fuel is added. Due to the combustion of fuel, the temperature rises at constant pressure. This process is shown on *T-s* diagram in Fig. 5.24.
   Heat supplied:

$$Q_{3-4} = \left(m_a + m_f\right)c_{pg}T_4 - m_a c_{pa}T_3$$

where

   m_a = mass flow rate of air, kg/s,
   m_f = mass flow rate of fuel supplied, kg/s,
   c_{pg} = specific heat at constant pressure of gases,
   c_{pa} = specific heat at constant pressure of air.

4.   **Turbine: Process 4–5, adiabatic isentropic expansion in a turbine**

The gases at high temperature and pressure are supplied to the turbine where they expand partially isentropically. The turbine is designed to produce power which is just sufficient to drive the compressor, fuel pump, and other auxiliaries apparatus. This process is shown in *T-s* diagram in Fig. 5.24.
   Turbine work:

$$W_{4-5} = \left(m_a + m_f\right)c_{pg}(T_4 - T_5)$$
$$= W_{2-3}, \text{compressor work}$$
$$= m_a c_{pa}(T_3 - T_2)$$

6.   **Nozzle: Process 5–6, adiabatic isentropic expansion in a nozzle**

After the gases leave the turbine, they expand isentropically further in the nozzle and there occurs a conversion the energy of hot gases into kinetic energy. The gases exit from the nozzle with a velocity that is much greater than the velocity of aircraft. Due to an increase in velocity of nozzle exit, a reaction or thrust in opposite direction is produced and this propels the aircraft. This method of propulsion is best suited to aircraft travelling at or above 800 km/hour (i.e., best suited for subsonic speed, 0.60 $< M < 1$).

Applying steady flow energy equation,

$$h_5 + \frac{C_5^2}{2} + gz_5 + q_{5-6} = h_6 + \frac{C_j^2}{2g} + gz_6 + w_{5-6} \qquad (5.19)$$

Nozzle can be analysed with following assumptions:

(i)   Heat transfer: $q_{5-6} = 0$ for adaiabatic flow,
(ii)   Wok interaction: $w_{5-6} = 0$ always for nozzle, and
(iii) Change in kinetic and potential energies are neglected.

Equation (5.19) is reduced to

$$h_5 + \frac{C_5^2}{2} = h_6 + \frac{C_j^2}{2}$$

or

$$\frac{C_j^2}{2} = (h_5 - h_6) + \frac{C_5^2}{2}$$

or

$$C_j = \sqrt{2(h_5 - h_6) + C_5^2}$$
$$= \sqrt{2c_{pg}(T_5 - T_6) + C_5^2} \text{ m/s}$$

where
   c_{pg} = specific heat of gases is in J/kgK,
   T_5 and T_6 are in °C or K,
   C_5 = Velocity at inlet of nozzle is in m/s,
   $C_j = \sqrt{2000c_{pg}(T_5 - T_6) + C_5^2}$ m/s.

where c_{pg} is in kJ/kgK.

If C_5 and very small as compared to the velocity C_j and C_5 is neglected

$$\therefore C_j = \sqrt{2000c_{pg}(T_5 - T_6)}\,\text{m/s}$$

Thrust or Propulsive force:

$$F = (m_a + m_f)C_j - m_a C_a$$

Propulsive power:

$$P = F \cdot C_a = \left[(m_a + m_f)C_j - m_a C_a\right]C_a$$

$$\text{Power produced by jet engine} = \frac{(m_a + m_f)C_j^2}{2} - \frac{m_a C_a^2}{2}$$

Propulsive efficiency:

$$\eta_P = \frac{\text{Propulsive power}}{\text{Power produced by the jet engine}}$$

$$\eta_P = \frac{\left[(m_a + m_g)C_j - m_a C_a\right]C_a}{\frac{(m_a + m_f)C_j^2}{2} - \frac{m_a C_a^2}{2}}$$

If mass flow rate of fuel m_f is neglected

$$\therefore \eta_p = \frac{2C_a}{C_j + C_a}$$

Overall efficiency:

$$\eta_o = \frac{\text{Propulsive power}}{\text{Heat energy released per second by the combustion of fuel}}$$

$$\eta_o = \frac{\left[(m_a + m_f)C_j - m_a C_a\right]C_a}{m_f \, \text{C.V.}}$$

$$= \frac{\left[\left(\frac{m_a}{m_f} + 1\right)C_j - \frac{m_a}{m_f}C_a\right]C_a}{\text{C.V}}$$

where $C.V. =$ Calorific value of fuel.

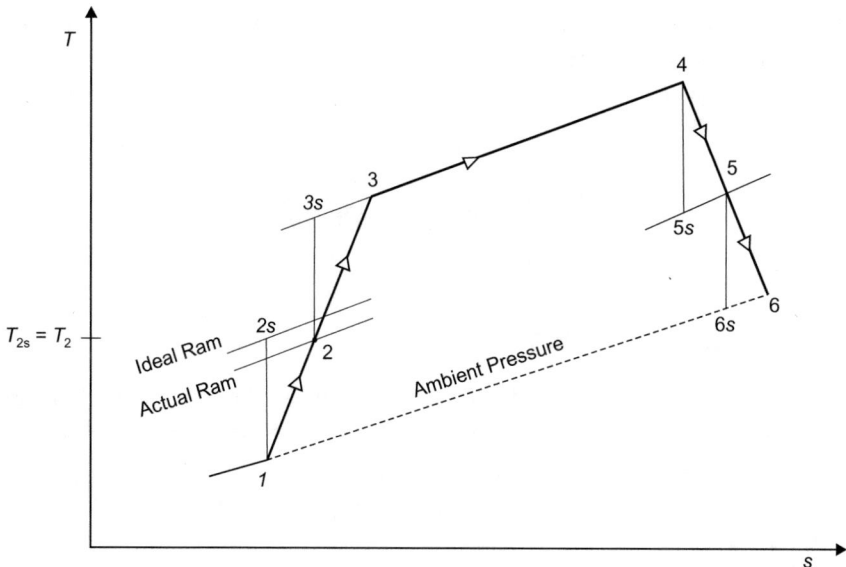

Fig. 5.25 Actual cycle

5.11.1 Actual Cycle

In an actual cycle the diffuser, the compressor, the turbine, and the nozzle does not work as adiabatic isentropic. Thus some losses do occur due to internal friction. The actual cycle is shown on T- diagram in Fig. 5.25 and the various processes are discussed below.

1. **Diffuser: Process 1–2**

The high velocity of air equal to the speed of aircraft is entered in the diffuser from the ambient at pressure p_1 and temperature T_1. Due to internal friction, the pressure increases less than the adiabatic isentropic process at the same temperature and enthalpy as shown in Fig. 5.25.

For process 1–2 s,

$$\frac{T_{2s}}{T_1} = 1 + \frac{(\gamma - 1)}{2} M^2$$

also

$$\frac{T_{2s}}{T_1} = \frac{T_2}{T_1} \qquad \therefore T_{2s} = T_2$$

$$\therefore \quad \frac{T_2}{T_1} = 1 + \frac{(\gamma - 1)}{2} M^2$$

Ram Efficiency: η_R. It is defined as the ratio of the actual increase in pressure to the isentropic increase in pressure.

Mathematically,

Ram efficiency:

$$\eta_R = \frac{\text{Actual increase in pressure}}{\text{Isentropic increase in pressure}}$$

$$= \frac{p_2 - p_1}{p_{2s} - p_1}$$

2.   **Compressor: Process 2–3**

The rum air at state 2 is entered in the compressor where the temperature increases higher than the adiabatic isentropic process at the same pressure as shown in Fig. 5.25.

Compressor Efficiency: η_C. It is defined as the ratio of isentropic increase in temperature to the actual increase in temperature in the compressor.

Mathematically,

Compressor efficiency:

$$\eta_C = \frac{\text{Isentropic increase in temperature}}{\text{Actual increase in temperature}}$$

$$= \frac{T_{3s} - T_2}{T_3 - T_2} \tag{5.20}$$

The compressor efficiency is also called the isentropic efficiency.

For isentropic process 2–3 s,

$$\frac{T_{3s}}{T_2} = \left(\frac{p_3}{p_2}\right)^{\frac{\gamma-1}{\gamma}}$$

or

$$T_{3s} = T_2 \left(\frac{p_3}{p_2}\right)^{\frac{\gamma-1}{\gamma}}$$

For given values of T_2, p_2, and p_3, we can determine the value of T_{3s}. By substituting the value of T_{3s} in Eq. (5.20), we can find out the value of T_3 at a given value of η_C.

Now, compressor input work,

$$W_{2-3} = m_a(h_3 - h_2)$$
$$= m_a c_{pa}(T_3 - T_2)$$

3. **Combustion Chamber**: **Process 3–4, heat addition at constant pressure**

In combustion chamber, heat addition due to combustion of fuel with air at constant pressure. This process is shown on T–s diagram in Fig. 5.25.

Heat addition per second:

$$Q_{3-4} = (m_a + m_\rho)c_{pg}T - m_a c_{pa}T_3$$

4. **Turbine**: **Process 4–5**

Due to internal friction, the gas turbine does not work as adiabatic isentropic. The temperature decreases in the actual process 4–5 lower than the adiabatic isentropic process at the same pressure as shown in Fig. 5.25.

Turbine Efficiency: η_T. It is defined as the ratio of the actual decrease in temperature to the isentropic decrease in temperature.

Mathematically,

Turbine efficiency:

$$\eta_T = \frac{\text{Actual decrease in temperature}}{\text{Isentropic decrease in temperature}}$$

$$\eta_T = \frac{T_4 - T_5}{T_4 - T_{5s}} \tag{5.21}$$

The turbine efficiency is also called the isentropic efficiency.

For isentropic process 4–5 s,

$$\frac{T_4}{T_{5s}} = \left(\frac{p_4}{p_5}\right)^{\frac{\gamma-1}{\gamma}}$$

or

$$T_{5s} = T_4 \left(\frac{p_5}{p_4}\right)^{\frac{\gamma-1}{\gamma}}$$

For given values of T_4, p_4, and p_5, we can determine the value of T_{5s}. By substituting the value of T_{5s} in Eq. (5.21), we can find out the value of T_5 at a given value of η_T.

Now, Turbine output work:

$$W_{4-5} = \left(m_a + m_\rho\right) c_{pg} (T_4 - T_5)$$

5.   Nozzle: Process 5–6

Due to internal friction, the expansion of gases is not adiabatic isentropic. The temperature decreases in the actual process 5–6 lower than the adiabatic isentropic process at the same pressure as shown in Fig. 5.25.

Nozzle Efficiency: η_N. It is defined as the ratio of actual drop in temperature to isentropic drop in temperature.

Mathematically,

Nozzle Efficiency:

$$\eta_N = \frac{\text{Actual drop in temperature}}{\text{Isentropic drop in temperature}}$$
$$= \frac{T_5 - T_6}{T_5 - T_{6s}} \tag{5.22}$$

For isentropic process 5–6 s,

$$\frac{T_5}{T_{6s}} = \left(\frac{p_5}{p_6}\right)^{\frac{\gamma-1}{\gamma}}$$

For given values of T_5, p_5, and p_6, we can determine the value of T_{6s}. By substituting the value of T_{6s} in Eq. (5.22), we can find out the value of T_6 at the given value of η_N.

Velocity of Jet:

$$C_j = \sqrt{2000 c_{pg} (T_5 - T_6)} \, \text{m/s}$$

where

$c_{pg} = $ specific heat at constant pressure is in kJ/kgK,
T_5, T_6 are in °C or K,
$C_5 \approx 0$, the velocity at nozzle inlet is neglected.

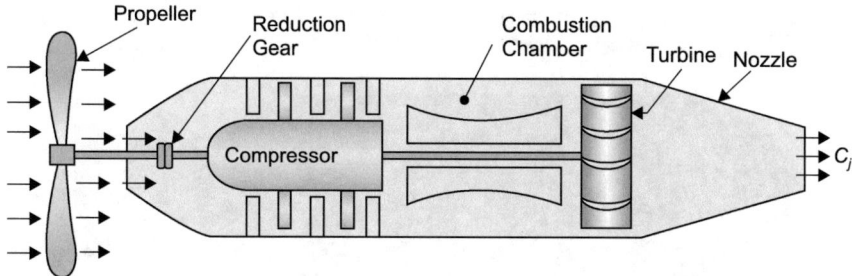

Fig. 5.26 Turbo prop engine

5.12 Turbo Prop

The turbo prop engine consists of the following main components (Fig. 5.26):

1. Propeller,
2. Diffuser,
3. Compressor,
4. Combustion chamber,
5. Turbine, and
6. Nozzle.

The diffuser, compressor, combustion chamber, turbine, and nozzle are the same function as in the turbojet engine. But in turbo prop engine, the turbine produces more power than it does in turbo jet engine because the turbine provides power to both the compressor and the propeller. In turbo prop engine, nearly 80% to 90% of the propulsion power is generated by the propeller and only a small propulsion power 10 to 20%, depending upon the flight velocity, from the exhaust nozzle. The reduction gear must be placed between the turbine shaft and the propeller to enable the propeller to operate efficiently because the gas turbine works at very speed and the propeller requires relatively low speed. The propulsive is provided by dual momentum. (i) The propeller increases the air momentum, and (ii) The overall engine, from diffuser to nozzle, provides an internal momentum increase.

The sum of the above two thrusts is the total thrust developed by the turbo prop. The turbo prop engine is best suited to aircraft travelling at a speed below 800 km/hr.

5.13 Ramjet

Ramjet engine does not require a compressor and turbine. It consists of a diffuser (used for compression), a combustion chamber, and a nozzle.

The compression of air takes place due to the ram effect at high flight Mach numbers. Therefore, the compressor and turbine are not required.

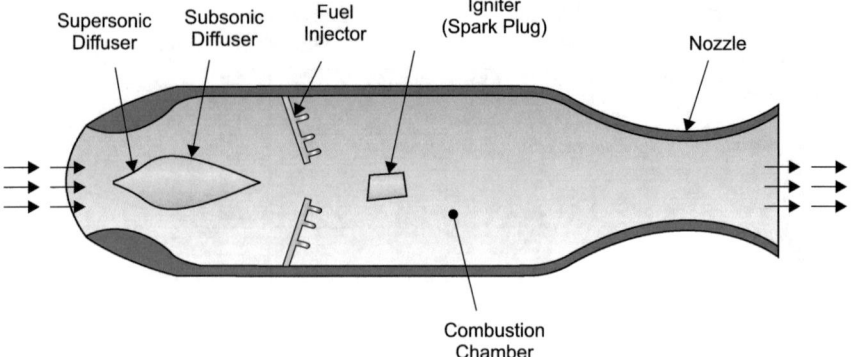

Fig. 5.27 Ram Jet

The best performance of ramjet engine has obtained a flight above Mach number 2 (two times the speed of sound). In a ramjet, the air is slowed down to about Mach number 0.2, fuel is added to the air and burned at this slow velocity and the combustion gases are expanded and accelerated in a nozzle. Due to the absent of the compressor and turbine, the ram jet engine cannot be started while it is at rest. So, it needs a launching device (Fig. 5.27).

Advantages:

(i)   No moving parts,
(ii)   Light in weight, and
(iii)   Wide variety of fuel may be used.

Disadvantages:

(i)   It cannot be started of its own. It has to be accelerated to a certain flight velocity by launching device.
(ii)   The fuel consumption is too large at low and moderate speed.

5.14   Rocket Engines–Non-breathing Engine

A rocket engine does not use atmospheric air for combustion. It contains its own oxygen tanks for fuel combustion. Propellant is the common name given to fuel and oxidizer. The rocket engine is independent of altitude and speed.

Classification of Rockets:

1.   Rocket engines may be classified into two categories according to propellant.

(i)   Liquid propellant rocket engines,
(ii)   Solid propellant rocket engines.

A liquid propellant rocket engine using liquid oxygen and liquid hydrocarbon fuel is shown in Fig. 5.28. In this system, the liquid oxidizer and fuel are stored in the supply tanks at low pressure in order to required a very light supply tank and forced into the combustion chamber at high pressure by the fuel and oxidizer pumps. The power required for driving the pumps is supplied by a gas turbine. The gases are formed in the combustion chamber by the fuel and the oxidizer for the gas turbine.

The rocket motor consists of an injection plate, a combustion chamber, and a discharge nozzle. The function of the injection plate is to receive the liquid oxidizer and fuel from the pumps through the valves and direct them in liquid streams so that they mix up with one another and burn in the combustion chamber. When the fuel burns in the combustion chamber, very high pressure and temperature gases are produced which are expanded in the nozzle to produce a high supersonic exit velocity of about 1500 to 3000 m/s.

Desirable properties for Rocket propellant:

1. High calorific value,
2. High density,
3. Ease of storage and handling,
4. Availability,
5. No chemical reaction with tanks, piping, valves, pumps, and injection nozzles,
6. Easily ignitable.

Application of Rocket:

1. For satellites,
2. For space ships,
3. Long-range artillery, and
4. Lethal weapons.

Problem 5.1: A gas turbine operates on a Brayton cycle with inlet conditions as 100 kPa, 20 °C, and a pressure ratio of 8. The temperature at turbine entry is 900 °C. Estimate the cycle efficiency.

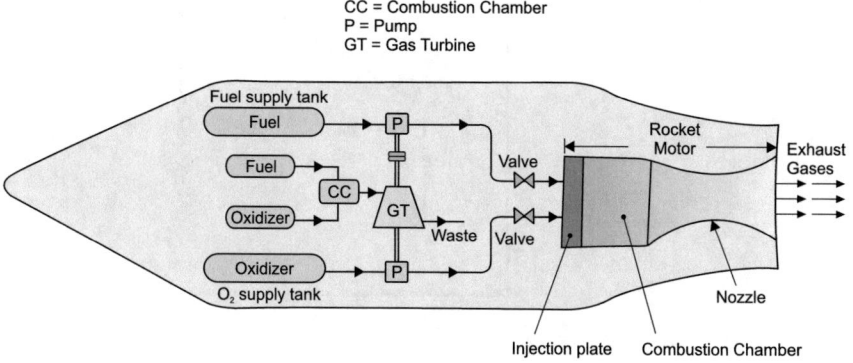

Fig. 5.28 Main component of a liquid propellant rocket engine

Solution: Given data (Fig. 5.29):

$p_1 = 100$ kPa.

$T_1 = 20\,°\text{C} = (20 + 273)\,\text{K} = 293\,\text{K}.$

$r_p = 8.$

$T_3 = 900\,°\text{C} = (900 + 273)\,\text{K} = 927\,\text{K}.$

For process 1–2,

$$\frac{T_2}{T_1} = \left(\frac{p_2}{p_1}\right)^{\frac{\gamma-1}{\gamma}}$$

$$\frac{T_2}{293} = (8)^{\frac{1.4-1}{1.4}} = (8)^{0.285} = 1.8$$

or

$$T_2 = 293 \times 1.8 = 523.4\,\text{K}$$

For process 3–4,

$$\frac{T_3}{T_4} = \left(\frac{p_3}{p_4}\right)^{\frac{\gamma-1}{\gamma}}$$

$$\frac{927}{T_4} = (8)^{\frac{1.4-1}{1.4}} = (8)^{0.285} = 1.8$$

or

Fig. 5.29 T-s diagram for Problem 5.1

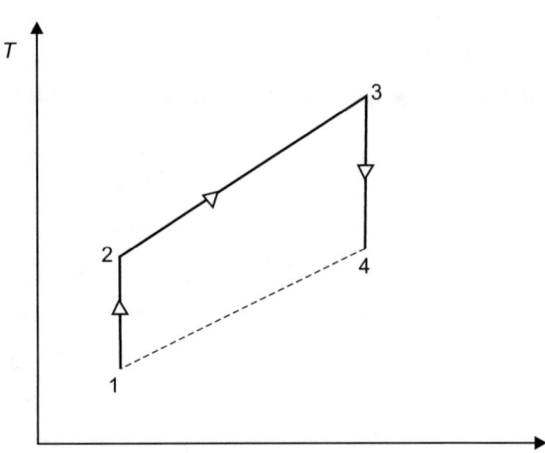

$$T_4 = \frac{927}{1.8} = 515 \text{ K}$$

Compressor input work: $W_C = mc_p(T_2 - T_1)$.
Turbine output work: $W_T = mc_p(T_3 - T_4)$.
Net output work:
$$W_{net} = W_T - W_C$$
$$= mc_p(T_{13} - T_4 - T_2 + T_1)^.$$
Heat supplied: $Q_{2-3} = mc_p(T_3 - T_2)$.
Cycle efficiency:

$$\eta = \frac{W_{net}}{Q_{2-3}}$$
$$= \frac{mc_p(T_3 - T_4 - T_2 + T_1)}{mc_p(T_3 - T_2)}$$
$$= \frac{T_3 - T_4 - T_2 + T_1}{T_3 - T_2} = \frac{927 - 515 - 523.4 + 293}{927 - 523.4}$$
$$= 0.4499 = 44.99\% = \mathbf{45\%}$$

or

$$\eta = 1 - \frac{1}{r_p^{(\gamma-1)/\gamma}}$$
$$= 1 - \frac{1}{(8)^{(1.4-1)/1.4}} = 1 - \frac{1}{8^{0.285}}$$
$$= 1 - \frac{1}{1.8} = 1 - 0.55 = 0.45 = 45\%$$

Problem 5.2: Air enters the compressor of a gas turbine at 100 kPa and 25 °C. For a pressure ratio of 6 and a maximum temperature of 850 °C, determine (i) the temperature of the air at the exit of the compresor, (ii) the back pressure ratio, (iii) the thermal efficiency.

Solution: Given data:
At inlet of the compressor.
Temperature: $T_1 = 25 \text{ °C} = (25 + 273) \text{ K} = 298 \text{ K}$.
Pressure: $p_1 = 100 \text{ kPa}$.
Pressure ratio: $r_p = 6$.
Maximum temperature: $T_3 = 850 \text{ °C} = (850 + 273) \text{ K} = 1123 \text{ K}$.

(i)   Temperature of air at exit of the compressor: T_2

For process 1–2 (Fig. 5.30):

$$\frac{T_2}{T_1} = \left(\frac{p_2}{p_1}\right)^{\frac{\gamma-1}{\gamma}}$$

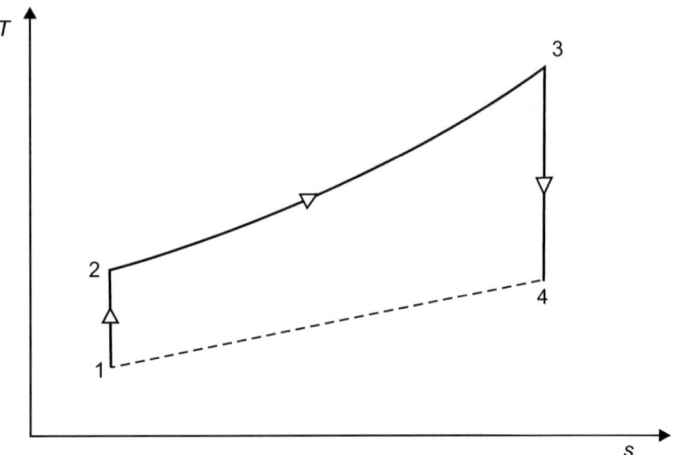

Fig. 5.30 T-s diagram for Problem 5.2

$$\frac{T_2}{T_1} = r_p^{(\gamma-1)/\gamma}$$

$$\frac{T_2}{298} = 6^{(1.4-1)/1.4} = 0.285$$

or

$$T_2 = 298 \times 6^{0.285} = \textbf{496.58 K}$$

(ii)   Back pressure ratio: BPR

For process 3–4:

$$\frac{T_3}{T_4} = \left(\frac{p_2}{p_1}\right)^{\frac{\gamma-1}{\gamma}}$$

$$\frac{T_3}{T_4} = r_p^{(\gamma-1)/\gamma}$$

$$\frac{1123}{T_4} = 6^{(1.4-1)/1.4} = 6^{0.285} = 1.666$$

or $T_4 = \frac{1123}{1666} = 670.06$ K.
Compressor work: $w_c = c_p(T_2 - T_1) = 1.005(496.58 - 298) = 199.57$ kJ/kg.
Turbine work: $w_T = c_p(T_3 - T_4)$
$$= 1.005(1123 - 670.06) = 455.20 \text{ kJ/kg}^{\cdot}$$

Back Pressure Ratio:

$$BPR = \frac{\text{compressor work} : w_c}{\text{turbine work} : w_T}$$

$$= \frac{w_c}{w_T} = \frac{199.57}{455.20} = 0.4384 = \mathbf{43.84\%}$$

(iii) **Thermal Efficiency:** η_{th}

$$\eta_{th} = 1 - \frac{1}{r_p^{(\gamma-1)/\gamma}}$$

$$= 1 - \frac{1}{6^{(1.4-1)/1.4}} = 1 - \frac{1}{6^{0.285}}$$

$$= 1 - 0.6 = 0.4 = \mathbf{40\%}$$

Problem 5.3: A gas turbine power plant is to produce 800 kW of power by compressing atmospheric air at 20 °C and 800 kPa. If the maximum temperature is 800 °C determine the minimum mass flow rate of the air.

Solution: Given data:
   Power:

$$P = 800 \ \text{kW}$$
$$T_1 = 20 \ °C = (20 + 273) \ K = 293 \ K$$
$$p_2 = 800 \ \text{kPa}$$

At inlet of turbine:

$$T_3 = 800 \ °C = (800 + 273) \ K = 1073 \ K$$

The minimum mass flow rate of the air means the minimum compressor input work and maximum plant output work. For maximum output work, (Fig. 5.31).

$$\frac{T_3}{T_1} = \left(\frac{p_2}{p_1}\right)^{2(\gamma-1)/\gamma} = (r_p)^{2(\gamma-1)/\gamma}$$

$$\frac{1073}{293} = r_p^{2(1.4-1)/1.4}$$

$$3.66 = r_p^{0.5714}$$

or

$$r_p = (3.66)^{1/0.5714} = (3.66)^{1.75} = 9.68$$

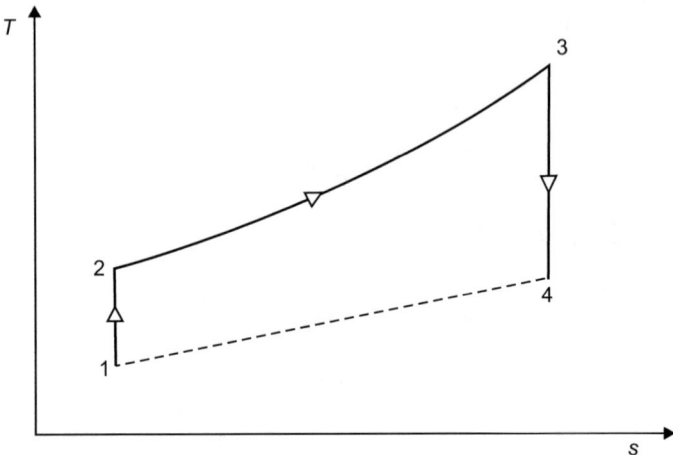

Fig. 5.31   T-s diagram for Problem 5.3

For process 1–2,

$$\frac{T_2}{T_1} = \left(r_p\right)^{(\gamma-1)/\gamma}$$

$$\frac{T_2}{293} = (9.68)^{(1.4-1)/1.4} = (9.68)^{0.2857} = 1.91$$

or

$$T_2 = 293 \times 1.91 = 559.63 \text{ K}$$

For process 3–4,

$$\frac{T_3}{T_4} = \left(r_p\right)^{(\gamma-1)/\gamma} = (9.68)^{(1.4-1)/1.4} = (9368)^{0.2857}$$

$$\frac{1076}{T_4} = 1.91$$

or

$$T_4 = \frac{1073}{1.91} = 561.78 \text{ K}$$

Net power output:

$$P = \text{Turbine output power} - \text{Compressor input power}$$

$$P = mc_p(T_3 - T_4) - mc_p(T_2 - T_1)$$
$$P = mc_p[T_3 - T_4 - T_2 + T_1]$$
$$800 = m \times 1.005[1073 - 561.78 - 559.63 + 293]$$
$$800 = m \times 1005 \times 244.59$$

or

$$m = \mathbf{3.24\ kg/s}$$

Problem 5.4: Determine the compressor outlet pressure that will result in maximum work output for a Brayton cycle in which the compressor inlet air conditions are 20 °C and 100 kPa and the maximum temperature is 1000 °C.

Solution: Given data:

$$T_1 = 20\ ^\circ\text{C} = (20 + 273)\ \text{K} = 293\ \text{K}$$
$$p_1 = 100\ \text{kPa}$$
$$T_3 = 1000\ ^\circ\text{C} = (1000 + 273)\ \text{K} = 1273\ \text{K}$$

For maximum work output condition (Fig. 5.32),

$$\frac{T_3}{T_1} = \left(\frac{p_2}{p_1}\right)^{\frac{2(\gamma-1)}{\gamma}}$$

$$\frac{1273}{293} = \left(\frac{p_2}{100}\right)^{\frac{2(1.4-1)}{1.4}}$$

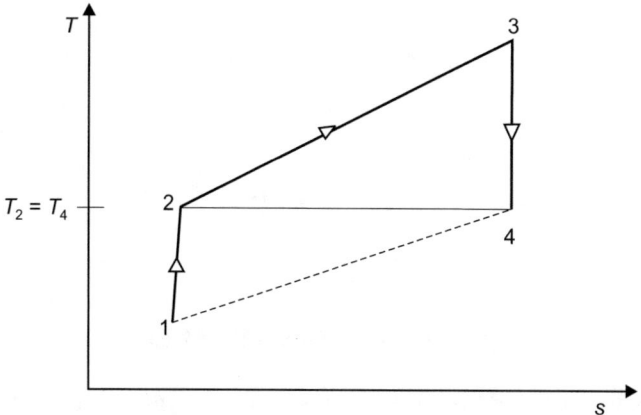

Fig. 5.32   T-s diagram for Problem 5.4

$$4.34 = \left(\frac{p_2}{100}\right)^{0.571}$$

or

$$\frac{p_2}{100} = (4.37)^{1.75} = 13.208$$

or

$$p_2 = 13.208 \times 100 = 1320.8 \text{ kPa}$$

or

$$T_2 = \sqrt{T_1 T_3} \quad \text{for maximum work output.}$$

$$T_2 = \sqrt{293 \times 1273} = 610.72 \text{ K.}$$

For process 1–2,

$$\frac{T_2}{T_1} = \left(\frac{p_2}{p_1}\right)^{\frac{\gamma-1}{\gamma}}$$

$$\frac{610.72}{293} = \left(\frac{p_2}{100}\right)^{\frac{1.4-1}{1.4}}$$

$$2.08 = \left(\frac{p_2}{100}\right)^{0.2857}$$

or

$$\frac{p_2}{100} = (2.08)^{\frac{1}{0.2857}} = (2.08)^{3.5}$$

or

$$\frac{p_2}{100} = 12.978$$

or

$$p_2 = 12978 \times 100 = \mathbf{1297.8 \text{ kPa}}$$

Problem 5.5: A gas turbine plant operates on the Brayton cycle between $T_{\min} = 300$ K and $T_{\max} = 1073$ K. Find the corresponding cycle efficiency. How does this

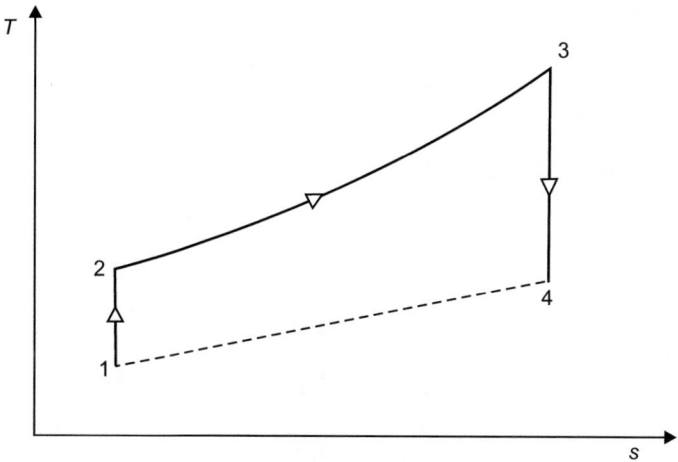

Fig. 5.33 T-s diagram for Problem 5.5

efficiency compare with the Carnot cycle efficiency operating between the same two temperatures.

Solution: Given data (Fig. 5.33):

Minimum temperature: $T_1 = T_{min} = 300$ K.
Maximum temperature: $T_3 = T_{max} = 1073$ K.
For maximum work done condition:

$$\frac{T_3}{T_1} = (r_p)^{\frac{2(\gamma-1)}{\gamma}}$$

$$\frac{1073}{300} = (r_p)^{\frac{2(1.4-1)}{1.4}}$$

$$3.576 = r_p^{0.5714}$$

or

$$r_p = (3.576)^{1.75} = 9.3$$

$$T_2 = \sqrt{T_1 T_3} \quad \text{for maximum work done}$$
$$= \sqrt{300 \times 1073} = 567.36 \text{ K}$$

Maximum specific work output:

$$w_{max} = c_p \left[\sqrt{T_1} - \sqrt{T_3} \right]^2$$

Heat supplied:

$$q_{2-3} = c_p(T_3 - T_2)$$

∴ **Maximum thermal efficiency**:

$$
\begin{aligned}
\eta_{max} &= \frac{w_{max}}{q_{2-3}} \\
&= \frac{c_p\left(\sqrt{T_1} - \sqrt{T_3}\right)^2}{c_p(T_3 - T_2)} \\
&= \frac{\left(\sqrt{T_1} - \sqrt{T_3}\right)^2}{T_3 - T_2} \\
&= \frac{\left(\sqrt{300} - \sqrt{1073}\right)^2}{1073 - 567.36} \\
&= \frac{238.27}{505.64} = 0.4712 = \mathbf{47.12\%}
\end{aligned}
$$

Thermal efficiency:

$$
\begin{aligned}
\eta_{Brayton} &= 1 - \frac{1}{r_p^{(\gamma-1)/\gamma}} \\
&= 1 - \frac{1}{(9.3)^{(1.4-1)/1.4}} \\
&= 1 - \frac{1}{9.3^{0.2857}} = 1 - \frac{1}{1.89} \\
&= 1 - 0.5291 = 0.4709 = \mathbf{47.09\%}
\end{aligned}
$$

Carnot efficiency:

$$
\begin{aligned}
\eta_{Canot} &= 1 - \frac{T_{min}}{T_{max}} = 1 - \frac{T_1}{T_3} \\
&= 1 - \frac{30}{1073} = 1 - 0.2795 = 0.7205 = \mathbf{72.05\%}
\end{aligned}
$$

Problem 5.6: An open-cycle gas turbine plant works between the pressure range of 1 bar and 6 bar and temperature range of 300 K and 1023 K. The calorific value of the fuel used in 43000 kJ/kg. Find the air-fuel (A/F) ratio and the thermal efficiency of the plant. Assume the compression and expansion are isentropic and pressure losses are neglected.

Solution: Given data (Fig. 5.34):

$p_1 = 1$ bar.

Fig. 5.34 T-s diagram for
Problem 5.6

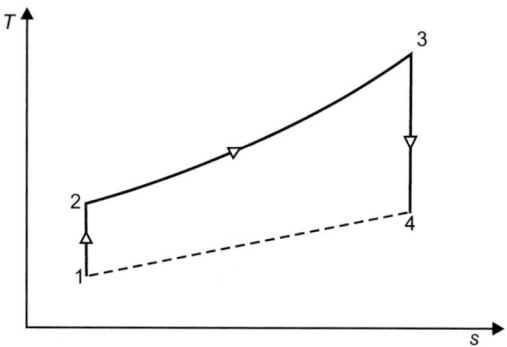

$p_2 = 6$ bar.
$T_1 = 300$ K.
$T_3 = 1023$ K.
Calorific value of the fuel: CV = 43000 kJ/kg.
For adiabatic process 1–2,

$$\frac{T_2}{T_1} = \left(\frac{p_2}{p_1}\right)^{\frac{\gamma-1}{\gamma}}$$

$$\frac{T_2}{300} = \left(\frac{6}{1}\right)^{\frac{1.4-1}{1.4}} = (6)^{0.285}$$

$$T_2 = 300 \times 6^{0.285} = 499.91 \text{ K}$$

For adiabatic process 3–4,

$$\frac{T_3}{T_4} = \left(\frac{p_2}{p_1}\right)^{\frac{\gamma-1}{\gamma}}$$

$$\frac{1023}{T_4} = \left(\frac{6}{1}\right)^{\frac{1.4-1}{1.4}} = (6)^{0.2857} = 1.668$$

or

$$T_4 = \frac{1023}{1.668} = 613.31 \text{ K}$$

Heat supplied:

$$Q_{2-3} = (m_a + m_f)c_{pg}T_3 - m_a c_{pa} T_2$$

Also

$$Q_{2-3} = m_f \times CV$$

$$\therefore \qquad (m_a + m_f)c_{pg}T_3 - m_a c_{pa}T_2 = m_f \times CV$$

$$\left(\frac{m_a}{m_f} + 1\right)c_{pg}T_3 - \frac{m_a}{m_f}c_{pa}T_2 = CV$$

$$\left(\frac{m_a}{m_f} + 1\right) \times 1.005 \times 1025 - \frac{m_a}{m_f} \times 1.005 \times 499.91 = 43000$$

[Note that if the properties of gases and not given, then we will assume the properties of gases same as the properties of air, i.e., $c_{pg} = c_{pa} = 1.005$ kJ/kgK, $\gamma = 1.4$]

$$\left(\frac{m_a}{m_f} + 1\right) \times 1030.12 - 502.40\frac{m_a}{m_f} = 43000$$

$$1030.12\frac{m_a}{m_f} + 1030.12 - 502.40\frac{m_a}{m_f} = 43000$$

$$527.72\frac{m_a}{m_f} = 41969.88$$

or

$$\frac{m_a}{m_f} = \mathbf{79.53}$$

Thermal efficiency:

$$\eta_{th} = 1 - \frac{1}{r_p^{(\gamma-1)/\gamma}} \qquad \text{only for ideal cycle}$$

$$= 1 - \frac{1}{(6)^{(1.4-1)/1.4}} = 1 - \frac{1}{6^{0.2857}}$$

$$= 1 - 0.5993 = 0.4006 = \mathbf{40.06\%}$$

Problem 5.7: Air enters the compressor of a gas turbine power plant operating on Brayton cycle at 1 bar, 27 °C. The pressure ratio of the cycle is 6. If $W_T = 2.5 \, W_C$, where W_T and W_C are the turbine and compressor work respectively, determine the maximum temperature and the cycle efficiency.

Solution: Given data:

$$p_1 = 1 \text{ bar}$$
$$T_1 = 27\,°C = (27 + 273)\,K = 300\,K$$

Pressure ratio:

$$r_p = 6$$
$$W_T = 2.5 W_C$$

For process 1–2,

$$\frac{T_2}{T_1} = \left(\frac{p_2}{p_1}\right)^{\frac{\gamma-1}{\gamma}}$$

$$\frac{T_2}{300} = (6)^{\frac{1.4-1}{1.4}}$$

$$T_2 = 300 \times (6)^{0.2857} = 500.54\,K$$

Compressor input work:

$$W_C = m\,c_p(T_2 - T_1)$$
$$= 1 \times 1.005(500.54 - 300) = 201.54 \text{ kJ/kg}$$

Cycle efficiency (Fig. 5.35):

$$\eta_{th} = 1 - \frac{1}{r_p^{(\gamma-1)/\gamma}} = 1 - \frac{1}{(6)^{(1.4-1)/1.4}}$$
$$= 1 - 0.5993 = 0.4007 = \mathbf{40.07\%}$$

Fig. 5.35 T-s diagram for Problem 5.7

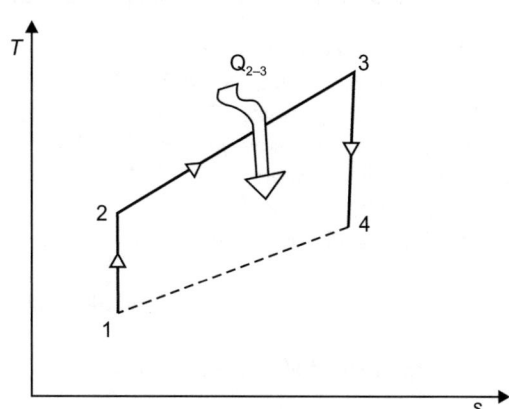

also

$$\eta_{th} = \frac{W_{net}}{Q_{2-3}}$$

$$0.4007 = \frac{W_T - W_C}{Q_{2-3}}$$

$$0.4007 = \frac{2.5W_C - W_C}{Q_{2-3}}$$

$$0.4007 = \frac{1.5W_C}{Q_{2-3}}$$

$$0.4007 = \frac{1.5 \times 201.54}{Q_{2-3}}$$

or $Q_{2-3} = 754.45$ kJ/kg.

also $Q_{2-3} = mc_p(T_3 - T_2)$.
$\therefore 754.45 = 1 \times 1005(T_3 - 500.54)$.

or $T_5 - 500.54 = 750.69$.
   or $T_5 = \mathbf{1251.23}$ K.

Problem 5.8: In a gas turbine plant the minimum and maximum temperature are 25 °C and 850 °C, respectively. If the compressor and turbine efficiencies are 85% and 88%, respectively, determine for maximum work output.

(i)   the maximum work output,
(ii)   the net work output per unit mass flow rate, and
(iii) the thermal efficiency of the plant.

Assume for both compressor and turbine $\gamma = 1.4$ and $c_p = 1.005$ kJ/kgK.

Solution: Given data: (refer Fig 5.36)
   Minimum temperature: $T_1 = 25\ ^\circ C = (25 + 273)$ K $= 298$ K.
   Maximum temperature: $T_3 = 850\ ^\circ C = (850 + 273)$ K $= 1123$ K.
   Compressor efficiency: $\eta_C = 85\% = 0.85$.
   Turbine efficiency: $\eta_T = 88\% = 0.88$.
   Adiabatic index: $\gamma = 1.4$
   Specific heat at constant pressure:

$$c_P = 1.005 \text{ kJ/kgK}$$

For maximum work output condition

$$\frac{T_2}{T_1}\eta_C\eta_T = r_p\frac{2(\gamma - 1)}{\gamma}$$

Fig. 5.36 T-s diagram for Problem 5.8

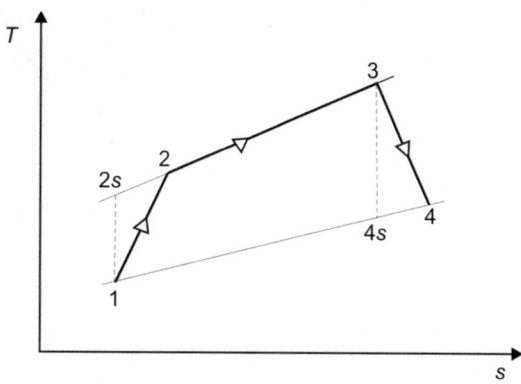

$$\frac{1123}{298} \times 0.85 \times 0.88 = r_p \frac{2(1.4-1)}{1.4}$$

$$2.818 = r_p^{0.5714}$$

or

$$r_p = 2.818^{1.75} = 6.129$$

For isentropic process 1–2 s

$$\frac{T_{2s}}{T_1} = \left(\frac{p_2}{p_1}\right)^{\frac{\gamma-1}{\gamma}}$$

$$\frac{T_{2s}}{T_1} = (r_p)^{(\gamma-1)\gamma_1}$$

$$\frac{T_{2s}}{298} = (6.129)^{\frac{1.4-1}{1.4}}$$

or

$$\frac{T_{2s}}{298} = (6.129)^{0.2857}$$

or

$$T_{2s} = 500.23 \text{ K}$$

$$\eta_C = \frac{T_{2s} - T_1}{T_2 - T_1}$$

$$0.85 = \frac{500.23 - 298}{T_2 - 298}$$

or

$$0.85 = \frac{202.23}{T_2 - 298}$$

or

$$T_2 - 2298 = \frac{202.23}{0.85} = 237.91$$

or

$$T_2 = 535.91 \text{ K}$$

For isentropic process 3–4 s

$$\frac{T_3}{T_{4s}} = \left(\frac{p_2}{p_1}\right)^{\frac{\gamma-1}{\gamma}}$$

$$\frac{1123}{T_{4s}} = (6.129)^{\frac{1.4-1}{1.4}}$$

$$\frac{1123}{T_{4s}} = (6.129)^{0.2857}$$

$$\frac{1123}{T_{4s}} = 1.678$$

or

$$T_{4s} = \frac{1123}{T_{4s}} = 669.24 \text{ K}$$

$$\eta_T = \frac{T_3 - T_4}{T_3 - T_{4s}}$$

$$0.88 = \frac{1123 - T_4}{1123 - 669.24}$$

$$0.88 = \frac{1123 - T_4}{453.76}$$

or

$$453.76 \times 0.88 = 1123 - T_4$$
$$399.30 = 1123 - T_4$$

or

$$T_4 = 723.7 \text{ K}$$

(i) Pressure ratio:

$$r_p = \mathbf{6.129}$$

(ii) Net work output per unit mass flow rate: w_{net}

Compressor input work: $w_C = c_p(T_2 - T_1) = 1.005(535.91 - 298) = 239.09$ kJ/kg.
  Turbine output work: $w_T = c_p(T_3 - T_4) = 1.005(1123 - 723.7) = 401.29$ kJ/kg.
  Net work output: $w_{\text{net}} = w_T - w_C = 401.29 - 239.09 = 162.2$ kJ/kg.

(iii) Thermal efficiency: η_{th}

$$\eta_{\text{th}} = \frac{\text{net work output} : w_{\text{net}}}{\text{heat supplied} : q_{2-3}}$$
$$= \frac{w_{\text{net}}}{q_{3-4}} = \frac{w_{\text{net}}}{c_p(T_3 - T_2)}$$
$$= \frac{162.2}{1.005(1123 - 535.91)} = \frac{162.2}{590.02} = \mathbf{27.49\%}$$

Problem 5.9: In an air standard gas turbine engine, air at a temperature of 15 °C and pressure of 1.01 bar enters the compressor which is compressed through a pressure ratio of 5. Air enters the turbine at a temperature of 815 °C and expands to an original pressure of 1.01 bar. Assume that turbine is working on an ideal Brayton cycle. Take $\gamma = 1.4$, $c_p = 1.005$ kJ/kgK. Determine the thermal efficiency of the cycle. In the above example if the isentropic efficiency of the compressor is 83% and turbine efficiency is 92%. Determine revised thermal efficiency.

Solution: Given data:

For Ideal Brayton Cycle:

$$T_1 = 15 \text{ °C} = 288 \text{ K}$$
$$p_1 = 1.01 \text{ bar}$$

Pressure ratio (Fig. 5.37):

$$r_p = 5$$

Fig. 5.37 T-s diagram
of Ideal Brayton cycle for
Problem 5.9

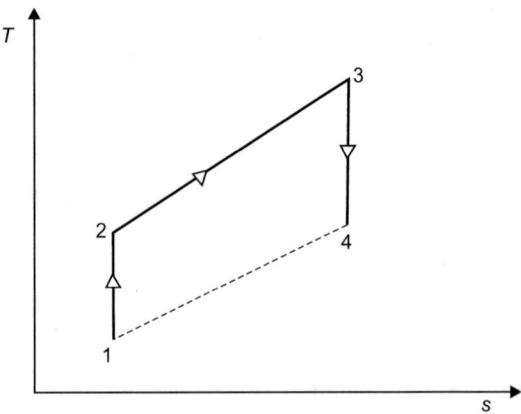

Fig. 5.37 T-s diagram of Ideal Brayton cycle for Problem 5.9

$$T_3 = 815\ ^\circ C = (815 + 273)\ K = 1088\ K$$
$$\gamma = 1.4$$
$$c_p = 1.005\ kJ/kgK$$

Thermal Efficiency:

$$\eta_{th} = 1 - \frac{1}{r_p^{\frac{(\gamma-1)}{\gamma}}}$$
$$= 1 - \frac{1}{5^{\frac{(1.4-1)}{1.4}}} = 1 - \frac{1}{5^{0.285}} = 1 - 0.6321$$
$$= 0.3679 = \mathbf{36.79\%}$$

For Actual Cycle: Isentropic efficiency of the compressor:

$$\eta_C = 83\% = 0.83$$

Isentropic efficiency of the turbine:

$$\eta_T = 92\% = 0.92$$

For adiabatic process 1–2 s,

$$\frac{T_{2s}}{T_1} = \left(\frac{p_2}{p_1}\right)^{\frac{\gamma-1}{\gamma}}$$

$$\frac{T_2}{288} = (5)^{\frac{1.4-1}{1.4}} = 5^{0.285}$$

$$T_2 = 288 \times 5^{0.285} = 455.61 \ \text{K}$$

$$\eta_C = \frac{T_{2s} - T_1}{T_2 - T_1}$$

$$0.83 = \frac{455.61 - 288}{T_2 - 288}$$

$$0.83 = \frac{167.61}{T_2 - 288}$$

or

$$T_2 - 288 = \frac{167.61}{0.83} = 201.94$$

$$T_2 = 201.94 + 288 = 489.94 \ \text{K}$$

For adiabatic process 3–4 s (Fig. 5.38)

$$\frac{T_3}{T_{4s}} = \left(\frac{p_2}{p_1}\right)^{\frac{\gamma-1}{\gamma}} = 5^{(1.4-1)/1.4} = 5^{0.285} = 1.582$$

$$\frac{1088}{T_{4s}} = 1.582$$

or

$$T_{4s} = \frac{1088}{1.582} = 687.73 \ \text{K}$$

Fig. 5.38 T-s diagram of actual Brayton cycle for Problem 5.9

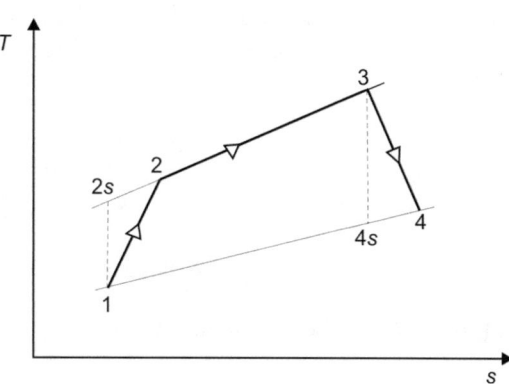

$$\eta_T = \frac{T_3 - T_4}{T_3 - T_{4s}}$$

$$0.92 = \frac{1088 - T_4}{1088 - 687.73}$$

$$0.92 = \frac{1088 - T_4}{400.27}$$

or

$$0.92 \times 400.27 = 1088 - T_4$$

$$368.24 = 1088 - T_4$$

or

$$T_4 = 719.76 \text{ K}$$

For unit mass flow rate
Turbine work: $w_T = c_p(T_3 - T_4) = 1.005(1088 - 719.76) = 370.08$ kJ/kg.
Compressor work: $w_C = c_p(T_2 - T_1) = 1.005(489.94 - 288) = 202.95$ kJ/kg.
Net work output: $w_{net} = w_T - w_C = 370.08 - 202.95 = 167.13$ kJ/kg.

Heta supplied:
$$q_s = q_{2-3} = c_p(T_3 - T_2) = 1.005(1088 - 489.94)$$
$$= 601.05 \text{ kJ/kg}$$

Thermal Efficiency:
$$\eta'_{th} = \frac{\text{net work output}: w_{net}}{\text{heat supplied}: q_s}$$
$$= \frac{167.13}{601.05} = 0.2780 = \mathbf{27.80\%}$$

Problem 5.10: Air enters the compressor of a gas turbine at 1 bar and 293 K. The pressure after compression is 6 bar. The air–fuel ratio is 60: 1. The isentropic efficiency of compressor and turbine is 0.84 and 0.88, respectively. Determine the power developed by the unit and overall efficiency.

Take:

$$c_{pa} = 1.005 \text{ kJ/kgK}$$
$$_{pg} = 1.11 \text{ kJ/kgK}$$
$$m_a = 5 \text{ kg/s}$$

Heating value of fuel $= 40$ MJ/kg

Solution: Given data: (refer Fig. 5.39)

$$p_1 = 1 \text{ bar}$$
$$T_1 = 293 \text{ K}$$

Pressure ratio:

$$r_p = \frac{p_2}{p_1} = 6 \text{ bar}$$

Air–fuel ratio

$$\frac{A}{F} = 60$$

Isentropic efficiency of compressor:

$$\eta_C = 0.84$$

Isentropic efficiency of turbine:

$$\eta_T = 0.88$$
$$c_{pa} = 1.005 \text{ kJ/kgK}$$
$$c_{pg} = 1.11 \text{ kJ/kgK}$$
$$m_a = 5 \text{ kg/s}$$

Heating value of fuel:

$$\text{H.V} = 40 \text{ MJ/kg} = 40000 \text{ kJ/kg}$$

Now,

Fig. 5.39 T-s diagram for Problem 5.10

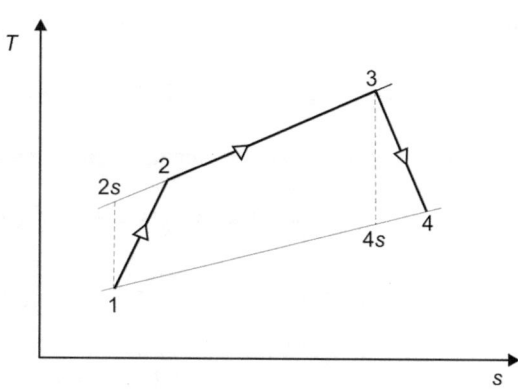

$$\frac{A}{F} = \frac{m_a}{m_f}$$

or

$$m_f = m_a \times \frac{F}{A} = 5 \times \frac{1}{60} = 0.0833 \text{ kg/s}$$

For Adiabatic Process: 1–2 s

$$\frac{T_{2s}}{T_1} = \left(\frac{p_2}{p_1}\right)^{\frac{\gamma-1}{\gamma}}$$

$$\frac{T_{2s}}{293} = (6)^{\frac{1.4-1}{1.4}} = (6)^{0.2857}$$

or

$$T_{2s} = (6)^{0.2857} \times 297 = 488.86 \text{ K}$$

$$\eta_C = \frac{T_{2s} - T_1}{T_2 - T_1}$$

$$0.84 = \frac{488.86 - 293}{T_2 - 293}$$

or

$$T_2 - 293 = \frac{195.86}{0.84} = 233.166$$

or

$$T_2 = 526.17 \text{ K}$$

Applying energy balance equation to combustion chamber, we get

$$\left(m_a + m_f\right)c_{pg}T_3 - m_a c_{pa}T_2 = m_f \times \text{HV}$$

$$(5 + 0.0833) \times 1.11 \times T_3 - 5 \times 1.005 \times 586.17 = 0.0833 \times 40000$$

$$5.64 \ T_3 - 2644 = 3332$$

or

$$5.64 \, T_3 = 5976$$

or

$$T_3 = 1059.57 \text{ K}$$

For abiabatic isentropic process 3–4 s

$$\frac{T_3}{T_{4s}} = \left(\frac{p_2}{p_1}\right)^{\frac{\gamma-1}{\gamma}} = (6)^{\frac{1.4-1}{1.4}} = 6^{0.2857} = 1.668$$

$$\frac{T_3}{T_{4s}} = 1.668$$

$$T_{4s} = \frac{T_3}{1.668} = \frac{1059.57}{1.668} = 635.23 \text{ K}$$

$$\eta_T = \frac{T_3 - T_4}{T_3 - T_{4s}}$$

$$0.88 = \frac{1059.57 - T_4}{1059.357 - 635.23}$$

$$0.88 = \frac{1059.57 - T_4}{424.34}$$

or

$$1059.57 - T_4 = 0.88 \times 424.34 = 373.419$$

or

$$T_4 = 686.15 \text{ K}$$

Power developed by turbine:

$$\begin{aligned} P_T &= \left(m_a + m_f\right)c_{pg}(T_3 - T_4) \\ &= (5 + 0.0833) \times 1.11 \times (1059.57 - 686.15) \\ &= 2107 \text{ kW} \end{aligned}$$

Power required to drive the compressor: P_C

$$\begin{aligned} P_C &= m_a c_{pg}(T_2 - T_1) \\ &= 5 \times 1.005(526.17 - 293) = 1171.67 \text{ kW} \end{aligned}$$

Power developed by the unit:

$$P = P_T - P_C = 2107 - 1171.67 = \textbf{935.33 kW}$$

Thermal efficiency:

$$\eta_{th} = \frac{P}{m_f \times HV}$$

$$= \frac{935.33}{0.0833 \times 40000} = 0.2807 = \textbf{28.07\%}$$

Overall efficiency:

$$\eta_0 = \eta_{th} \times \eta_m \quad | \because \text{ Assume mechanical efficiency}: \eta_m = 0.95$$
$$= 0.2807 \times 0.95 = 0.2666 = \textbf{26.66\%}$$

Problem 5.11: A gas turbine plant working on a Brayton cycle with a regenerator of 70% effectiveness. The air inlet temperature and pressure is 30 °C and 0.2 MPa. The pressure ratio is 7 and the maximum cycle temperature 1000 °C. Assuming turbine and compressor efficiency as 90% and 80% respectively, find an increase in efficiency due to regenerator.

Solution: Given data:

Effectiveness of regenerator:

$$\in = 70\% = 0.70$$

At inlet:

$$T_1 = 30\ °C = (30 + 273)\ K = 303\ K$$
$$p_1 = 0.2\ MPa = 2\ bar$$

Pressure ratio:

$$r_p = \frac{p_2}{p_1} = 7$$

Maximum cycle temperature $= 1000\ °C = (1000 + 273)\ K = 1273\ K$

Turbine efficiency:

$$\eta_T = 90\% = 0.90$$

Compressor efficiency:

$$\eta_C = 80\% = 0.80$$

Fig. 5.40 T-s diagram of cycle without regenerator for Problem 5.11

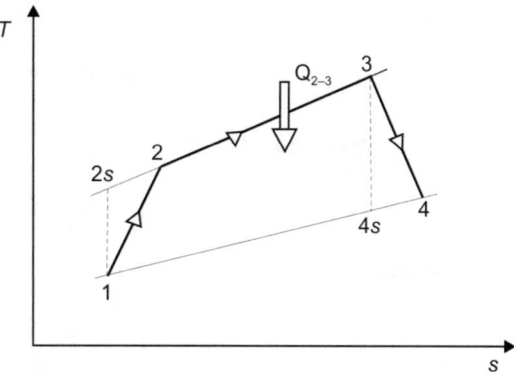

Case 1: Cycle without regenerator:

For process 1–2 s (Fig. 5.40),

$$\frac{T_{2s}}{T_1} = \left(\frac{p_2}{p_1}\right)^{\frac{\gamma-1}{\gamma}}$$

$$\frac{T_{2s}}{T_1} = (7)^{\frac{1.4-1}{1.4}}$$

$$\frac{T_{2s}}{303} = (7)^{0.285} = 1.74$$

or

$$T_{2s} = 527.22 \,°\text{C}$$

Compressor efficiency:

$$\eta_C = \frac{T_{2s} - T_1}{T_2 - T_1}$$

$$0.80 = \frac{527.22 - 303}{T_2 - 303}$$

$$0.80 = \frac{224.22}{T_2 - 303}$$

or

$$T_2 - 303 = \frac{224.22}{0.80} = 280.275$$

or

$$T_2 = 583.275 \text{ K}$$

Maximum cycle temperature:

$$T_3 = 1273 \text{ K}$$

For process 3–4 s,

$$\frac{T_3}{T_{4s}} = \left(\frac{p_2}{p_1}\right)^{\frac{\gamma-1}{\gamma}}$$

$$\frac{1273}{T_{4s}} = (7)^{\frac{1.4-1}{1.4}} = 7^{0.285} = 1.74$$

or

$$T_{4s} = \frac{1273}{1.74} = 731.60 \text{ K}$$

Turbine efficiency:

$$\eta_T = \frac{T_3 - T_4}{T_3 - T_{4s}}$$

$$0.90 = \frac{1273 - T_4}{1273 - 731.60}$$

$$0.90 = \frac{1273 - T_4}{541.4}$$

or

$$0.90 \times 541.4 = 1273 - T_4$$
$$487.26 = 1273 - T_4$$

or

$$T_4 = 1273 - 487.26 = 785.74 \text{ K}$$

Turbine work:

$$w_T = c_p(T_3 - T_4)$$
$$= 1.005(1273 - 785.74) = 489.69 \text{ kJ/kg}$$

Compressor work:

$$w_C = c_p(T_2 - T_1)$$
$$= 1.005(583.275 - 303) = 281.67 \text{ kJ/kg}$$

Network output:

$$w_{\text{net}} = \text{Turbine work} - \text{Compressor work}$$
$$= w_T - w_C$$
$$= 489.69 - 281.67 = 208.02 \text{ kJ/kg}$$

Heat supplied:

$$q_{2-3} = c_p(T_3 - T_2)$$
$$= 1.005(1273 - 583.275) = 693.17 \text{ kJ/kg}$$

Thermal efficiency:

$$\eta_1 = \frac{w_{\text{net}}}{q_{2-3}} = \frac{208.02}{693.17} = 0.30$$

Case–II: Cycle with regenerator:

$$T_1 = 303 \text{ K}$$
$$T_2 = 583.275 \text{ K}$$

Maximum cycle temperature (Fig. 5.41):

Fig. 5.41 T-s diagram of cycle with generator for Problem 5.11

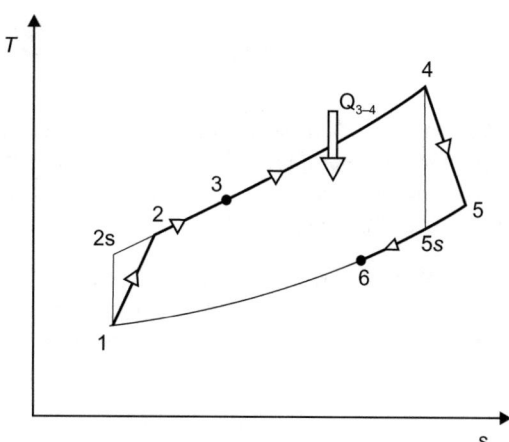

$$T_4 = 12763 \text{ K}$$
$$T_5 = 785.74 \text{ K}$$

Effectiveness:

$$\in = \frac{T_3 - T_2}{T_5 - T_2}$$

$$0.70 = \frac{T_3 - 583.273}{785.74 - 583.273}$$

$$0.70 = \frac{T_3 - 583.273}{202.467}$$

or

$$0.70 \times 202.467 = T_3 - 583.273$$
$$141.723 = T_3 - 583.273$$

or $T_3 = 724.99 \approx 725$ K.
   Heat supplied:

$$q_{3-4} = c_p(T_4 - T_3)$$
$$= 1.005(1273 - 725) = 550.74 \text{ kJ/kg}$$

Thermal efficiency:

$$\eta_2 = \frac{w_{net}}{q_{3-4}} = \frac{208.02}{550.74} = 0.3777$$

$$\text{Increase in efficiency} = \frac{\eta_2 - \eta_1}{\eta_1} = \frac{0.3777 - 0.30}{0.30} = 0.259 = \mathbf{25.9\%}$$

Problem 5.12: A small gas turbine plant has an output of 1 MW at a maximum to minimum temperature ratio of 5 and a pressure ratio of 8. The isentropic efficiencies of the compressor and turbine are 82% and 85% respectively. The compressor draws air at 27 °C. Determine

(i)   the mass flow rate through the turbine,
(ii)   the thermal efficiency of the plant, and
(iii)   the efficiencies of the reversible Brayton cycle and the Carnot's cycle between the same temperatures.

   Assume the working fluid in cycle is air.

Solution: Given data:

Power output:

$$P = 1 \text{ MW} = 1000 \text{ kW}$$

$$\frac{T_3}{T_1} = 5$$

Pressure ratio:

$$r_p = \frac{p_2}{p_1} = 8$$

Isentropic efficiency of the compressor:

$$\eta_C = 82\% = 0.82$$

Isentropic efficiency of the turbine (Fig. 5.42):

$$\eta_T = 85\% = 0.85$$
$$T_1 = 27\,^\circ\text{C} = (27 \quad 273) \text{ K} = 300 \text{ K}$$

For isentropic process 1–2 s,

$$\frac{T_{2s}}{T_1} = \left(\frac{p_2}{p_1}\right)^{\frac{\gamma-1}{\gamma}}$$

$$\frac{T_{2s}}{T_1} = r_p^{\frac{\gamma-1}{\gamma}}$$

$$\frac{T_{2s}}{300} = (8)^{\frac{1.4-1}{1.4}} = 8^{0.2857}$$

Fig. 5.42 T-s diagram for Problem 5.12

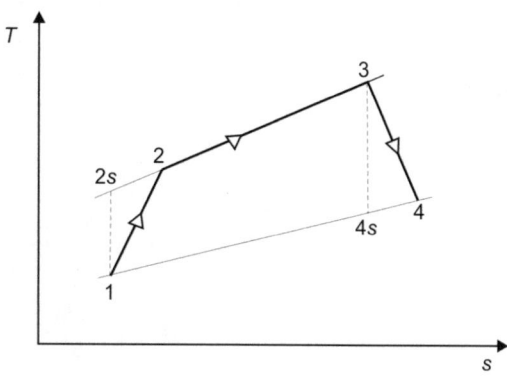

or

$$T_{2s} = 300 \times 8^{0.2857} = 543.41 \text{ K}$$

$$\eta_C = \frac{T_{2s} - T_1}{T_2 - T_1}$$

$$0.82 = \frac{543.41 - 300}{T_2 - 300}$$

$$0.82 = \frac{243.41}{T_2 - 300}$$

or $T_2 - 300 = \frac{243.41}{0.82} = 296.84$.

or

$$T_2 = 596.84 \text{ K}$$

$$\frac{T_3}{T_1} = 5 \text{ (Given)}$$

$$T_3 = 5T_1 = 5 \times 300 = 1500 \text{ K}$$

For isentropic process 3–4 s,

$$\frac{T_3}{T_{4s}} = \left(\frac{p_2}{p_1}\right)^{\frac{\gamma-1}{\gamma}}$$

$$\frac{1500}{T_{4s}} = (8)^{\frac{1.4-1}{1.4}} = 80.2857$$

$$\frac{1500}{T_{4s}} = 1.81$$

or

$$T_{4s} = 828.72 \text{ K}$$

$$\eta_T = \frac{T_3 - T_4}{T_3 - T_{4s}}$$

$$0.85 = \frac{1500 - T_4}{1500 - 828.72}$$

$$0.85 = \frac{1500 - T_4}{671.28}$$

$$0.85 \times 671.28 = 1500 - T_4$$
$$570.58 = 1500 - T_4$$

or

$$T_4 = 929.42 \text{ K}$$

(i) Mass flow rate through the turbine: m

Compressor input work:

$$w_C = c_p(T_2 - T_1)$$
$$= 1.005(896.84 - 300) = 298.32 \text{ kJ/kg}$$

Turbine output work:

$$w_T = c_p(T_3 - T_4)$$
$$= 1.005(1500 - 929.42) = 573.43 \text{ kJ/kg}$$

Network output:

$$w_{\text{net}} = w_T - w_C$$
$$= 573.43 - 298.32 = 25.11 \text{ kJ/kg}$$

Net power output:

$$P = m \; w_{\text{net}}$$
$$P = m \times 275.11 \text{ kW}$$

where

$$m = \text{mass flow rate in kg/s}$$

also

$$P = 1000 \text{ kW}$$

$$\therefore 1000 = m \times 275.11$$

or

$$m = 3.63 \text{ kg/s}$$

(ii) Thermal efficiency: η_{th}

$$\eta_{th} = \frac{\text{net work output} : w_{net}}{\text{heat supplied} : q_{2-3}}$$

$$= \frac{w_{net}}{c_p(T_3 - T_2)} = \frac{275.11}{1.005(1500 - 596.84)}$$

$$= \frac{275.11}{907.67} = 0.3030 = \mathbf{30.30\%}$$

(iii) Efficiency of the reversible Brayton cycle 1–2 s to 3–4 s,

$$\eta_{Brayton} = 1 - \frac{1}{r_p^{(\gamma-1)/\gamma}}$$

$$= 1 - \frac{1}{(8)^{\frac{1.4-1}{1.4}}} = 1 - \frac{1}{8^{0.2857}}$$

$$= 1 - 0.5520 = 0.448 = \mathbf{44.80\%}$$

Carnot efficiency:

$$\eta_{Carnot} = 1 - \frac{T_{min}}{T_{max}} = 1 - \frac{T_1}{T_3}$$

$$= 1 - \frac{300}{1500} = 1 - 0.2 = 0.8 = \mathbf{80\%}$$

Problem 5.13. Find the required air–fuel ratio in gas turbine plant whose turbine and compressor efficiencies are 85% and 80%, respectively. Maximum cycle temperature is 875 °C. The working fluid can be taken as air ($c_p = 1$ kJ/kgK, $\gamma = 1.4$) which enters the compressor at 100 kPa and 27 °C. The pressure ratio is 4. The fuel has a calorific value of 42,000 kJ/kg. There is a loss of 10% of calorific value in the combustion chamber. Also, find the cycle efficiency.

Solution: Given data:
Turbine efficiency: $\eta_T = 85\% = 0.85$.
Compressor efficiency: $\eta_C = 80\% = 0.80$.
Maximum temperature: $T_3 = 875 \text{ °C} = (875 + 273) \text{ K} = 1148 \text{ K}$

$$c_{pa} = 1 \text{ kJ/kgK}$$
$$\gamma = 1.4$$
$$p_1 = 100 \text{ kPa}$$
$$T_1 = 27 \text{ °C} = (27 + 273) \text{ K} = 300 \text{ K}$$

Pressure ratio: $r_p = 4$.
Calorific value: $CV = 42000$ kJ/kg.
Calorific value process in the combustion chamber $= 10\%$
For process 1–2 s,

$$\frac{T_{2s}}{T_1} = \left(\frac{p_2}{p_1}\right)^{\frac{\gamma-1}{\gamma}}$$

$$\frac{T_{2s}}{300} = (4)^{\frac{1.4-1}{1.4}} = (4)^{0.2857}$$

or

$$T_{2s} = 300 \times (4)^{0.2857} = 445.78 \text{ K}$$

Compressor efficiency (Fig. 5.43):

$$\eta_C = \frac{T_{2s} - T_1}{T_2 - T_1}$$

$$0.80 = \frac{445.78 - 300}{T_2 - 300}$$

or

$$T_2 - 300 = \frac{145.78}{0.8} = 182.22$$

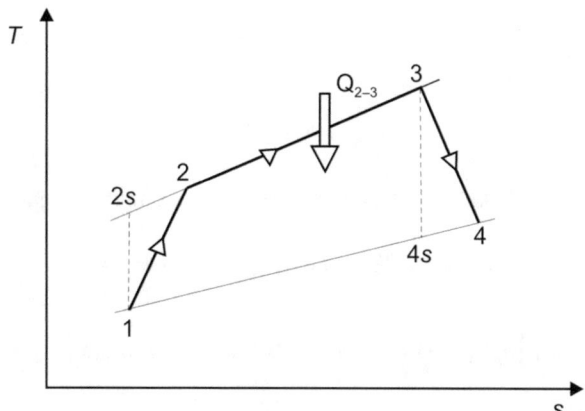

Fig. 5.43 T-s diagram for Problem 5.13

or

$$T_2 = 482.22 \text{ K}$$

For process 3–4 s,

$$\frac{T_3}{T_{4s}} = \left(\frac{p_2}{p_1}\right)^{\frac{\gamma-1}{\gamma}}$$

$$\frac{1148}{T_{4s}} = (4)^{\frac{1.4-1}{1.4}} = (4)^{0.2857} = 1.485$$

$$\frac{1148}{T_{4s}} = 1.485$$

or

$$T_{4s} = 773.06 \text{ K}$$

Turbine efficiency:

$$\eta_T = \frac{T_3 - T_4}{T_3 - T_{4s}}$$

$$0.85 = \frac{1148 - T_4}{1148 - 773.06}$$

or

$$318.67 = 1148 - T_4$$

or

$$T_4 = 829.33 \text{ K}$$

Applying energy balance equation to combustion chamber, we get

$$\left(m_a + m_f\right)c_{pg}T_3 - m_a c_{pa}T_2 = 0.9 \times m_f \times CV$$

$$\left(\frac{m_a}{m_f} + 1\right)c_p T_3 - \frac{m_a}{m_f}c_{pa}T_2 = 0.9 \times CV \qquad \text{Assume, } c_{pg} = c_{pa}$$

$$\left(\frac{m_a}{m_f} + 1\right)1 \times 1148 - \frac{m_a}{m_f} \times 1 \times 482.22 = 0.9 \times 42000$$

$$1148\frac{m_a}{m_f} + 1148 - 482.22\frac{m_a}{m_f} = 37800$$

$$665.78\frac{m_a}{m_f} = 36652$$

or

$$\frac{m_a}{m_f} = 55.05$$

Cycle efficiency:

$$\eta = \frac{W_{net}}{\text{Heat supplied}}$$

$$= \frac{W_T - W_C}{Q_{2-3}}$$

$$= \frac{(m_a + m_f)c_{pa}(T_3 - T_4) - m_a c_{pa}(T_2 - T_1)}{m_f \times 0.90 \times 42000}$$

$$= \frac{\left(\frac{m_a}{m_f} + 1\right) \times 1(1148 - 829.33) - \frac{m_a}{m_f} \times 1(482.22 - 300)}{37800}$$

$$= \frac{(55.05 + 1) \times 318.67 - 55.05 \times 182.22}{37800}$$

$$= 0.2071 = \mathbf{20.71\%}$$

Problem 5.14: A gas turbine plant consists of two turbines. One turbine to drive the compressor and other turbines to develop power output and both are having their own combustion chambers which are served by air directly from the compressor. Air enters the compressor at 1 bar and 288 K and is compressed to 8 bar with an isentropic efficiency of 76%. Due to heat added in the combustion chamber, the inlet temperature of the gas to both turbines is 900 °C. The isentropic efficiency of each turbine is 86% and the mass flow rate of air at the compressor is 23 kg/s. The calorific value of fuel is 42000 kJ/kg. Calculate the output of the plant and the thermal efficiency if mechanical efficiency is 95% and generator efficiency is 96%. Take c_p = 1.005 kJ/kgK and $\gamma = 1.4$ for air and $c_{pg} = 1.128$ kJ/kgK and $\gamma = 1.34$ for gases.

Solution: Given data:
For compressor (Fig. 5.44):

$$p_1 = 1 \text{ bar}$$
$$T_1 = 288 \text{ K}$$
$$p_2 = 8 \text{ bar}$$

Isentropic efficiency: $\eta_C = 76\%$

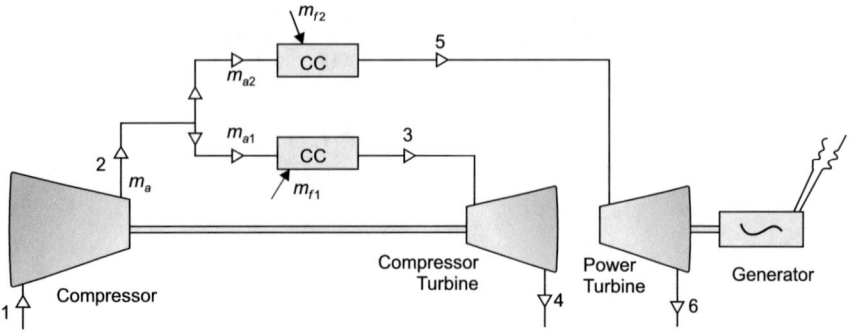

Fig. 5.44 Schematic for Problem 5.14

Mass flow rate of air: $m_a = 23$ kg/s.
For process 1–2 s (Fig. 5.45),

$$\frac{T_{2s}}{T_1} = \left(\frac{p_2}{p_1}\right)^{\frac{\gamma-1}{\gamma}}$$

$$\frac{T_{2s}}{288} = \left(\frac{8}{1}\right)^{\frac{1.4-1}{1.4}} = (8)^{0.2857}$$

or

$$T_{2s} = 288 \times (8)^{0.2857} = 521.68 \text{ K}$$

$$\eta_c = \frac{T_{2s} - T_1}{T_2 - T_1}$$

$$0.76 = \frac{521.68 - 288}{T_2 - 288} = \frac{233.68}{T_2 - 288}$$

Fig. 5.45 *T-s* diagram for
Problem 5.14

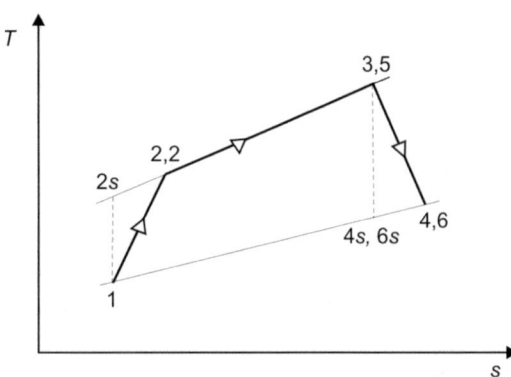

or

$$T_2 - 288 = \frac{233.68}{0.76} = 307.47$$

or

$$T_2 = 595.47 \text{ K}$$

For turbine:
Inlet temperature:

$$T_3 = T_5 = 900 \text{ °C} = (900 + 273) \text{ K} = 1173 \text{ K}$$
$$\eta_T = 86\% = 0.86$$

For process 3–4 s,

$$\frac{T_3}{T_{4s}} = \left(\frac{p_2}{p_1}\right)^{\frac{\gamma-1}{\gamma}}$$

$$\frac{1173}{T_{4s}} = (8)^{\frac{1.34-1}{1.34}} = (8)^{0.2537} \quad \gamma = 1.34 \text{ for gases}$$

or

$$T_{4s} = 692.12 \text{ } K$$
$$\eta_T = \frac{T_3 - T_4}{T_3 - T_{4s}}$$
$$0.86 = \frac{1173 - T_4}{1173 - 692.12} = \frac{1173 - T_4}{480.88}$$

or

$$0.86 \times 480.88 = 1173 - T_4$$

or

$$T_4 = 759.45 \text{ K}$$
$$= T_6$$

Calorific value of fuel: $CV = 42000 \text{ kJ/kg}$.
Mechanical efficiency: $\eta_m = 95\% = 0.95$.
Generator efficiency: $\eta_G = 96\% = 0.96$.

Applying energy balance equation to both combustion chambers, we get

$$\left(m_a + m_f\right)c_{pg}T_3 - m_a c_p T_2 = m_f C V$$

$$\left(23 + m_f\right) \times 1.128 \times 1173 - 23 \times 1.005 \times 595.47 = m_f \times 42000$$

$$\left(23 + m_p\right) \times 1323.14 - 13764.28 = 42000\, m_f$$
$$30432.22 + 1323.14\, m_f - 13764.28 = 42000\, m_f$$
$$16667.94 = 40676.86\, m_f$$

or

$$m_f = 0.4097 \text{ kg/s}$$

Let

$$m_{f1} = \text{Fuel supplied in C.C.} - \text{I}$$
$$m_{a1} = \text{Air supplied in C.C.} - \text{I}$$
$$m_{f2} = \text{Fuel supplied in C.C.} - \text{II}$$
$$m_{a2} = \text{Air supplied in C.C.} - \text{II}$$

Applying energy balance equation in C.C.-I, we get

$$\left(m_{a1} + m_{f1}\right)c_{pg}T_3 - m_{a1}c_{pa}T_2 = m_{f1}C.V$$

$$\left(\frac{m_{a1}}{m_{f1}} + 1\right)c_{pg}T_3 - \frac{m_{a1}}{m_{f1}}c_{pa}T_2 = C.V.$$

$$\left(\frac{m_{a1}}{m_{f1}} + 1\right)1.123 \times 1173 - \frac{m_{a1}}{m_{f1}} \times 1.005 \times 595.47 = 42000$$

$$\left(\frac{m_{a1}}{m_{f1}} + 1\right)1323.14 - \frac{m_{a1}}{m_{f1}} \times 598.44 = 42000$$

$$1323.14\frac{m_{a1}}{m_{f1}} + 1323.14 - 598.44\frac{m_{a1}}{m_{f1}} = 42000$$

$$724.7\frac{m_{a1}}{m_{f1}} = 40676.86$$

or

$$\frac{m_{a1}}{m_{f1}} = 56.129$$

Now, the power required to drive the compressor is equal to the power developed by compressor turbine.

$$m_a c_{pa}(T_2 - T_1) = (m_{a1} + m_{f1})c_{pg}(T_3 - T_4)$$

$$\therefore \quad 23 \times 1.005(595.47 - 288) = m_{f1}\left(\frac{m_{a1}}{m_{f1}} + 1\right)c_{pg}(T_3 - T_4)$$

$$23 \times 1.005(595.47 - 288) = m_{f1}(56.129 + 1) \times 1.128 \times (1173 - 759.45)$$

$$7107.169 = m_{f1} \times 26649.78$$

Or $m_{f1} = 0.266$ kg/s.
Fuel supplied in C.C.-II:

$$m_{f2} = m_f - m_{f1} = 0.4097 - 0.266 = 0.1437 \text{ kg/s}$$

$\therefore$ Air supplied in C.C.-I:

$$m_{a1} = 56.129 m_{f1} = 56.125 \times 0.266 = 14.92 \text{ kg/s}$$

Air supplied in C.C.-II:

$$m_{a2} = m_a - m_{a1} = 23 - 14.92 = 8.08 \text{ kg/s}$$

Power output by power turbine:

$$W_{\text{net}} = (m_{a2} + m_{f2})c_{pg}(T_5 - T_6)$$
$$= (8.08 + 0.1437) \times 1.128(173 - 759.45) = 3836.22 \text{ kW}$$

Thermal efficiency:

$$\eta_{\text{th}} = \frac{W_{\text{net}}}{m_f \times C.V.} = \frac{3836.22}{0.4097 \times 42000} = 0.2229 = 22.29\%$$

Plant efficiency:

$$\eta_{\text{th}} = \eta_{\text{th}}\eta_m\eta_G$$
$$= 0.2229 \times 0.95 \times 0.96 = 0.2032$$

also

$$\eta_P = \frac{\text{Output of the plant}}{\text{Input to the plant}}$$

$$\therefore 0.2032 = \frac{\text{Output of the plant}}{m_f \times C.V.}$$

or

$$\text{Output of the plant} = 0.2032 \times 0.4097 \times 42000$$
$$= 3496.54 \text{ kW} = \textbf{3.496 MW}$$

Problem 5.15: In a gas turbine plant working on the Brayton cycle the air at inlet is at 27 °C, 1 bar. The pressure ratio is 6.25 and the maximum temperature is 800 °C. The turbine and compressor efficiencies are each 80%. Find (i) the turbine exhaust temperature, (ii) the compressor work per kg of air, (iii) heat supplied per kg of air, (iv) the turbine work per kg of air, and (v) the cycle efficiency.

Solution: Given data:

$$T_1 = 27 \text{ °C} = (273 + 27) \text{ K} = 300 \text{ K}$$
$$p_1 = 1 \text{ bar}$$

Pressure ratio: $r_p = 6.25$.
Maximum temperature (Fig. 5.46):

$$T_3 = 800 \text{ °C} = (273 + 800) \ K = 1073 \ K$$
$$\eta_T = 0.8$$
$$\eta_C = 0.8$$

For ideal process 1–2 s,

$$\frac{T_{2s}}{T_1} = \left(r_p\right)^{\frac{\gamma-1}{\gamma}}$$

Fig. 5.46 T-s diagram for Problem 5.15

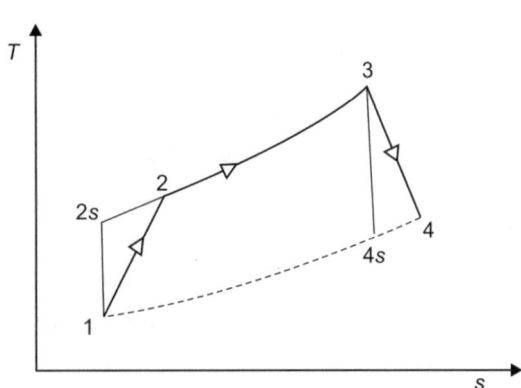

$$\frac{T_{2s}}{300} = (6.25)^{\frac{1.4-1}{1.4}} = (6.25)^{0.2857}$$

$$T_{2s} = 300 \times (6.25)^{0.2857} = 506.41 \text{ K}$$

$$\eta_C = \frac{T_{2s} - T_1}{T_2 - T_1}$$

$$0.80 = \frac{506.41 - 300}{T_2 - 300}$$

or

$$T_2 - 300 = \frac{206.41}{0.80} = 258.01$$

or

$$T_2 = 558.01 \text{ K}$$

For ideal process 3–4 s,

$$\frac{T_3}{T_{4s}} = (r_p)^{\frac{\gamma-1}{\gamma}}$$

$$\frac{1073}{T_{4s}} = (6.25)^{\frac{1.4-1}{1.4}} = (6.25)^{0.2857}$$

or

$$T_{4s} = 635.64 \text{ K}$$

$$\eta_T = \frac{T_3 - T_4}{T_3 - T_{4s}}$$

$$0.80 = \frac{1073 - T_4}{1073 - 635.64}$$

$$0.80 = \frac{1073 - T_4}{437.36}$$

or

$$0.80 \times 437.36 = 1073 - T_4$$

or

$$349.88 = 1073 - T_4$$

or

$$T_4 = 723.12 \text{ K}$$

(i)   Turbine exhaust temperature: $T_4 = \textbf{723.12 K}$
(ii) Compressor work per kg of air: w_{1-2}

$$w_{1-2} = c_p(T_2 - T_1) = 1.005(5588.01 - 300) = \textbf{259.30 kJ/kg}$$

(iii) Heat supplied per kg of air: q_{2-3}

$$q_{2-3} = c_p(T_3 - T_2) = 1.005(1073 - 558.01) = \textbf{517.56 kJ/kg}$$

(iv) Turbine work per kg of air: w_{3-4}

$$w_{3-4} = c_p(T_3 - T_4) = 1.005(1073 - 723.12) = \textbf{351.62 kJ/kg}$$

(v)   Cycle efficiency:

$$\eta_{th} = \frac{w_{\text{net}}}{q_{2-3}} = \frac{w_{3-4} - w_{1-2}}{q_{2-3}}$$
$$= \frac{351.62 - 259.30}{517.56} = 0.1783 = \textbf{17.83\%}$$

Problem 5.16: A simple open-cycle gas turbine has a compressor turbine and a free power turbine. It develops an electrical power output of 250 MW. The cycle takes in air at 1 bar and 288 K. The compressor ratio is 14. The turbine inlet temperature is 1500 K. The isentropic efficiencies of the compressor and turbine are 0.86 and 0.89, respectively. The mechanical efficiency of each shaft is 0.98. Combustion efficiency is 0.98, while combustion pressure loss is 0.03 bar. The alternator efficiency is 0.98. Take the calorific value of fuel equal to 42,000 kJ/kg.
    Take:

$$c_{pa} = 1.005 \text{ kJ/kgK}$$
$$c_{pg} = 1.15 \text{ kJ/kgK}$$

Calculate the following:

(i)   Air–fuel ratio,
(ii)   Specific work output,
(iii) Specific fuel consumption,
(iv)   Mass flow rate of air, and
(v)   Cycle thermal efficiency.

Solution: Given data:
    Electrical power output: $P_E = 250 \text{ MW} = 250 \times 10^3 \text{ kW}$.

At inlet of compressor: $p_1 = 1$ bar

$$T_1 = 288 \text{ K}$$

Compressor ratio: $r_p = \frac{p_2}{p_1} = 14$

$$p_2 = p_1 \times 14 = 1 \times 14 = 14 \text{ bar}$$

Turbine inlet temperature: $T_3 = 1500$ K (Fig. 5.47).

$$\eta_C = 0.86$$
$$\eta_T = 0.89$$

Mechanical efficiency each shaft: $\eta_m = 0.98$.
Combustion efficiency: $\eta_{\text{comb}} = 0.98$.
Pressure loss in combustion chamber:

$$\Delta p = 0.03 \text{ bar}$$

Alternator efficiency: $\eta_A = 0.98$.
Calorific value of fuel:

$$CV = 42000 \text{ kJ/kg}$$
$$c_{pa} = 1.005 \text{ kJ/kgK}$$
$$c_{pg} = 1.15 \text{ kJ/kgK}$$

For ideal process 1–2 s,

Fig. 5.47 T-s diagram for
Problem 5.16

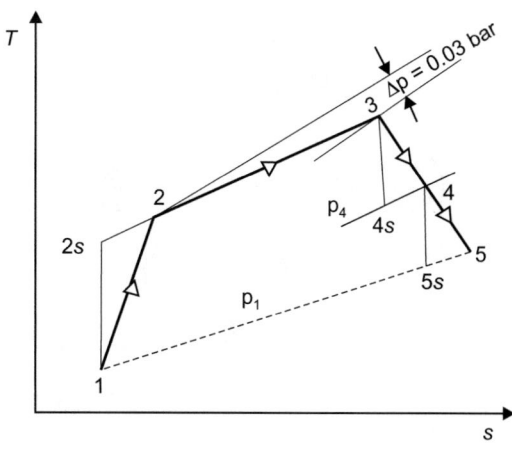

$$\frac{T_{2s}}{T_1} = \left(\frac{p_2}{p_1}\right)^{\frac{\gamma-1}{\gamma}}$$

$$\frac{T_{2s}}{288} = (14)^{\frac{1.4-1}{1.4}} = 14^{0.2857}$$

or

$$T_{2s} = 288 \times 14^{0.2857} = 612.12 \text{ K}$$

$$T_2 = \frac{T_{2s} - T_1}{T_2 - T_1}$$

$$0.86 = \frac{612.12 - 288}{T_2 - 288}$$

or

$$T_2 - 288 = \frac{324.12}{0.86} = 376.88$$

or

$$T_2 = 664.88 \text{ K}$$

Pressure at point 3,

$$p_3 = p_2 - \Delta p = 14 - 0.03 = 13.97 \text{ bar}$$

Applying energy balance equation is combustion chamber, we get

$$(m_a + m_f)c_{pg}T_3 - m_a c_{pa}T_2 = \eta_{com} \times m_f \times CV$$

$$\left(\frac{m_a}{m_f} + 1\right)c_{pg}T_3 - \frac{m_a}{m_f}c_{pa}T_2 = \eta_{com}CV$$

$$\left(\frac{m_a}{m_f} + 1\right) \times 1.15 \times 1500 - \frac{m_a}{m_f} \times 1.005 \times 664.88 = 0.98 \times 42000$$

$$1725\left(\frac{m_a}{m_f} + 1\right) - 668.20\frac{m_a}{m_f} = 41160$$

$$1725\frac{m_a}{m_f} + 1725 - 668.20\frac{m_a}{m_f} = 41160$$

$$1056.8\frac{m_a}{m_f} = 39435$$

or

$$\frac{m_a}{m_f} = \textbf{37.31}$$

Work required to run the compressor = Work produced by compressor turbine

$$\frac{m_a c_{pa}(T_2 - T_1)}{\eta_m} = (m_a + m_f)c_{pg}(T_3 - T_4)\eta_m$$

$$\frac{m_a c_{pa}(T_2 - T_1)}{\eta_m} = m_a\left(1 + \frac{m_f}{m_a}\right)c_{pg}(T_3 - T_4)\eta_m$$

$$\frac{c_{pa}(T_2 - T_1)}{\eta_m} = \left(1 + \frac{m_f}{m_a}\right)c_{pg}(T_3 - T_4)\eta_m$$

$$\frac{1.005(664.88 - 288)}{0.98} = \left(1 + \frac{1}{37.31}\right) \times 1.15(1500 - T_4) \times 0.98$$

$$386.49 = 1735.50 - 1.157T_4$$

or

$$T_4 = 1165.95 \text{ K}$$

$$\eta_T = \frac{T_3 - T_4}{T_3 - T_{4s}} \qquad \text{for compressor turbine}$$

$$0.89 = \frac{1500 - 1165.95}{1500 - T_{4s}} = \frac{334.05}{1500 - T_{4s}}$$

or

$$1500 - T_{4s} = \frac{334.05}{0.89} = 375.33$$

or

$$T_{4s} = 1124.67 \text{ K}$$

For process 3–4 s,

$$\frac{T_3}{T_{4s}} = \left(\frac{p_3}{p_4}\right)^{(\gamma-1)/\gamma}$$

$$\frac{1500}{1124.67} = \left(\frac{13.97}{p_4}\right)^{(1.4-1)/1.4}$$

or $\frac{13.97}{p_4} = (1.333)^{3.5}$.

or $p_4 = 5.10$ bar.

For process 4–5 s,

$$\frac{T_4}{T_{5s}} = \left(\frac{p_4}{p_1}\right)^{\frac{\gamma-1}{\gamma}}$$

$$\frac{1165.95}{T_{5s}} = \left(\frac{5.10}{1}\right)^{\frac{1.4-1}{1.4}} = (5.10)^{0.285} = 1.59$$

or

$$T_{5s} = 733.30 \text{ K}$$

$$\eta_T = \frac{T_4 - T_5}{T_4 - T_{5s}} \qquad \text{for power turbine}$$

$$0.89 = \frac{1165.95 - T_5}{1165.95 - 733.30} = \frac{1165.95 - T_5}{432.65}$$

or

$$0.89 \times 432.65 = 1165.96 - T_5$$

or $T_5 = 780.90$ K.

Alternator efficiency:

$$\eta_A = \frac{\text{Electrical power} : P_E}{\text{Shaft power} : P_s}$$

$$0.98 = \frac{250 \times 10^3}{P_s}$$

or $P_s = 255.10 \times 10^3$ kW.

Mechanical efficiency:

$$\eta_m = \frac{\text{Shaft power} : P_s}{\text{Power output of power turbine} : P}$$

$$0.98 = \frac{255.10 \times 10^3}{P}$$

or $P = 260.30 \times 10^3$ kW.

also $P = (m_a + m_\rho c_{pg}(T_4 - T_5)$

$$\therefore 260.30 \times 10^3 = m_f \left(\frac{m_a}{m_f} + 1 \right) c_{pg}(T_4 - T_5)$$

$$\therefore 260.30 \times 10^3 = m_f(37.31 + 1) \times 1.15(1165.95 - 782.90)$$

or

$$m_f = 15.34 \text{ kg/s}$$

$$\therefore m_a = 37.31 \times 15.34 = 572.33 \text{ kg/s}$$

(i)   Air–fuel ratio: $\frac{m_a}{m_f} = \mathbf{37.31}$

(ii)   Specific work output: $w_{4-5} = \frac{P}{m_a + m_f} = \frac{260.30 \times 10^3}{572.33 + 15.34} = 442.93 \text{ kJ/kg}$

(iii)   Specific fuel consumption $= \frac{m_f}{P} = \frac{3600 m_f}{P} \text{ kg/kWh}$
$$= \frac{3600 \times 15.34}{260.30 \times 10^3} = \mathbf{0.2121 \text{ kg/kWh}}$$

(iv)   Mass flow rate of air: $m_a = \mathbf{572.33 \text{ kg/s}}$

(v)   Cycle thermal efficiency: $\eta_{th} = \dfrac{P}{m_f \times CV \times \eta_{comb}} = \dfrac{260.30 \times 10^3}{15.34 \times 42000 \times 0.98}$
$$= 0.4122 = 4.22\%$$

Problem 5.17: In an air-standard regenerative gas turbine cycle, the pressure ratio is 5. Air enters the compressor at 1 bar, 300 K, and leaves at 490 K. The maximum temperature in the cycle is 1000 K. If the effectiveness of the regenerator and the adiabatic efficiency of the turbine are 80%, find the cycle efficiency. Assume for air $\gamma = 1.4$ and $c_p = 1.005$ kJ/kgK. Show the cycle on T-s diagram.

Solution: Given data for regenerative gas turbine cycle:
Pressure ratio: $r_p = 5$.
At compressor inlet: $p_1 = 1$ bar

$$T_1 = 300 \text{ K}$$

At compressor exit: $T_2 = 490$ K.
Maximum temperature in cycle: $T_4 = 1000$ K.
Effectiveness: $\epsilon = 80\% = 0.80$.
Adiabatic efficiency of turbine: $\eta_T = 80\% = 0.80$.
For process 1–2, adiabatic isentropic compression (Fig. 5.48)

$$\frac{T_2}{T_1} = (r_p)^{\frac{\gamma - 1}{\gamma}}$$

Fig. 5.48 T-s diagram for
Problem 5.17

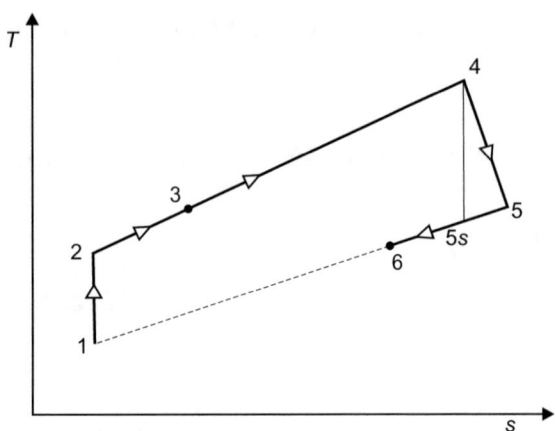

$$\frac{T_2}{300} = (5)^{\frac{1.4-1}{1.4}} = (5)^{0.2857}$$

$$T_2 = 300 \times (5)^{0.2857} = 475.13 \text{ K}$$

Assume process 4–5 s, adiabatic isentropic expansion,

$$\frac{T_4}{T_{5s}} = \left(r_p\right)^{\frac{\gamma-1}{\gamma}}$$

$$\frac{1000}{T_{5s}} = (5)^{\frac{1.4-1}{1.4}} = (5)^{0.2857}$$

or

$$T_{5s} = \frac{1000}{(5)^{0.2857}} = 631.40 \text{ K}$$

$$\eta_T = \frac{T_4 - T_5}{T_4 - T_{5s}}$$

$$0.80 = \frac{1000 - T_5}{1000 - 631.40}$$

$$0.80 = \frac{1000 - T_5}{368.6}$$

or

$$1000 - T_5 = 0.80 \times 368.6$$

$$1000 - T_5 = 294.88$$

or

$$T_5 = 705.12 \text{ K}$$

Effectiveness:

$$\in = \frac{T_3 - T_2}{T_5 - T_2}$$

$$\therefore 0.80 = \frac{T_3 - 475.13}{705.12 - 475.13}$$

$$0.80 = \frac{T_3 - 475.13}{229.99}$$

or

$$0.80 \times 229.99 = T_3 - 475.13$$
$$183.99 = T_3 - 475.13$$

or

$$T_3 = 659.12 \text{ K}$$

Specific compressor work:

$$w_{1-2} = c_p(T_2 - T_1)$$
$$= 1.005(490 - 300) = 190.95 \text{ kJ/kg}$$

Specific turbine work:

$$w_{4-5} = c_p(T_4 - T_5)$$
$$= 1.005(1000 - 705.12) = 296.35 \text{ kJ/kg}$$

Net specific work output:

$$w_{\text{net}} = w_{4-5} - w_{1-2}$$
$$= 296.35 - 190.95 = 105.4 \text{ kJ/kg}$$

Specific heat supplied:

$$q_{3-4} = c_p(T_4 - T_3)$$
$$= 1.005(1000 - 659.12) = 342.58 \text{ kJ/kg}$$

Cycle efficiency:

$$\eta_{th} = \frac{w_{net}}{q_{3-4}} = \frac{105.4}{342.58} = 0.3076 = \mathbf{30.76\%}$$

Problem 5.18: In a compound gas turbine, the air from the compressor passes through a heat exchanger heated by the exhaust gases from the L.P. turbine, and then into the combustion chamber. The H.P. turbine drives the compressor only. The exhaust from the high-pressure turbine passes through the reheater to the L.P. turbine which is coupled to the generator.

The following data refer to the plant:
Pressure ratio of compressor: 4
Isentropic efficiency of compressor: 0.86.
Isentropic efficiency of H.P. turbine: 0.84.
Isentropic efficiency of L.P. turbine: 0.80.
Mechanical efficiency of drive to compressor: 0.92.
Effectiveness of heat exchanger: 0.70.
Temperature of gases entering H.P. turbine: 660 °C.
Temperature of gases entering L.P. turbine: 625 °C.
For air, $c_p = 1.005$ kJ/kgK, $\gamma = 1.4$
For gases, $c_p = 1.15$ kJ/kgK, $\gamma = 1.33$.
Ambient conditions: 100 kPa, 15 °C.
Determine:

(i)   The pressure of the gases entering L.P. turbine,
(ii)   The net specific power, and
(iii) The overall efficiency.

Solution: Given data:
Pressure ratio (Figs. 5.49 and 5.50):

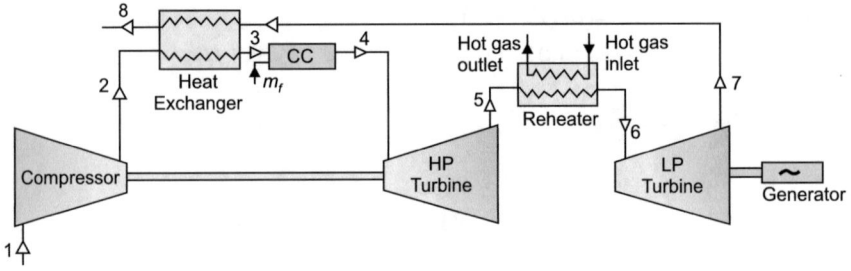

Fig. 5.49 Schematic for Problem 5.18

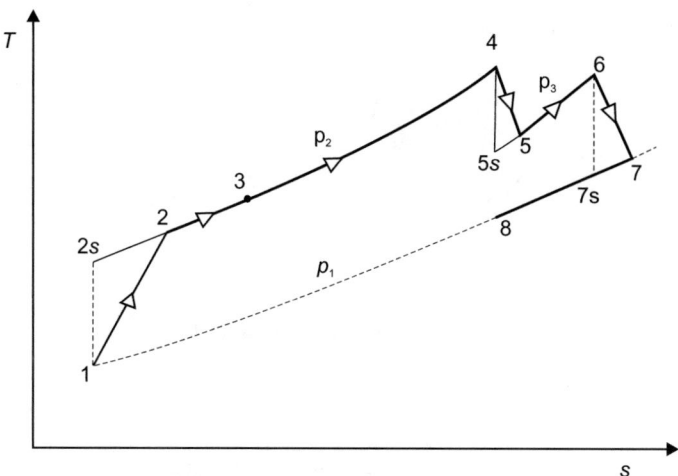

Fig. 5.50 T-s diagram for Problem 5.18

$$r_p = \frac{p_2}{p_1} = 4$$

$$\eta_C = 0.86$$

$$\eta_{T1} = 0.84$$

$$\eta_{T2} = 0.80$$

$$\eta_m = 0.92$$

$$\in = 0.70$$

$$T_4 = 660\ ^\circ\text{C} = (660 + 273)\ \text{K} = 933\ \text{K}$$

$$T_6 = 625\ ^\circ\text{C} = (625 + 273)\ \text{K} = 898\ \text{K}$$

For air:

$$c_{pa} = 1.005\ \text{kJ/kgK}$$

$$\gamma_a = 1.4$$

For gases:

$$c_{pa} = 1.15\ \text{kJ/kgK}$$

$$\gamma_g = 1.33$$

Ambient conditions:

$$p_1 = 100\ \text{kPa}$$

$$T_1 = 15\ ^\circ\text{C} = (15 + 273)\ \text{K} = 288\ \text{K}$$

For ideal reference process 1–2 s,

$$\frac{T_{2s}}{T_1} = \left(\frac{p_2}{p_1}\right)^{\frac{\gamma_a - 1}{\gamma_a}}$$

$$\frac{T_{2s}}{288} = (4)^{\frac{1.4-1}{1.4}} = (4)^{0.2857} = 1.48$$

or

$$T_{2s} = 288 \times 1.48 = 426.24 \text{ K}$$

$$\eta_C = \frac{T_{2s} - T_1}{T_2 - T_1}$$

$$0.86 = \frac{426.24 - 288}{T_2 - 288} = \frac{138.24}{T_2 - 288}$$

or

$$T_2 - 288 = \frac{138.24}{0.86} = 160.74$$

or

$$T_2 = 160.74 + 288 = 448.74 \text{ K}$$

Mechanical efficiency of compressor,

$$\eta_m = \frac{\text{Compressor work} : w_{1-2}}{\text{H.P.turbine work} : w_{4-5}}$$

$$\eta_m = \frac{w_{1-2}}{w_{4-5}}$$

$$0.92 = \frac{c_{pa}(T_2 - T_1)}{c_{pg}(T_4 - T_5)}$$

$$0.92 = \frac{1.005(448.74 - 288)}{1.15(933 - T_5)}$$

or

$$0.92 \times 1.15(933 - T_5) = 161.54$$
$$1.058(933 - T_5) = 161.54$$

or

$$933 - T_5 = \frac{161.54}{1.054} = 152.68$$

or

$$T_5 = 933 - 152.68 = 780.32 \text{ K}$$

Isectropic efficiency of H.P. turbine:

$$\eta_{T1} = \frac{T_4 - T_5}{T_4 - T_{5s}}$$
$$0.84 = \frac{933 - 780.32}{933 - T_{5s}}$$
$$0.84 = \frac{152.68}{933 - T_{5s}}$$

or

$$933 - T_{5s} = \frac{152.68}{0.84} = 181.76$$

or

$$T_{5s} = 751.24 \text{ K}$$

$$\frac{p_2}{p_1} = 14$$
$$\frac{p_2}{100} = 4$$

or

$$p_2 = 400 \text{ kPa}$$

For ideal process 4–5 s,

$$\frac{T_4}{T_{5s}} = \left(\frac{p_2}{p_3}\right)^{\frac{\gamma_g - 1}{\gamma_g}}$$
$$\frac{933}{751.24} = \left(\frac{400}{p_3}\right)^{\frac{1.33-1}{1.33}}$$
$$1.24 = \left(\frac{400}{p_3}\right)^{0.248}$$

or

$$(1.24)^{1/0.248} = \frac{400}{p_3}$$
$$2.38 = \frac{400}{p_3}$$

or

$$p_3 = \frac{400}{2.38} = \mathbf{168.06 \ kPa}$$

For ideal reference process 6–7 s,

$$\frac{T_6}{T_{7s}} = \left(\frac{p_3}{p_1}\right)^{\frac{\gamma_g - 1}{\gamma_g}}$$
$$\frac{898}{T_{7s}} = \left(\frac{168.06}{100}\right)^{\frac{1.33-1}{1.33}} = (1.68)^{0.248} = 1.137$$
$$\frac{898}{T_{7s}} = 1.137$$

or

$$T_{7s} = \frac{898}{1.137} = 789.79 \ K$$

Isentropic efficiency of L.P. turbine:

$$\eta_{T2} = \frac{T_6 - T_7}{T_6 - T_{7s}}$$
$$0.80 = \frac{898 - T_7}{898 - 789.79}$$
$$0.80 = \frac{898 - T_7}{108.21}$$

or

$$108.21 \times 0.80 = 898 - T_7$$
$$86.56 = 898 - T_7$$

or

$$T_7 = 898 - 86.56$$

$$= 811.44 \text{ K}$$

Net specific power = specific work output by L.P.trurbine

$$= w_{6-7}$$
$$= c_{pg}(T_6 - T_7) = 1.15(898 - 811.44) = \textbf{99.54 kJ/kg}$$

Effectiveness of heat exchanger,

$$\in = \frac{c_{pa}(T_3 - T_2)}{c_{pa}(T_7 - T_2)}$$

$$0.70 = \frac{1.005 \times (T_3 - 448.74)}{1.15 \times (811.44 - 448.74)}$$

$$\frac{0.70 \times 1.15}{1.005} = \frac{T_3 - 448.74}{362.7}$$

$$0.80 = \frac{T_3 - 448.74}{362.7}$$

or

$$362.7 \times 0.80 = T_3 - 448.74$$
$$290.16 = T_3 - 448.74$$

or

$$T_3 = 290.16 + 448.74 = 738.9 \text{ K}$$

Net specific heat supplied :

$$q_s = \text{specific heat supplied in combustion chamber}$$
$$+ \text{ specific heat supplied in reheater}$$

$$q_s = q_{3-4} + q_{5-6}$$
$$= c_{pg}(T_4 - T_3) + c_{pg}(T_6 - T_5)$$
$$= c_{pg}[T_4 - T_3 + T_6 - T_5]$$
$$= 1.15 \times [933 - 738.9 + 898 - 78032] = 358.54 \text{ kJ/kg}$$

Overall efficiency:

$$\eta_o = \frac{\text{Net specific power}}{\text{Net specific heat supplied}}$$

$$= \frac{w_{6-7}}{q_s} = \frac{99.54}{358.54} = 0.2776 = \textbf{27.76\%}$$

Ans:

(i)   The pressure of the gases entering L.P. turbine = **168.06 kPa**,
(ii)   The net specific power = **99.54 kJ/kg**,
(iii)   The overall efficiency = **27.76%**.

Problem 5.19: A closed-cycle gas turbine using argon as the working fluid has two-stage compression with perfect intercooling. The overall pressure ratio is 9 and the pressure ratio in each stage is equal. Each stage has an isentropic efficiency of 85%. The turbine is also two-stage will equal pressure ratio with interstage reheat to the original temperature. Each turbine stage has an isentropic efficiency of 90%. The turbine inlet temperature is 1100 K and the compressor inlet is 27 °C. Determine:

(i)   The work done per kg of fluid flow,
(ii)   The work ratio, and
(iii)   The overall efficiency.

Take c_p and γ for argon 0.5207 kJ/kgK and 1.667, respectively.

Solution: Given data:
Overall pressure ratio: $\frac{p_3}{p_1} = 9$.
Pressure ratio in each stage is equal (Fig. 5.51),

i.e.,

$$\frac{p_2}{p_1} = \frac{p_3}{p_2} = \sqrt{\frac{p_3}{p_1}} = \sqrt{9} = 3$$

Isentropic efficiency of each compressor: $\eta_C = 85\% = 0.85$.
Two-stage turbine with equal pressure ratio, i.e., $\frac{p_3}{p_2} = \frac{p_2}{p_1} = 3$ (Fig. 5.52).
Isentropic efficiency of each turbine:

$$\eta_T = 90\% = 0.90$$

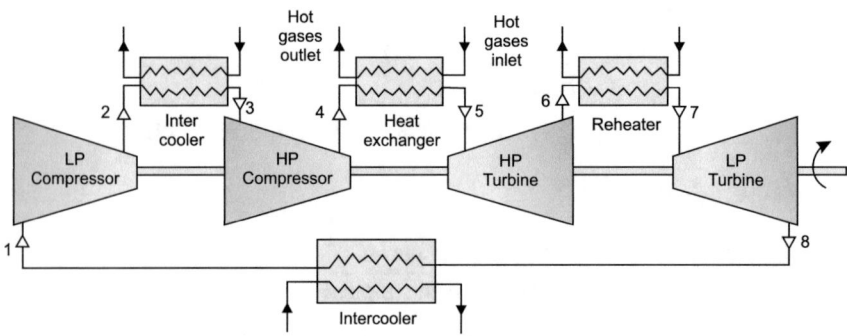

Fig. 5.51 Schematic for Problem 5.19

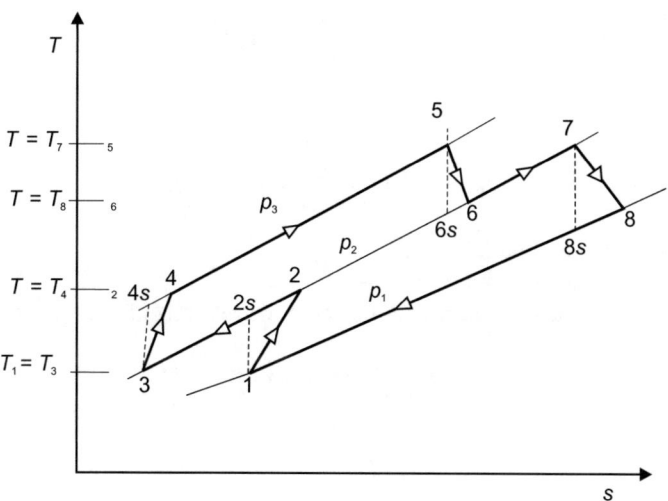

Fig. 5.52 T-s diagram for Problem 5.19

$$T_5 = T_7 = 1100 \text{ K}$$
$$T_1 = 27 \text{ °C} = (27 + 273) \text{ K} = 300 \text{ K}$$
$$c_p = 0.5207 \text{ kJ/kgK}$$
$$\gamma = 1.667$$

For ideal reference process 1–2 s,

$$\frac{T_{2s}}{T_1} = \left(\frac{p_2}{p_1}\right)^{\frac{\gamma-1}{\gamma}}$$
$$\frac{T_{2s}}{300} = (3)^{\frac{1}{1.667-1}} = (3)^{0.4} = 1.55$$

or

$$T_{2s} = 300 \times 1.55 = 465 \text{ K}$$

$$\eta_C = \frac{T_{2s} - T_1}{T_2 - T_1} \qquad \text{for L.P.compressor}$$
$$0.85 = \frac{465 - 300}{T_2 - 300} = \frac{165}{T_2 - 300}$$

or

$$T_2 - 300 = \frac{165}{0.85} = 194.11$$

or

$$T_2 = 194.11 + 300 = 494.11 \text{ K}$$

For perfect intercooling, $T_1 = T_3$.
For same compression ratio, $\frac{p_2}{p_1} = \frac{p_3}{p_2}$.
For same isentropic efficiency for L.P. compressor and H.P. compressor

$$T_2 = T_4 = 494.11 \text{ K}$$

and

$$\begin{aligned}
w_{1-2} &= w_{3-4} \\
&= c_p(T_2 - T_1) \\
&= 0.5207 \times (494.11 - 300) = 101.07 \text{ kJ/kg}
\end{aligned}$$

For ideal process 5–6 s in H.P. turbine,

$$\frac{T_5}{T_{6s}} = \left(\frac{p_3}{p_2}\right)^{\frac{\gamma-1}{\gamma}}$$
$$\frac{1100}{T_{6s}} = (3)^{\frac{1.667-1}{1.667}} = (3)^{0.4} = 1.55$$

or

$$T_{6s} = \frac{1100}{1.55} = 709.67 \text{ K}$$

$$\eta_T = \frac{T_5 - T_6}{T_5 - T_{6s}} \quad \text{for H.P.turbine}$$
$$0.90 = \frac{1100 - T_6}{1100 - 709.67} = \frac{1100 - T_6}{390.33}$$

or

$$0.90 \times 390.33 = 1100 - T_6$$
$$351.29 = 1100 - T_6$$

or

$$T_6 = 1100 - 351.29 = 748.71 \text{ K}$$

For same,

$$T_5 = T_7 = 1100 \text{ K}$$

For same pressure ratio, $\frac{p_2}{p_1} = \frac{p_3}{p_2}$.
For same isentropic efficiency for L.P. turbine and H.P. turbine
Therefore,

$$
\begin{aligned}
w_{5-6} &= w_{7-8} \\
&= c_p(T_5 - T_6) \\
&= 0.5207 \times (100 - 748.71) \\
&= 182.91 \text{ kJ/kg}
\end{aligned}
$$

Net heat supplied:

$$
\begin{aligned}
q_s &= \text{Heat supplied in heat exchanger} + \text{heat supplied in reheater} \\
&= q_{4-5} + q_{6-7} \\
&= c_p(T_5 - T_4) + (T_7 - T_6) \\
&= c_p(T_5 - T_4 + T_7 - T_6) \\
&= 0.5207(1100 - 494.11 + 1100 - 748.71) \\
&= 498.40 \text{ kJ/kg}
\end{aligned}
$$

(i)
$$
\begin{aligned}
\text{Net work done} &= w_{5-6} + w_{7-8} - w_{1-2} - w_{3-4} \\
&= 2w_{5-6} - 2w_{1-2} \qquad |\because w_{7-8} = w_{5-6}, w_{3-4} = w_{1-2} \\
&= 2 \times 182.91 - 2 \times 101.07 \\
&= \mathbf{163.68 \text{ kJ/kg}}
\end{aligned}
$$

(ii) Work ratio:

$$
\begin{aligned}
WR &= \frac{\text{Net work done}}{\text{Net turbine work}} \\
&= \frac{163.68}{w_{5-6} + w_{7-8}} \\
&= \frac{163.68}{2 \times w_{5-6}} = \frac{163.68}{2 \times 182.91} = \mathbf{0.447}
\end{aligned}
$$

(iii) Overall cycle efficiency:

$$\eta_o = \frac{\text{Net work output}}{\text{Net sheat supplied}}$$

$$= \frac{163.68}{498.40} = 0.3284 = \mathbf{32.84\%}$$

Problem 5.20: A gas turbine power plant works between the pressure of 1 kg/cm^2 and 5 kg/cm^2 and temperature of 285 K and 1100 K. The intercooler cools the air at 2.3 kg/cm^2 to 285 K before the air is sent to the second stage compressor. The compressed air from the second-stage compressor passes through a generator, whose effectiveness is 0.72, and then through the combustion chamber. The heated air is then expanded in a high-pressure turbine to 2.3 kg/cm^3 and is then reheated to 1100 K. The air is finally expanded in the low-pressure turbine to 1 kg/cm^2. Assuming the compressor and turbine efficiencies to be 85%, determine: (i) the ratio of compressor work to be turbine work, (ii) power developed for an air flow of 3 kg/s, (iii) the thermal efficiency of the cycle, (iv) heat rejected for second to the atmosphere. Assume that all the components are mounted on the same shaft. Sketch the flow diagram of the turbine and represent the process on T-s plane. Also assume $c_p = 0.24$ kJ/kgK and ratio of specific heats to be 1.4.

Solution: Given data (Figs. 5.53 and 5.54):

$$p_1 = 1 \text{ kg/cm}^2$$
$$p_3 = 5 \text{ kg/cm}^2$$
$$T_1 = 285 \text{ K}$$
$$T_6 = 1100 \text{ K}$$
$$p_2 = 2.3 \text{ kg/cm}^2$$
$$T_3 = T_1 = 285 \text{ K}$$
$$\epsilon = 0.72$$
$$T_8 = T_6 = 1100 \text{ K}$$
$$\eta_c = 0.85$$
$$\eta_T = 0.85$$
$$m_a = 3 \text{ kg/s}$$

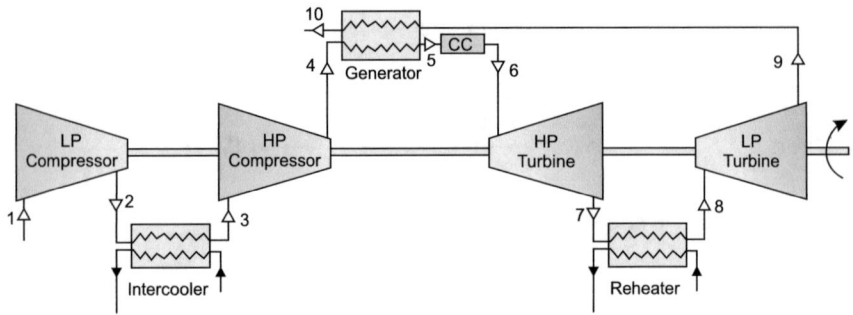

Fig. 5.53 Schematic Flow diagram of gas turbine power plant for Problem 5.20

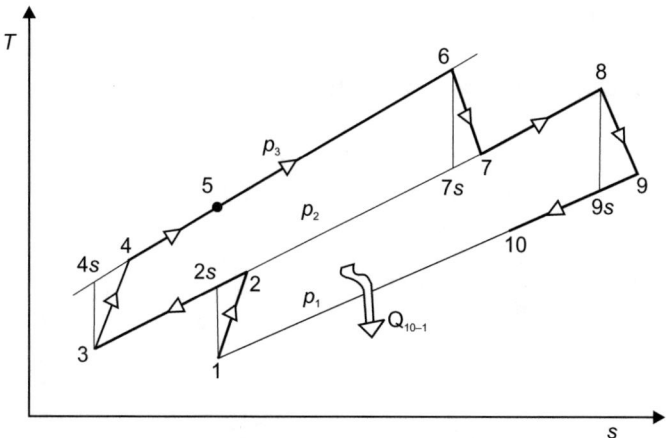

Fig. 5.54 *T-s* diagram for Problem 5.20

$$c_p = 0.24 \text{ kJ/kgK}$$
$$\gamma = 1.4$$

For isentropic process 1–2 s (assumed)

$$\frac{T_{2s}}{T_1} = \left(\frac{p_2}{p_1}\right)^{\frac{\gamma-1}{\gamma}} = \left(\frac{2.3}{1}\right)^{(1.4-1)/1.4} = (2.3)^{0.2857} = 1.268$$

$$\frac{T_{2s}}{285} = 1.268$$

or

$$T_{2s} = 1.268 \times 285 = 361.38 \text{ K}$$

$$\eta_C = \frac{T_{2s} - T_1}{T_2 - T_1} \quad \text{for low pressure compressor}$$

$$0.85 = \frac{361.38 - 285}{T_2 - 285} = \frac{76.38}{T_2 - 285}$$

or

$$T_2 - 285 = \frac{76.38}{0.85} = 89.85$$

or

$$T_2 = 89.85 + 285 = 374.85 \text{ K}$$

For isentropic process 3–4 s (assumed)

$$\frac{T_{4s}}{T_3} = \left(\frac{p_3}{p_2}\right)^{(\gamma-1)/\gamma} = \left(\frac{5}{2.3}\right)^{(1.4-1)/1.4}$$
$$= (217)^{0.2857} = 1.247$$

$$\frac{T_{4s}}{285} = 1.247$$

or

$$T_{4s} = 1.247 \times 285 = 355.39 \text{ K}$$

$$\eta_C = \frac{T_{4s} - T_3}{T_4 - T_3} \quad \text{for high pressure compressor}$$

$$0.85 = \frac{355.39 - 285}{T_4 - 285} = \frac{70.39}{T_4 - 285}$$

or

$$T_4 - 285 = \frac{70.39}{0.85} = 82.81$$
$$T_4 = 82.81 + 285 = 367.81 \text{ K}$$

For isentropic process 6–7 s (assumed)

$$\frac{T_6}{T_{7s}} = \left(\frac{p_3}{p_2}\right)^{(\gamma-1)/\gamma} = \left(\frac{5}{2.3}\right)^{(1.4-1)/1.4} = (2.17)^{0.2857} = 1.247$$

$$\frac{1100}{T_{7s}} = 1.247$$

$$T_{7s} = \frac{1100}{1.247} = 882.11 \text{ K}$$

$$\eta_T = \frac{T_6 - T_7}{T_6 - T_{7s}} \quad \text{for high pressure compressor}$$

$$0.85 = \frac{1000 - T_7}{1000 - 882.11} = \frac{1100 - T_7}{217.89}$$

or

$$217.89 \times 0.85 = 1100 - T_7$$
$$185.20 = 1100 - T_7$$

or

$$T_7 = 1100 - 185.20 = 914.8 \text{ K}$$

For isentropic process 8–9 s (assumed)

$$\frac{T_8}{T_{9s}} = \left(\frac{p_2}{p_1}\right)^{(\gamma-1)/\gamma} = \left(\frac{2.3}{1}\right)^{(1.4-1)/1.4} = (2.3)^{0.2857} = 1.268$$

$$\frac{1100}{T_{9s}} = 1.268$$

or

$$T_{9s} = \frac{1100}{1.268} = 867.50 \text{ K}$$

$$\eta_T = \frac{T_8 - T_9}{T_8 - T_{9s}} \quad \text{for low pressure compressor}$$

$$0.85 = \frac{1100 - T_9}{1100 - 867.50} = \frac{1100 - T_9}{232.5}$$

or

$$0.85 \times 232.5 = 1100 - T_9$$
$$197.62 = 1100 - T_9$$

or

$$T_9 = 1100 - 197.62 = 902.38 \text{ K}$$

Effectiveness of a generator:

$$\in = \frac{T_5 - T_4}{T_9 - T_4} = \frac{T_5 - 367.81}{902.38 - 367.81} = \frac{T_5 - 367.81}{534.57}$$

$$0.72 = \frac{T_5 - 367.81}{534.57}$$

or

$$0.72 \times 534.57 = T_5 - 367.81$$
$$384.89 = T_5 - 367.81$$

or

$$T_5 = 752.7 \text{ K}$$

Assume working fluid in the cycle is air.
Applying energy balance equation for generator.

$$\text{Heat gained by cold air} = \text{Heat lost by hot air}$$
$$m_a c_p (T_5 - T_4) = m_{a^c p}(T_9 - T_{10})$$

or

$$T_5 - T_4 = T_9 - T_{10}$$
$$752.7 - 367.81 = 902.38 - T_{10}$$

or

$$T_{10} = 517.49 \text{ K}$$

Compressor work:

$$\begin{aligned} W_{1-2} &= mc_p(T_2 - T_1) \qquad \text{for LP Compressor} \\ &= 3 \times 0.24 \times (374.85 - 285) \\ &= 65.41 \text{ kW} \end{aligned}$$

Compressor work:

$$\begin{aligned} W_{3-4} &= mc_p(T_4 - T_3) \qquad \text{for HP Compressor} \\ &= 3 \times 0.24 \times (367.81 - 285) \\ &= 59.62 \text{ kW} \end{aligned}$$

Net compressor work:

$$\begin{aligned} W_{C,net} &= W_{1-2} + W_{3-4} \\ &= 65.41 + 59.62 = 125.03 \text{ kW} \end{aligned}$$

Net heat supplied:

$$\begin{aligned} Q_s &= \text{Heat supplied in combustion chamber} + \text{Heat supplied in reheater} \\ &= m_a c_p(T_6 - T_5) + m_a c_p(T_8 - T_7) \\ &= m_a c_p(T_7 - T_5 + T_8 - T_7) \\ &= 3 \times 0.24 \times (914.8 - 752.7 + 1100 - 914.8) \\ &= 250.05 \text{ kW} \end{aligned}$$

Net turbine work:

$$W_{T,net} = \text{High pressure turbine work} + \text{Low pressure turbine work}$$
$$= m_a c_p(T_6 - T_7) + m_a c_p(T_8 - T_9)$$
$$= m_a c_p(T_6 - T_7 + T_8 - T_9)$$
$$= 3 \times 0.24 \times (1100 - 914.8 + 1100 - 902.38)$$
$$= 275.63 \text{ kW}$$

Heat rejected per second to the atmosphere,

$$Q_{10-1} = m_a c_p(T_{10} - T_1)$$
$$= 3 \times 0.24 \times (517.49 - 285)$$
$$= \mathbf{167.39 \text{ kW}}$$

(i) Ratio of compressor work to the turbine work,

$$\frac{W_{C,net}}{W_{T,net}} = \frac{125.03}{275.63} = \mathbf{0.4536}$$

(ii) Power developed for an air:
$$P_{net} = W_{T,net} - W_{C,net}$$
$$= 275.63 - 125.03 = \mathbf{150.6 \text{ kW}}$$

(iii) Thermal efficiency of the cycle,

$$\eta_{th} = \frac{\text{Net powe developed}}{\text{Net the rate of heat supplied}}$$
$$= \frac{P_{net}}{Q_s} = \frac{150.6}{250.05} = 0.6022$$
$$= \mathbf{60.22\%}$$

(iv) Heat rejected per second to the atmosphere:

$$Q_{9-10} = \mathbf{167.39 \text{ kW}}$$

Problem 5.21: A turbojet unit is flying at a speed of 268 m/s at an altitude where the ambient conditions are 0.20 bar and 220 K. The air enters an ideal diffuser and leaves the combustor at 1350 K and 1 bar. The fuel supplied has a heating value of 43,000 kJ/kg. Assume all compression and expansion processes to be isentropic. Determine

(i) The air–fuel ratio,
(ii) The specific thrust, and
(iii) The propulsive efficiency.

Take c_p and γ for the compression process 1.005 kJ/kgK and 1.4, for combustion and expansion processes, 1.102 kJ/kgK and 1.33, respectively.

Solution: Given data (Fig. 5.55):

$$V_a = 268 \text{ m/s}$$
$$p_1 = 0.20 \text{ bar}$$
$$T_1 = 220 \text{ K}$$
$$T_4 = 1350 \text{ K}$$
$$p_3 = p_4 = 1 \text{ bar}$$

Heating value:

$$HV = 43000 \text{ kJ/kg}$$
$$c_{pa} = 1.005 \text{ kJ/kgK}$$
$$\gamma_1 = 1.4$$
$$c_{pg} = 1.102 \text{ kJ/kgK}$$
$$\gamma_2 = 1.33$$

For diffuser:

$$h_1 + \frac{V_a^2}{2} = h_2 \qquad \text{Neglecting exit velocity } V_2$$
$$c_{pg} T_1 + \frac{C_a^2}{2} = c_{pa} T_2$$

Fig. 5.55 T-s diagram for Problem 5.21

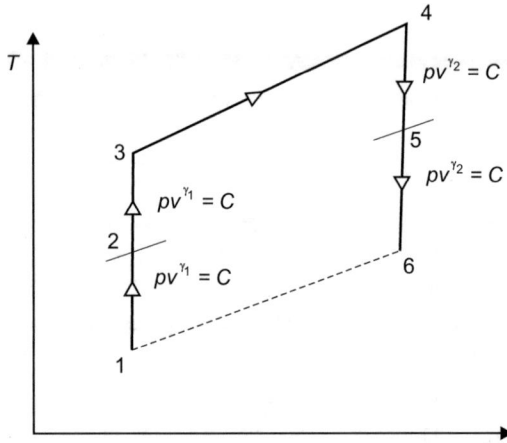

$$c_{pa}T_1 + \frac{C_a^2}{2000} = c_{pa}T_2$$

where

c_{pa} is in kJ/kgK

T_1 and T_2 are in °C or K

C_a is in m/s

$$1.005 \times 220 + \frac{(268)^2}{2000} = 1.005 \times T_2$$
$$221.1 + 35.91 = 1.005\,T_2$$
$$257.01 = 1.005\,T_2$$

or

$$T_2 = 255.73 \text{ K}$$

$$\frac{T_2}{T_1} = \left(\frac{p_2}{p_1}\right)^{\frac{\gamma_1-1}{\gamma_1}}$$
$$\frac{255.73}{220} = \left(\frac{p_2}{0.20}\right)^{\frac{1.4-1}{1.4}}$$
$$1.16 = \left(\frac{p_2}{0.20}\right)^{0.2857}$$

or

$$\frac{p_2}{0.20} = (1.16)^{3.5}$$

or

$$p_2 = 0.20 \times (1.16)^{3.5} = 0.3362 \text{ bar}$$

For compressor: Process 2–3, adiabatic isentropic compression

$$\left(\frac{T_3}{T_2}\right) = \left(\frac{p_3}{p_2}\right)^{\frac{\gamma_1-1}{\gamma_1}}$$
$$\frac{T_3}{255.73} = \left(\frac{1}{0.3362}\right)^{\frac{1.4-1}{1.4}} = (2.97)^{0.2857} = 1.364$$

or

$$T_3 = 1.364 \times 255.73 = 348.81 \text{ K}$$

For combustion chamber:
Applying energy balance equation, we get

$$(m_a + m_f)c_{pg}T_4 - m_a c_{pg}T_3 = m_f \times HV$$

$$\left(\frac{m_a}{m_f} + 1\right) \times 1.102 \times 1350 - \frac{m_a}{m_f} \times 1.005 \times 348.81 = 43000$$

$$\left(\frac{m_a}{m_f} + 1\right) \times 1487.7 - 35055\frac{m_a}{m_f} = 43000$$

$$1487.7\frac{m_a}{m_f} + 1487.7 - 350.55\frac{m_a}{m_f} = 43000$$

$$1137.15\frac{m_a}{m_f} = 41512.3$$

or

$$\frac{m_a}{m_f} = \mathbf{36.50}$$

For turbine: Process 4–5, adiabatic isentropic expansion,

$$\text{Turbine work} = \text{Compressor work}$$

$$(m_a + m_f)c_{pg}(T_4 - T_5) = m_a c_{pa}(T_3 - T_2)$$

$$\left(1 + \frac{m_f}{m_a}\right)c_{pg}(T_4 - T_5) = c_{pa}(T_3 - T_2)$$

$$\left(1 + \frac{1}{36.50}\right) \times 1.102(1350 - T_5) = 1.005(348.81 - 255.73)$$

$$1.132(1350 - T_5) = 93.54$$

$$1350 - T_5 = 82.63$$

or

$$T_5 = 1267.37 \text{ K}$$

$$\frac{T_4}{T_5} = \left(\frac{p_4}{p_5}\right)^{\frac{\gamma_2 - 1}{\gamma_2}}$$

$$\frac{1350}{1267.37} = \left(\frac{1}{p_5}\right)^{\frac{1.33-1}{1.33}} = \left(\frac{1}{p_5}\right)^{0.248}$$

or

$$(1.065)^{4.03} = \frac{1}{p_5}$$

or

$$p_5 = 0.7758 \text{ bar}$$

For nozzle: Process 5–6 adiabatic isentropic

$$\frac{T_5}{T_6} = \left(\frac{p_5}{p_6}\right)^{\frac{\gamma_2-1}{\gamma_2}}$$

$$\frac{1267.37}{T_6} = \left(\frac{0.7758}{0.20}\right)^{\frac{1.33-1}{1.33}} \qquad \because p_6 = p_1$$

$$\frac{1267.36}{T_6} = 1.399$$

or

$$T_6 = \frac{1267.36}{1.399} = 905.90 \text{ K}$$

$$h_5 = h_6 + \frac{C_j^2}{2000} \qquad \text{Neglecting inlet velocity } V_5$$

$$c_{pg}T_5 = c_{pg}T_6 + \frac{C_j^2}{2000}$$

$$1.102 \times 1267.37 = 1.102 \times 905.90 + \frac{C_j^2}{2000}$$

$$1396.64 = 998.30 + \frac{C_j^2}{2000}$$

or

$$\frac{C_j^2}{2000} = 398.34$$

or

$$C_J = 892.56 \text{ m/s}$$

(i) Air–fuel ratio: $\frac{m_a}{m_f} = \mathbf{36.50}$

$$\text{Specific thrust} = \left(1 + \frac{m_f}{m_a}\right)C_j - C_a$$

(ii)
$$= \left(1 + \frac{1}{36.50}\right) \times 892.56 - 268$$

$$= \mathbf{649.01 \ N/kg/s \ of \ air}$$

(iii)   Propulsive efficiency: $\eta_P = \frac{2C_a}{C_j + C_a} = \frac{2 \times 268}{892.56 + 268} = 0.4618 = \mathbf{46.18\%}$

Summary

1.   **Gas Turbine**: A gas turbine is a type of rotodynamic machine that converts the thermal energy of gas into mechanical work.

2.   **Classification of Gas Turbines**: Gas turbines are classified into two categories are

    (i)   Closed-cycle gas turbine,
    (ii)   Open-cycle gas turbine.

3.   **Closed-Cycle Gas Turbine**: It works on Joule's cycle.
    Main parts:

    (i)   Compressor,
    (ii)   Heat exchanger,
    (iii)   Turbine, and
    (iv)   Intercooler.

4.   **Open-Cycle Gas Turbine**: It works on Brayton's cycle.
    Main parts:

    (i)   Compressor,
    (ii)   Combustion chamber, and
    (iii)   Turbine,

    Thermal efficiency:

$$\eta_{\text{th}} = 1 - \frac{1}{r_p^{(\gamma-1)/\gamma}}$$

    where

$$r_p = \text{pressure ratio,}$$
$$\gamma = \text{adiabatic index.}$$

5.   **Back work ratio: BWR**

$$\text{BWR} = \frac{\text{Compressor work} : W_C}{\text{Turbine work} : W_T}$$

$$\text{BWR} = \frac{W_C}{W_T}$$

6. **Condition for maximum work output in the ideal Brayton cycle**

(i) $\dfrac{T_{max}}{T_{min}} = \dfrac{T_3}{T_1} = r_p^{2(\gamma-1)\gamma}$

(ii) $T_2 = T_4 = \sqrt{T_1 T_3}$

(iii) $w_{max} = c_p\left(\sqrt{T_1} - \sqrt{T_3}\right)^2$

7. **Compressor efficiency**: η_C

$$\eta_C = \frac{(\Delta T)_{isentropic}}{(\Delta T)_{actual}}$$

$$\eta_C = \frac{T_{2s} - T_1}{T_2 - T_1}$$

The compressor efficiency is also called the isentropic efficiency of the compressor.

8. **Turbine efficiency**: η_T

$$\eta_T = \frac{(\Delta T)_{actual}}{(\Delta T)_{isentropic}}$$

$$\eta_T = \frac{T_3 - T_4}{T_3 - T_{4s}}$$

The turbine efficiency is also called the isentropic efficiency.

9. **Maximum work output condition for the actual cycle**

$$\frac{T_{max}}{T_{min}}\eta_C\eta_T = \frac{T_3}{T_1}\eta_C\eta_T = r_p^{2(\gamma-1)\gamma}$$

$$\sqrt{\frac{T_{max}}{T_{min}}}\eta_C\eta_T = \sqrt{\frac{T_3}{T_1}}\eta_C\eta_T = r_p^{(\gamma-1)\gamma}$$

10. **Refinement of open cycle gas turbine**

The following processes are used for refinement of the cycle.

(i)   Regeneration,

(ii)   Reheat, and

(iii) Intercooling.

11. **Cycle with regeneration**:

⇩Heat supplied,

W_{net} = constant, and

⇧Thermal efficiency.

12. **Cycle with reheat**:

⇧W_{net}

⇧Heat supplied, and
⇩Thermal efficiency.

13. **Cycle with intercooling:**

⇧W_{net}
⇧Heat supplied, and
⇧Thermal efficiency.

14. **Cycle with regeneration, reheat, and intercooling:**

⇧W
⇧Heat supplied, and
⇧Thermal efficiency.

15. **Jet propulsion:** The word 'propulsion' means moving forward, the propulsion produced due to action of jet is called het propulsion.

16. **Propulsive force, propulsive efficiency, and thermal efficiency,**

(i)   Propulsive force: $F = m_a(C_J - C_a)$.

where m_f is neglected because $m_a \gg m_f$.

m_a =mass of air enter to the aircraft, m/s,
C_a =Absolute velocity of aircraft,
C_j =Relative velocity of the jet.

(ii)   Propulsive power: $P = m_a(C_j - C_a)C_a$.

(iii)   Propulsive efficiency $\eta_P = \frac{2C_a}{C_j+C_a} = \frac{2\sigma}{1+\sigma}$.

where $\sigma = \frac{C_a}{C_j}$.

17. **Classification of jet propulsion engines.** Jet propulsion engines may be classified into two categories:
   1.   Air-breathing engine
      (i)   Turbo jet,
      (ii)   Turbo prop, and
      (iii)   Pulse jet.
   2.   Non-air-breathing engine.
      (i)   Rocket.

18. **Turbo Jet:** The turbo jet consists the following main components:
   1.   Diffuser,
   2.   Compressor,
   3.   Combustion chamber,
   4.   Turbine, and
   5.   Nozzle.

19. **Turbo Prop:** The turbo prop consists the following main components:
   1.   Propeller,
   2.   Diffuser,
   3.   Compressor,

      4.   Combustion chamber,

      5.   Turbine, and

      6.   Nozzle.

20.   **Rocket Engine**: A rocket engine does no use atmospheric air for combustion. It contains own oxygen tanks for fuel combustion. Propellant is the common name given to fuel and oxidiser. The rocket engine is independent altitude and speed.

Assignment–1

1.   What is the difference between Joule's cycle and Brayton's cycle?

2.   Sketch 'Brayton cycle' on T-s and p–v planes.

3.   For Brayton cycle, derive the thermal efficiency.

$$\eta_{th} = 1 - \frac{1}{r_p^{(\gamma-1)/\gamma}}$$

where r_p =pressure ratio.

4.   Differentiate between closed- and open-cycle gas turbine.

5.   Prove that for maximum work output of the ideal Brayton cycle

$$\frac{T_{max}}{T_{min}} = r_p \frac{2(\gamma - 1)}{\gamma}$$

where

$$T_{max} = \text{maximum temperature in the cycle}$$
$$T_{min} = \text{minimum temperature in the cycle}$$
$$r_p = \text{pressure ratio.}$$

6.   Obtain an expression for the specific work output of a gas turbine unit in terms of pressure ratio, isentropic efficiencies of the compressor and turbine and the maximum and minimum temperature, T_3 and T_1. Hence, show that the pressure ratio r_p for maximum power of given by $r_p = \left(\eta_T \eta_C \frac{T_3}{T_1}\right)^{\gamma/2(\gamma-1)}$

7.   Derive an expression for maximum work output per kg of air in terms of the maximum temperature (T_{max}) and minimum temperature (T_{min}).

8.   Discuss the advantages of constant pressure closed gas turbine.

9.   Derive an expression for efficiency of Brayton cycle in terms of pressure ratio and adiabatic exponent.

10.   What is the back work ratio? Why are the back work ratios relatively high in gas turbine?

11.   Discuss various techniques which can be used to enhance the thermal efficiency and specific work output of an open cycle gas turbine.

12.   Define the effectiveness of a regenerator used in open gas turbine cycles.

13. Draw the layout and T-s diagram of an open cycle gas turbine with heat exchanger, reheat, and intercooling.
14. What is the specific purpose of cooling of working fluid in an intercooler of the closed-cycle gas turbine?
15. The cycle with regeneration gives maximum thermal efficiency at low-pressure ratio. Why?
16. With the help of p-v and T-s diagrams, define the isentropic efficiencies of the compressor and turbine.
17. Explain the principle of jet propulsion.
18. What is propulsive efficiency?
19. With the help of neat diagram, explain the working of turbo prop jet engine.
20. Differentiate between turbojet engine and rocket engine.

Assignment–2

1. A simple Brayton cycle using air as working fluid has a pressure ratio of 3. The air enters the compressor at 1 bar and 25 °C. The maximum temperature in the cycle is 650 °C. Determine:

    (i)   Compressor work,
    (ii) Turbine work,
    (iii) Thermal efficiency,
    (iv) Work ratio, [**Ans.** (i) 110.50 kJ/kg, (ii) 250.03 kJ/kg, (iii) 27%, (iv) 0.5580].

2. A power plant operating an Brayton cycle has a pressure ratio of 6. The maximum and minimum temperature are 1000 °C and 20 °C, respectively. Determine:

    (i) The net work output, and
    (ii) The thermal efficiency of the cycle.

    Take $\gamma = 1.4$ and $c_p = 1.005$ kJ/kgK.[**Ans.** (i) 315.22 kJ/kg, (ii) 40%]
3. A simple Brayton cycle using air as working fluid has a pressure ratio of 10. The air enters the compressor at 17 °C and the turbine at 827 °C. Determine:

    (i)   The air temperature at the compressor exit,
    (ii) The back pressure ratio, and
    (iii) The thermal efficiency, [**Ans.** (i) 286.7 °C, (ii) 50.87%, (iii) 48.20%].

4. Air enters the compressor of a gas turbine at 27 °C and 100 kPa, where it is compressed isentropically to 700 kPa. Heat is supplied to air in the amount of 950 kJ/kg before it enters the turbine. If the isentropic efficiency of the turbine is 88%, determine,

    (i) The net work output, and
    (ii) The thermal efficiency.

    Assume $\gamma = 1.4$, $c_p = 1.005$ kJ/kgK.[**Ans.** (i) 329.64 kJ/kg, (ii) 34.69%].

Fig. 5.56 T-s diagram for
Problem 4

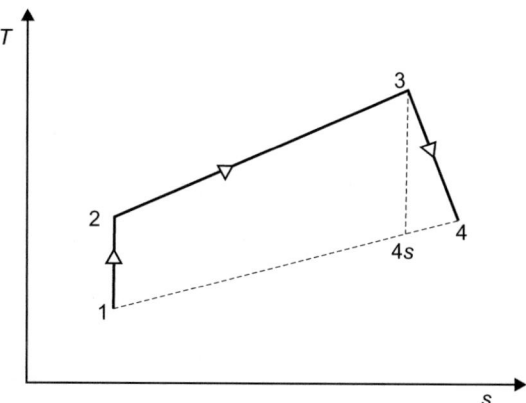

5. Repeat problem 4, if the gas turbine operates on a simple ideal Brayton cycle, determine the percentage increase in the thermal efficiency of the cycle.[**Ans.** 18.64%].

6. A simple Brayton cycle using air as the working fluid has a pressure ratio of 8. The minimum and maximum temperatures in the cycle are 27 and 827 °C. Assuming an isentropic efficiency of 80% for the compressor and 85% for the turbine. Determine:

    (i)   The air temperature at the turbine exit,
    (ii)   The net work output, and
    (iii) The thermal efficiency. [**Ans.** (i) 408.17 °C, (ii) 115.14 kJ/kg, (iii) 23.11%].

7. In a gas turbine plant the minimum and maximum temperatures are 20 °C and 786 °C, respectively. Determine for maximum work output:

    (i)   The pressure ratio,
    (ii)   The net work output in kJ/kg, and
    (iii) The thermal efficiency.

    Assume $\gamma = 1.4$ and $c_p = 1.005$ kJ/kg. [**Ans.** (i) 9.47, (ii) 239.12 kJ/kg, (iii) 47.39].

Hint:

    See Fig. 5.57.

    For maximum work output condition:

(a) $\frac{T_3}{T_1} = r_p^{2(\gamma-1)\gamma}$

(b) $T_2 = T_4 = \sqrt{T_1 T_3}$

(c) $w_{max} = c_p\left(\sqrt{T_1} - \sqrt{T_3}\right)^2$

8. In a gas turbine working on Brayton cycle air enters into the compressor at 1 bar and 27 °C. The pressure ratio is 8 and the maximum temperature is 900 °C. If

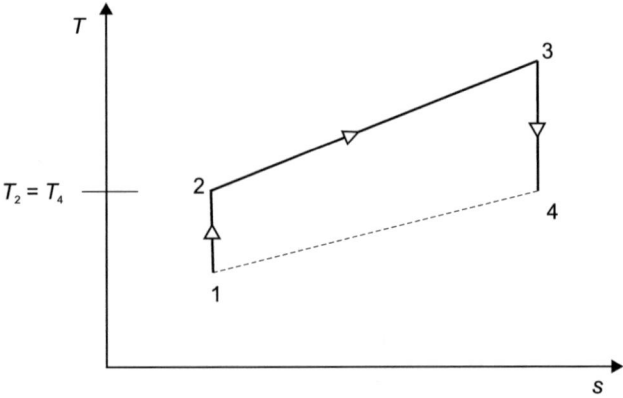

Fig. 5.57   T-s diagram of maximum work output for Problem 7

the turbine and compressor efficiencies are taken as 90% and 80%, respectively. Calculate the cycle efficiency. Assume $\gamma = 1.4$ and $c_p = 1.005$ kJ/kgK. [**Ans.** 29.77%]

9. Air enters the compressor of a gas turbine plant operating on Brayton cycle at 1 bar pressure and 27 °C temperature. The pressure ratio is 6 and the maximum cycle temperature is limited to 800 °C. If the compressor and turbine efficiencies are 85% and 88%, respectively. Make calculation for the net work output, cycle efficiency, and the work ratio. [**Ans.** $w_{net} = 143.11$ kJ/kg, $\eta = 26.50\%$, $WR = 0.3765$]

10. In a gas turbine plant, working on the Brayton cycle with a regenerator of 75% effectiveness. The air at the inlet to the compressor is at 1 bar, 30 °C, the pressure ratio is 6, and the maximum cycle temperature is 900 °C. If the turbine and compressor have each an efficiency of 80%, determine the percentage increase in the cycle efficiency due to regeneration. [**Ans.** 42.56%]

Chapter 6
Thermodynamic Property Relations

Nomenclature

The following is the nomenclature introduced in this chapter:

T	C or K	Temperature
p	kPa	Pressure
V	m^3	Volume
v	m^3/kg	Specific volume
H	kJ	Enthalpy
h	kJ/kg	Specific enthalpy
U	kJ	Internal energy
u	kJ/kg	Specific internal energy
S	kJ/K	Entropy
s	kJ/kgK	Specific entropy
G	kJ	Gibbs function
F	kJ	Helmholtz function
R	kJ/kgK	Gas constant
c_v	kJ/kgK	Specific heat at constant volume
c_p	kJ/kgK	Specific heat at constant pressure
β	1/°C	Volume expansivity
K_T	1/Pa	Isothermal compressibility
K_s	1/Pa	Adiabatic compressibility
h_{fg}	kJ/kg	Specific enthalpy of vaporization
h_{sg}	kJ/kg	Specific enthalpy of sublimation
s_{fg}	kJ/kgK	Specific entropy change during phase change occurs

© The Author(s) 2022
S. Kumar, *Thermal Engineering Volume 2*,
https://doi.org/10.1007/978-3-030-89216-6_6

6.1 Introduction

We know that three thermodynamic properties such as temperature (T), pressure (p), and volume (V) can be measured directly. However, some other thermodynamic properties such as enthalpy (H), internal energy (U), entropy (S), Gibbs function (G), and Helmholtz function (F) cannot be measured directly. Therefore, it is essential to develop the thermodynamic relations for the properties which can be measured directly. The thermodynamic property relations are derived with the help of the first law, the second law of thermodynamics, and the mathematical concepts of partial differentiation.

6.2 Some Mathematical Relations

1. **Exact (or perfect) differential**

Consider a variable z which is a continuous function of x and y.
   That is,

$$z = f(x, y).$$

According to the condition of exact differential

$$dz = \left(\frac{\partial z}{\partial x}\right)_y dx + \left(\frac{\partial z}{\partial y}\right)_x dy$$

$$dz = M dx + N dy$$

where

$$M = \left(\frac{\partial z}{\partial x}\right)_y$$

$= $ partial derivative of z w.r.t. x (at constant y variable)

$$N = \left(\frac{\partial z}{\partial y}\right)_x$$

$= $ partial derivative of z w.r.t. y (at constant x variable)

Differentiating M partially w.r.t. y, we get

$$\left(\frac{\partial M}{\partial y}\right)_x = \frac{\partial^2 z}{\partial x \partial y} \tag{6.1}$$

Differentiating N partially w.r.t. x, we get

$$\left(\frac{\partial M}{\partial y}\right)_x = \frac{\partial^2 z}{\partial y \partial x} \tag{6.2}$$

From Eqs. (6.1) and (6.2), we get

$$\left(\frac{\partial M}{\partial y}\right)_x = \left(\frac{\partial N}{\partial x}\right)_y \tag{6.3}$$

All $z\,(x,\,y)$ that obey Eq. (6.3) are point functions and are called path functions if they do not satisfy this equation.

2. **Cyclic Relation**

As z is a function of two independent variables x and y.
    That is,

$$z = f(x, y).$$

Similarly,

x is function of two independent variables y and z,	y is function of two independent variables x and z,
That is, $x = f(y,z)$ According to the condition of exact differential, we get	That is, $y = f(x,z)$ According to the condition of exact differential, we get
$dx = \left(\frac{\partial x}{\partial y}\right)_z dy + \left(\frac{\partial x}{\partial z}\right)_y dz$ (6.4)	$y = \left(\frac{\partial x}{\partial y}\right)_z dy + \left(\frac{\partial x}{\partial z}\right)_y dz$ (6.5)

Substituting the value of dy from Eq. (6.5) in Eq (6.4), we get

$$dx = \left(\frac{\partial x}{\partial y}\right)_z \left[\left(\frac{\partial y}{\partial x}\right)_z dx + \left(\frac{\partial y}{\partial z}\right)_x dz\right] + \left(\frac{\partial x}{\partial z}\right)_y dz$$

$$= \left(\frac{\partial x}{\partial y}\right)_z \left(\frac{\partial y}{\partial x}\right)_z dx + \left(\frac{\partial x}{\partial y}\right)_z \left(\frac{\partial y}{\partial z}\right)_x dz + \left(\frac{\partial x}{\partial z}\right)_y dz$$

$$dx = dx + \left[\left(\frac{\partial x}{\partial y}\right)_z \left(\frac{\partial y}{\partial z}\right)_x + \left(\frac{\partial x}{\partial z}\right)_y\right] dz$$

or

$$\left(\frac{\partial x}{\partial y}\right)_z \left(\frac{\partial y}{\partial z}\right)_x + \left(\frac{\partial x}{\partial y}\right)_y = 0$$

or

$$\left(\frac{\partial x}{\partial y}\right)_z \left(\frac{\partial y}{\partial z}\right)_x = -\left(\frac{\partial x}{\partial z}\right)_y$$

or

$$\left(\frac{\partial x}{\partial y}\right)_z \left(\frac{\partial y}{\partial z}\right)_x \left(\frac{\partial z}{\partial x}\right)_y = -1 \qquad (6.6)$$

Equation (6.6) is called the **cyclic relation**

Problem 6.1: Prove that for an ideal gas:

$$\left(\frac{\partial p}{\partial v}\right)_T \left(\frac{\partial v}{\partial T}\right)_p \left(\frac{\partial T}{\partial p}\right)_v = -1$$

Solution: We know that the equation of state for an ideal gas

$$pv = RT \qquad (6.7)$$

Equation (6.7) involves the three variables p, v, and T. Out of three variables, one variable can be taken as the dependent variable and the remaining two variables being the independent

$$p = \frac{RT}{v} \qquad (6.8)$$

Let

$$p = f(v, T)$$

Differentiating Eq. (6.8) w.r.t. v, at $T = C$, we get

$$\left(\frac{\partial p}{\partial v}\right)_T = -\frac{RT}{v^2} \qquad (6.9)$$

From Eq. (6.7)

$$v = \frac{RT}{p} \tag{6.10}$$

Let

$$v = f(T, p)$$

Differentiating Eq. (6.10) w.r.t. T, at $p = C$, we get

$$\left(\frac{\partial v}{\partial T}\right) = \frac{R}{p} \tag{6.11}$$

From Eq. (6.7)

$$T = \frac{pv}{R} \tag{6.12}$$

Let

$$T = f(p, v)$$

Differentiating Eq. (6.7) w.r.t. p, at $v = C$, we get

$$\left(\frac{\partial v}{\partial T}\right) = \frac{R}{p} \tag{6.13}$$

Multiplying Eqs. (6.9), (6.11), and (6.13) we get

$$\left(\frac{\partial p}{\partial v}\right)_T \times \left(\frac{\partial v}{\partial T}\right)_p \times \left(\frac{\partial T}{\partial p}\right)_v = -\frac{RT}{v^2} \times \frac{R}{p} \times \frac{v}{R}$$

$$\left(\frac{\partial p}{\partial v}\right)_T \left(\frac{\partial v}{\partial T}\right)_p \left(\frac{\partial T}{\partial p}\right)_v = -\frac{RT}{pv}$$

$$\left(\frac{\partial p}{\partial v}\right)_T \left(\frac{\partial v}{\partial T}\right)_p \left(\frac{\partial T}{\partial p}\right)_v = -1.$$

Hence proved.

6.3 Maxwell Relations

The equations which relate the partial derivatives of properties p, v, T, and s to each other for a compressible fluid are called the **Maxwell relations**. Four Maxwell relations are derived as following:

1. **First Maxwell relation:**

According to the first law of thermodynamic for a process:

$$dQ = dU + dW$$

$$TdS = dU + pdV \qquad \because \delta Q = TdS, \delta W = pdV$$

For a unit mass

$$Tds = du + pdv$$

or

$$du = Tds - pdv \tag{6.14}$$

The above Eq. (6.14) is in the form

$$dz = Mdx + Ndy$$

where

$$M = T$$

$$x = s$$

$$N = -p$$

and

$$y = v$$

$$\therefore$$

$$\left(\frac{\partial T}{\partial v}\right)_s = -\left(\frac{\partial p}{\partial s}\right)_v \qquad (6.15)$$

Equation (6.15) is known as the **first Maxwell relation.**

2. **Second Maxwell relation:**

By definition of enthalpy

$$H = U + pV$$

For a unit mass

$$h = u + pv \qquad (6.16)$$

On differentiating Eq. (6.16), we get

$$dh = du + pdv + vdp$$

Calling Eq. (6.14)

$$du = Tds - pdv$$

or

$$du + pdv = Tds$$

∴

$$dh = Tds + vdp \qquad (6.17)$$

The above Eq. (6.17) is in the form

$$dz = Mdx + Ndy$$

where

$$M = T$$

$$x = s$$

$$N = v$$

and

$$y = p$$

$\therefore$

$$\left(\frac{\partial T}{\partial p}\right)_s = \left(\frac{\partial v}{\partial s}\right)_p \tag{6.18}$$

Equation (6.18) is known as the **second Maxwell relation.**

3.   **Third Maxwell relation:**

By definition of Helmholtz function

$$F = U - TS$$

For a unit mass

$$f = u - Ts \tag{6.19}$$

On differentiating Eq. (6.19), we get

$$df = du - Tds - sdT$$

Calling Eq. (6.14)

$$du = Tds - pdv$$

or

$$du - Tds = -pdv$$

$\therefore$

$$df = -pdv - sdT \tag{6.20}$$

The above Eq. (6.20) is in the form

$$dz = M\,dx + N\,dy$$

where

$$M = -p$$

$$x = v$$

$$N = -s$$

and

$$y = T$$

$\therefore$

$$-\left(\frac{\partial p}{\partial T}\right)_v = -\left(\frac{\partial s}{\partial v}\right)_T$$

or

$$\left(\frac{\partial p}{\partial T}\right)_v = \left(\frac{\partial s}{\partial v}\right)_T \tag{6.21}$$

Equation (6.21) is known as the **third Maxwell relation.**

4. Fourth Maxwell relation:

By definition of Gibbs function

$$G = H - TS$$

For a unit mass

$$g = h - Ts \tag{6.22}$$

On differentiating Eq. (6.22), we get

$$dg = dh - T\,ds - s\,dT$$

Calling Eq. (6.17)

$$dh = T ds + v dp$$

or

$$dh - T ds = v dp$$

$\therefore$

$$dg = v dp - s dT \qquad (6.23)$$

The above Eq. (6.23) is in the form

$$dz = M dx + N dy$$

where

$$M = v$$

$$x = p$$

$$N = -s$$

and

$$y = T$$

$\therefore$

$$\left(\frac{\partial v}{\partial T}\right)_p = -\left(\frac{\partial s}{\partial p}\right)_T \qquad (6.24)$$

Equation (6.24) is known as the **fourth Maxwell relation.**

   Equation (6.15), (6.18), (6.21), and (6.24) are the four Maxwell relations. They are very useful relations only for simple compressible fluids. They are used to determine the entropy change, which cannot be measured directly, by simply measuring the change in properties p, v, and T.

   1st Maxwell relation:

$$\left(\frac{\partial T}{\partial v}\right)_s = -\left(\frac{\partial p}{\partial s}\right)_v$$

2nd Maxwell relation:

$$\left(\frac{\partial T}{\partial p}\right)_s = \left(\frac{\partial v}{\partial s}\right)_p$$

3rd Maxwell relation:

$$\left(\frac{\partial p}{\partial T}\right)_v = \left(\frac{\partial s}{\partial v}\right)_T$$

4th Maxwell relation:

$$\left(\frac{\partial v}{\partial T}\right)_p = -\left(\frac{\partial s}{\partial p}\right)_T$$

6.4   Volume Expansivity: β

It is defined as the ratio of volumetric strain to an increase in temperature at constant pressure.

Mathematically, it is defined as

$$\text{Volume expansivity} : \beta = \frac{\text{Volumertic strain: } \varepsilon_v}{\text{Increase in temperature: } dT}$$

$$\beta = \frac{\varepsilon_v}{dT} \quad \text{at} \quad p = c$$

where volumetric strain:

$$\varepsilon_v = \frac{dV}{V}$$

It is defined as the change in volume to the original volume.

$$\therefore$$

$$\beta = \frac{1}{V}\left(\frac{\partial V}{\partial T}\right)_p = \frac{1}{v}\left(\frac{\partial v}{\partial T}\right)_p$$

The volume expansivity is negative for some substances like liquid water between 0 and 4 °C.

6.5 Isothermal Compressibility: K_T

It is defined as the ratio of volumetric strain to an increase in pressure at a constant temperature.

Mathematically, it is defined as

$$\text{Isothermal compressibility} : K_T = \frac{\text{Volumetric strain} : \varepsilon_v}{\text{Increase in pressure} : dp}$$

$$K_T = \frac{\varepsilon_v}{dp} \quad \text{at} \quad T = c$$

where volumetric strain:

$$\varepsilon_v = \frac{dV}{V}$$

It is defined as the change in volume to the original volume.

$\therefore$

$$K_T = -\frac{1}{V}\left(\frac{\partial V}{\partial p}\right)_T = -\frac{1}{v}\left(\frac{\partial v}{\partial p}\right)_T$$

The *−ve* sign shows that the volume decreases as pressure increases.

6.6 Adiabatic Compressibility: K_S

It is defined as the ratio of volumetric strain to an increase in pressure at constant entropy.

Mathematical, it is defined as

$$\text{Adiabatic compressibility} : K_S = \frac{\text{Volumetric strain} : \varepsilon_v}{\text{Increase in pressure} : dp}$$

$$K_S = \frac{\varepsilon_v}{dp} \quad \text{at} \quad S = c$$

$$K_S = -\frac{1}{V}\left(\frac{\partial V}{\partial p}\right)_s$$

$$= -\frac{1}{v}\left(\frac{\partial v}{\partial p}\right)_s$$

The −ve sign shows that the volume decreases as pressure increases.

6.7 Ratio of Volumetric Expansivity and Isothermal Compressibility: $\frac{\beta}{K_T}$

We know that the volumetric expansivity:

$$\beta = \frac{1}{v}\left(\frac{\partial v}{\partial T}\right)_p$$

or

$$\left(\frac{\partial v}{\partial T}\right)_p = \beta v \qquad (6.25)$$

Also isothermal compressibility:

$$K_T = -\frac{1}{v}\left(\frac{\partial v}{\partial p}\right)_T$$

or

$$\left(\frac{\partial v}{\partial p}\right)_T = -K_T v \qquad (6.26)$$

We know that the cyclic relation of the equation of state for ideal gas

$$\left(\frac{\partial p}{\partial v}\right)_T\left(\frac{\partial v}{\partial T}\right)_p\left(\frac{\partial T}{\partial p}\right)_v = -1 \qquad (6.27)$$

Substituting the values of $\left(\frac{\partial v}{\partial T}\right)_p$ from Eq. (6.25) and $\left(\frac{\partial p}{\partial v}\right)_T$ from Eq. (6.26) in Eq. (6.27), we get

$$-\frac{1}{K_T v} \times \beta v \times \left(\frac{\partial T}{\partial p}\right)_v = -1$$

or

$$\frac{\beta}{K_T} \times \left(\frac{\partial T}{\partial p}\right)_v = 1$$

or

$$\frac{\beta}{K_T} = \left(\frac{\partial p}{\partial T}\right)_s$$

6.8   Change in Internal Energy

Let the specific internal energy u is a function of T and v.
   That is,

$$u = f(T, v)$$

In differential form, we can write

$$du = \left(\frac{\partial u}{\partial T}\right)_v dT + \left(\frac{\partial u}{\partial v}\right)_T dv$$

$$du = c_v dT + \left(\frac{\partial u}{\partial v}\right)_T dv \tag{6.28}$$

Call Eq. (6.14)

$$du = Tds - pdv$$

$$du = T\left(\frac{\partial s}{\partial v}\right)_T dv - pdv \tag{6.29}$$

Calling third maxwell relation, from Eq. (6.21) of Sect. 6.3,

$$\left(\frac{\partial p}{\partial T}\right)_v = \left(\frac{\partial s}{\partial v}\right)_T$$

or

$$\left(\frac{\partial s}{\partial v}\right)_T = \left(\frac{\partial p}{\partial T}\right)_v$$

Substituting .
$\left(\frac{\partial s}{\partial v}\right)_T = \left(\frac{\partial p}{\partial T}\right)_v$ in Eq. (6.29), we get

$$du = T\left(\frac{\partial p}{\partial T}\right)_v dv - pdv$$

$$du = \left[T\left(\frac{\partial p}{\partial T}\right)_v - p\right]dv$$

or

$$\left(\frac{\partial u}{\partial v}\right)_T = \left[T\left(\frac{\partial p}{\partial T}\right)_v - p\right] \tag{6.30}$$

Substituting the value of $\left(\frac{\partial u}{\partial v}\right)_T$ from Eq. (6.30) in Eq. (6.28), we get

$$du = c_v dT + \left[T\left(\frac{\partial p}{\partial T}\right)_v - p\right]dv \tag{6.31}$$

If the compressible system undergoes a change in state from (T_1, v_1) to (T_2, v_2) the change in specific internal energy is given by

$$u_2 - u_1 = \int_{T_1}^{T_2} c_v dt + \int_{v_1}^{v_2} \left[T\left(\frac{\partial p}{\partial T}\right)_v - p\right]dv$$

We know that the ratio of volumetric expansivity and isothermal compressibility:

$$\frac{\beta}{K_T} = \left(\frac{\partial p}{\partial T}\right)_v$$

or

$$\left(\frac{\partial p}{\partial T}\right)_v = \frac{\beta}{K_T}$$

Substituting the volume of $\left(\frac{\partial p}{\partial T}\right)_v = \frac{\beta}{K_T}$ in Eq. (6.31), we get

$$du = c_v dT + \left[\frac{T\beta}{K_T} - p\right]dv$$

6.8.1   Internal Energy of an Ideal Gas is a Function of Temperature Only

Calling Eq. (6.31)

$$du = c_v dT + \left[T \left(\frac{\partial p}{\partial T} \right)_T - p \right] dv \qquad (6.32)$$

The specific internal energy is function of T and v,
   i.e.,

$$u = f(T, v)$$

In differential form, we can write

$$du = \left(\frac{\partial u}{\partial T} \right)_v dT + \left(\frac{\partial u}{\partial v} \right)_T dv \qquad (6.33)$$

Comparing Eqs. (6.32) and (6.33), we get

$$\left(\frac{\partial u}{\partial v} \right)_T = T \left(\frac{\partial p}{\partial T} \right)_v - p \qquad (6.34)$$

We know that the equation of state for an ideal gas

$$pv = RT$$

or

$$p = \frac{RT}{v}$$

Differentiating w.r.t to T, at $v = C$

$$\left(\frac{\partial p}{\partial T} \right)_v = \frac{R}{v}$$

Substituting the value of $\left(\frac{\partial p}{\partial T} \right)_v = \frac{R}{v}$ in Eq. (6.34), we get

$$\left(\frac{\partial u}{\partial v} \right)_T = \frac{RT}{v} - p$$

$$\left(\frac{\partial u}{\partial v}\right)_T = p - p \quad \therefore pv = RT \text{ or } \frac{RT}{v} = p$$

$$\left(\frac{\partial u}{\partial v}\right)_T = 0$$

Thus, u unchanges when v changes at $T = c$

$$\left(\frac{\partial u}{\partial p}\right)_T \left(\frac{\partial p}{\partial v}\right)_T \left(\frac{\partial v}{\partial u}\right)_T = 1$$

or

$$\left(\frac{\partial u}{\partial p}\right)_T \left(\frac{\partial p}{\partial v}\right)_T = \left(\frac{\partial u}{\partial v}\right)_T$$

or

$$\left(\frac{\partial u}{\partial p}\right)_T \left(\frac{\partial p}{\partial v}\right)_T = 0 \quad \because \left(\frac{\partial u}{\partial v}\right)_T = 0$$

Since

$$\left(\frac{\partial u}{\partial p}\right)_T = 0, \left(\frac{\partial p}{\partial v}\right)_T = 0$$

Thus, u unchanges when p changes at $T = C$. So the internal energy of an ideal gas is function of temperature only.

That is,

$$u = f(T).$$

6.9 Change in Enthalpy

Let the specific enthalpy h is a function of T and p.

That is,

$$h = f(T, p)$$

In differential form, we can write

$$dh = \left(\frac{\partial h}{\partial T}\right)_p dT + \left(\frac{\partial h}{\partial p}\right)_T dp$$

$$dh = c_p dT + \left(\frac{\partial h}{\partial p}\right)_T dp \qquad (6.35)$$

Calling Eq. (6.17) of Sect. 6.3,

$$dh = Tds + vdp$$

$$dh = T\left(\frac{\partial s}{\partial p}\right)_T dp + vpd \qquad (6.36)$$

Calling the fourth Maxwell relation, from Eq. (6.24) of Sect. 6.3,

$$\left(\frac{\partial v}{\partial T}\right)_p = -\left(\frac{\partial s}{\partial p}\right)_T$$

or

$$\left(\frac{\partial s}{\partial p}\right)_T = -\left(\frac{\partial v}{\partial T}\right)_p$$

Substituting .
$\left(\frac{\partial s}{\partial p}\right)_T = -\left(\frac{\partial v}{\partial T}\right)_p$ is Eq. (6.36), we get

$$dh = -T\left(\frac{\partial v}{\partial T}\right)_p dp + vdp$$

or

$$dh = \left[-T\left(\frac{\partial v}{\partial T}\right)_p + v\right]dp$$

or

$$\left(\frac{\partial h}{\partial p}\right)_T = \left[-T\left(\frac{\partial v}{\partial T}\right)_p + v\right] \qquad (6.37)$$

Substituting the value of $\left(\frac{\partial h}{\partial p}\right)_T$ from Eq. (6.37) in Eq. (6.35), we get

$$dh = c_p dT + \left[-T\left(\frac{\partial v}{\partial T}\right)_p + v\right]dp$$

or

$$dh = c_p dT + \left[v - T\left(\frac{\partial v}{\partial T}\right)_p\right]dp \qquad (6.38)$$

If the compressible system undergoes a change in state from (T_1, p_1) to (T_2, p_2), the change in specific enthalpy is given by

$$h_2 - h_1 = \int_{T_1}^{T_2} c_p dT + \int_{p_1}^{p_2}\left[v - T\left(\frac{\partial v}{\partial T}\right)_p\right]dp$$

From Eq. (6.38)

$$dh = c_p dT + v\left[1 - \frac{T}{v}\left(\frac{\partial v}{\partial T}\right)_p\right]dp \qquad (6.39)$$

We know that the volume expansivity: β

$$\beta = \frac{1}{v}\left(\frac{\partial v}{\partial T}\right)_p$$

or

$$\frac{1}{v}\left(\frac{\partial v}{\partial T}\right)_p = \beta$$

$\therefore$ Eq. (6.39) becomes

$$dh = c_p dT + v[1 - \beta T]dp.$$

6.10   Change in Entropy

(i)   Calling Eq. (6.14) of Sect. 6.3,

$$du = Tds - pdv$$

or

$$Tds = du + pdv$$

$$ds = c_v \frac{dT}{T} + \frac{p}{T}dv$$

$$ds = c_v \frac{dT}{T} + \left(\frac{\partial p}{\partial T}\right)_v dv \tag{6.40}$$

For process 1–2

$$s_2 - s_1 = \int_{T_1}^{T_2} c_v \frac{dT}{T} + \int_{v_1}^{v_2} \left(\frac{\partial p}{\partial T}\right)_v dv \tag{6.41}$$

(ii)   Calling Eq. (6.17) of Sect. 6.3,

$$dh = Tds + vdp$$

or

$$Tds = dh - vdp$$

or

$$ds = c_p \frac{dT}{T} - \frac{v}{T}dp$$

For process 1–2

$$ds = c_p \frac{dT}{T} - \left(\frac{\partial v}{\partial T}\right)_p dp \tag{6.42}$$

$$s_2 - s_1 = \int_{T_1}^{T_2} c_p \frac{dT}{T} - \int_{p_1}^{p_2} \left(\frac{\partial v}{\partial T} \right) dp$$

6.11 *Tds* Relations

(a) According to first law of thermodynamics for a process

$$\delta Q = dU + p\,dV$$

For a unit mass

$$\delta q = du + p\,dv$$

$$T\,ds = c_v dT + p\,dv$$

$$T\,ds = c_v dT + T\frac{p}{T}\,\mathrm{dv}$$

$$T\,ds = c_v dT + T\left(\frac{\partial p}{\partial T} \right)_v d\boldsymbol{v} \qquad (6.43)$$

Calling third Maxwell relation, from Eq. (6.21) of Sect. 6.3

$$\left(\frac{\partial p}{\partial T} \right)_v = \left(\frac{\partial s}{\partial v} \right)_T$$

Substituting

$\left(\frac{\partial p}{\partial T} \right)_v = \left(\frac{\partial s}{\partial v} \right)_T$ in Eq. (6.43), we get

$$T\,ds = c_v dT + T\left(\frac{\partial s}{\partial v} \right)_T dv$$

$$T\,ds = c_v dT + T\left(\frac{\partial p}{\partial v} \right)_T \left(\frac{\partial s}{\partial p} \right)_T dv$$

$$T\,ds = c_v dT + T\left(\frac{\partial p}{\partial v} \right)_T \left(\frac{\partial s}{\partial p} \right)_T dv \qquad (6.44)$$

Calling fourth Maxwell relation, from Eq. (6.24) of Sect. 6.3,

$$\left(\frac{\partial v}{\partial T}\right)_p = -\left(\frac{\partial s}{\partial p}\right)_T$$

or

$$\left(\frac{\partial s}{\partial p}\right)_T = -\left(\frac{\partial v}{\partial T}\right)_p$$

Substituting.

$\left(\frac{\partial s}{\partial p}\right)_T = \left(\frac{\partial v}{\partial T}\right)_p$ in Eq. (6.44), we get.

∴

$$Tds = c_v dT - T - \left(\frac{\partial p}{\partial v}\right)_T \left(\frac{\partial v}{\partial T}\right)_p dv$$

$$Tds = c_v dT - \frac{1}{v}\left(\frac{\partial v}{\partial T}\right)_p \times v \left(\frac{\partial p}{\partial v}\right)_T Tdv$$

$$Tds = c_v dT + \frac{1}{v}\left(\frac{\partial v}{\partial T}\right)_p \times \left[\frac{1}{-\frac{1}{v}\left(\frac{\partial v}{\partial T}\right)_T}\right] Tdv$$

$$\boldsymbol{Tds = c_v dT + \frac{\beta}{K_T} Tdv} \tag{6.45}$$

where

$$\beta = \frac{1}{v}\left(\frac{\partial v}{\partial T}\right)_p , \text{ volume expansivity}$$

and

$$K_T = -\frac{1}{v}\left(\frac{\partial v}{\partial p}\right)_T , \text{ isothermal compressibility}$$

Equations (6.43) and (6.45) are known as first *Tds* equations

(b)   By definition of enthalpy

$$H = U + pV$$

For a unit mass

$$h = u + pv \tag{6.46}$$

On differentiating Eq. (6.46), we get

$$dh = du + pdv + vdp$$

Calling Eq. (6.14) of Sect. 6.3

$$du = Tds - pdv$$

or

$$du + pdv = Tds$$

$$\therefore$$

$$dh = Tds + vdp$$

or

$$Tds = dh - vdp$$

$$Tds = c_p dT - \left(\frac{\partial p}{\partial v}\right)_T$$

$$\boldsymbol{Tds = c_p dT - \left(\frac{\partial v}{\partial T}\right)_p Tdp} \tag{6.47}$$

$$Tds = c_p dT - \frac{v}{v}\left(\frac{\partial v}{\partial T}\right)_p Tdp$$

$$\boldsymbol{Tds = c_p dT - v\beta Tdp} \tag{6.48}$$

where

$$\frac{1}{v}\left(\frac{\partial v}{\partial T}\right)_p = \beta, \text{ volume expansivity}$$

Equations (6.47) and (6.48) are known as second *Tds* equations.

6.12   Difference Between c_p and c_v

According to first law of thermodynamic for a process

$$\delta Q = dU + pdV$$

For a unit mass

$$\delta q = du + pdv$$

$$Tds = c_v dT + pdv$$

$$Tds = c_v dT + \frac{Tp}{T} dv$$

$$\boldsymbol{Tds = c_v dT + T\left(\frac{\partial p}{\partial T}\right)_v dv}$$

The above equation is known as first Tds equation.

or

$$ds = c_v \frac{dT}{T} + \left(\frac{\partial p}{\partial T}\right)_v dv \qquad (6.49)$$

By definition of enthalpy

$$H = U + pV$$

For a unit mass

$$h = u + pv \qquad (6.50)$$

On differentiating Eq. (6.50), we get

$$dh = du + pdv + vdp$$

$$dh = T\, ds + vdp$$

or

$$Tds = dh - vdp$$

$$Tds = c_pdT - vdp$$

$$Tds = c_pdT - \frac{T}{T}vdp$$

$$\boldsymbol{Tds = c_pdT - T\left(\frac{\partial v}{\partial T}\right)_p dp}$$

The above equation is known as second *Tds* equation.

or

$$ds = c_p\frac{dT}{T} - \left(\frac{\partial v}{\partial T}\right)_p dp \qquad (6.51)$$

Eq. (6.51) = Eq. (6.49)

$$c_p\frac{dT}{T} - \left(\frac{\partial v}{\partial T}\right)_p dp = c_v\frac{dT}{T} + \left(\frac{\partial p}{\partial T}\right)_v dv$$

$$(c_p - c_v)\frac{dT}{T} = \left(\frac{\partial p}{\partial T}\right)_v dv + \left(\frac{\partial v}{\partial T}\right)_p dp$$

$$dT = \frac{T\left(\frac{\partial p}{\partial T}\right)_v}{c_p - c_v}dv + \frac{T\left(\frac{\partial v}{\partial T}\right)_p}{c_p - c_v}dp \qquad (6.52)$$

The temperature T is function of v and p,
    i.e.,

$$T = f(v, p)$$

In differential form, we can write

$$dT = \left(\frac{\partial T}{\partial v}\right)_p dv + \left(\frac{\partial T}{\partial p}\right)_v dp \qquad (6.53)$$

Comparing Eqs. (6.52) and (6.53), we get

$$\left(\frac{\partial T}{\partial v}\right)_p = \frac{T}{(c_p - c_v)}\left(\frac{\partial p}{\partial T}\right)_v$$

or

$$c_p - c_v = T \frac{\left(\frac{\partial p}{\partial T}\right)_v}{\left(\frac{\partial T}{\partial v}\right)_p}$$

$$c_p - c_v = T \left(\frac{\partial p}{\partial T}\right)_v \left(\frac{\partial v}{\partial T}\right)_p \tag{6.54}$$

and

$$\left(\frac{\partial T}{\partial p}\right)_v = \frac{T}{(c_p - c_v)} \left(\frac{\partial v}{\partial T}\right)_p$$

or

$$c_p - c_v = T \frac{\left(\frac{\partial v}{\partial T}\right)_p}{\left(\frac{\partial T}{\partial p}\right)_v}$$

$$c_p - c_v = T \left(\frac{\partial p}{\partial T}\right)_v \left(\frac{\partial v}{\partial T}\right)_p \tag{6.55}$$

We know that the cyclic relation,

$$\left(\frac{\partial p}{\partial v}\right)_T \left(\frac{\partial v}{\partial T}\right)_p \left(\frac{\partial T}{\partial p}\right)_v = -1$$

$$\left(\frac{\partial p}{\partial v}\right)_T \left(\frac{\partial v}{\partial T}\right)_p = -\frac{1}{\left(\frac{\partial T}{\partial p}\right)_v} = -\left(\frac{\partial p}{\partial T}\right)_v$$

or

$$\left(\frac{\partial p}{\partial T}\right)_v = -\left(\frac{\partial p}{\partial v}\right)_T \left(\frac{\partial v}{\partial T}\right)_p \tag{6.56}$$

Substituting the value of $\left(\frac{\partial p}{\partial T}\right)_v$ from Eq. (6.56) in Eq. (6.55), we get

$$c_p - c_v = -T \left(\frac{\partial p}{\partial v}\right)_T \left(\frac{\partial v}{\partial T}\right)_p \left(\frac{\partial v}{\partial T}\right)_p$$

$$c_p - c_v = -T \left(\frac{\partial p}{\partial v}\right)_T \left(\frac{\partial v}{\partial T}\right)_p^2 \tag{6.57}$$

$$c_p - c_v = T \frac{v^2 \left[\frac{1}{v^2} \left(\frac{\partial v}{\partial T} \right)_p^2 \right]}{v \left[-\frac{1}{v} \left(\frac{\partial v}{\partial p} \right)_T \right]}$$

$$c_p - c_v = \frac{T v \beta^2}{K_T} \tag{6.58}$$

where $\beta = \frac{1}{v} \left(\frac{\partial v}{\partial T} \right)_p$ volume expansivity and $K_T = -\frac{1}{v} \left(\frac{\partial v}{\partial p} \right)_T$ isothermal compressibility

Equation (6.57) gives the following information:

1. The difference between c_p and c_v is zero at absolute zero temperature for all substance (liquid, gas or solid), i.e., $c_p - c_v = 0$ at $T = 0$ K.
2. The difference between c_p and c_v is always $+ve$ because $\left(\frac{\partial v}{\partial T} \right)_p^2$ is always $+ve$ and $\left(\frac{\partial p}{\partial v} \right)_T$ is $-ve$ for all compressible fluids.
3. For a liquid and solid, $\left(\frac{\partial v}{\partial T} \right)_p$ is small and difference between c_p and c_v is small.

When $\left(\frac{\partial v}{\partial T} \right)_p = 0$ $c_p = c_v$ as true at the point of maximum density of water at 4 °C.

Problem 6.2: Prove that $c_p - c_v = R$ for an ideal gas.

Solution: Calling Eq. (6.57) of Sect. 6.12.

$$c_p - c_v = -T \left(\frac{\partial p}{\partial v} \right)_T \left(\frac{\partial v}{\partial T} \right)_p^2 \tag{6.59}$$

We know that the equation of state for an ideal gas

$$pv = RT$$

$$p = \frac{RT}{v} \tag{6.60}$$

Differentiating Eq. (6.59) w.r.t to v, at $T = C$, we get

$$\left(\frac{\partial p}{\partial v} \right)_T = -\frac{RT}{v^2} \tag{6.61}$$

and

$$v = \frac{RT}{p} \tag{6.62}$$

Differentiating Eq. (6.62) w.r.t to T, at $p = C$, we get

$$\left(\frac{\partial p}{\partial v}\right)_T = \frac{R}{p} \tag{6.63}$$

Substituting the values of $\left(\frac{\partial p}{\partial v}\right)_T$ from Eq. (6.61) and $\left(\frac{\partial v}{\partial T}\right)_p$ from Eq. (6.63) in Eq. (6.59), we get

$$c_p - c_v = -T \times \left(\frac{-RT}{v^2}\right) \times \left(\frac{R}{p}\right)^2$$

$$= T \times \frac{RT}{v^2} \times \frac{R^2}{p^2}$$

$$= R\frac{T^2 R^2}{v^2 p^2}$$

$$= R\left(\frac{RT}{pv}\right)^2 \quad \therefore pv = RT \quad \text{or} \frac{RT}{pv} = 1$$

$$\boldsymbol{c_p - c_v = R.}$$

Hence proved.

Problem 6.3: Prove that the ratio of specific heat is equal to the ratio of isothermal compressibility and adiabatic compressibility,

  i.e.,

$$\frac{c_p}{c_v} = \frac{K_T}{K_S}$$

Solution:

  The specific heat at constant pressure is given by

$$c_p = \left(\frac{\partial h}{\partial T}\right)_p = \left(\frac{\partial h}{\partial T} \times \frac{\partial s}{\partial s}\right)_p$$

$$= \left(\frac{\partial h}{\partial s}\right)_p \left(\frac{\partial s}{\partial T}\right)_p$$

$$= T\left(\frac{\partial s}{\partial T}\right)_p \quad \therefore \left(\frac{\partial h}{\partial s}\right)_p = T$$

$$c_p = T\left(\frac{\partial s}{\partial T}\right)_p \tag{6.64}$$

The specific heat at constant volume is given

$$c_v = \left(\frac{\partial u}{\partial T}\right)_v = \left(\frac{\partial u}{\partial T} \times \frac{\partial s}{\partial s}\right)_v$$

$$= \left(\frac{\partial u}{\partial s}\right)_v\left(\frac{\partial s}{\partial T}\right)_v$$

$$= T\left(\frac{\partial s}{\partial T}\right)_v \qquad \therefore \left(\frac{\partial u}{\partial s}\right)_v = T$$

$$c_v = T\left(\frac{\partial s}{\partial T}\right)_v \tag{6.65}$$

The ratio of specific heats:

$$\frac{c_p}{c_v} = \frac{T\left(\frac{\partial s}{\partial T}\right)_p}{T\left(\frac{\partial s}{\partial T}\right)_v}$$

$$\frac{c_p}{c_v} = \frac{\left(\frac{\partial s}{\partial T}\right)_p}{\left(\frac{\partial s}{\partial T}\right)_v} \tag{6.66}$$

where

$$\left(\frac{\partial s}{\partial T}\right)_p = \left(\frac{\partial s}{\partial T} \times \frac{\partial v}{\partial v}\right)_p$$

$$\left(\frac{\partial s}{\partial T}\right)_p = \left(\frac{\partial s}{\partial v}\right)_p\left(\frac{\partial v}{\partial T}\right)_p \tag{6.67}$$

Calling second Maxwell relation:

$$\left(\frac{\partial T}{\partial p}\right)_s = \left(\frac{\partial v}{\partial s}\right)_p$$

$$\left(\frac{\partial s}{\partial v}\right)_p = \left(\frac{\partial p}{\partial T}\right)_s$$

Calling fourth Maxwell relation:

$$\left(\frac{\partial v}{\partial T}\right)_p = -\left(\frac{\partial s}{\partial p}\right)_T$$

or

Eq. (6.67) becomes

$$\left(\frac{\partial s}{\partial T}\right)_p = -\left(\frac{\partial p}{\partial T}\right)_s\left(\frac{\partial s}{\partial p}\right)_T \tag{6.68}$$

and

$$\left(\frac{\partial s}{\partial T}\right)_v = \left(\frac{\partial s}{\partial T} \times \frac{\partial p}{\partial p}\right)_v$$

$$\left(\frac{\partial s}{\partial T}\right)_v = \left(\frac{\partial s}{\partial p}\right)_v \left(\frac{\partial p}{\partial T}\right)_v \qquad (6.69)$$

Calling first Maxwell relation:

$$\left(\frac{\partial T}{\partial v}\right)_s = \left(\frac{\partial p}{\partial s}\right)_v$$

or

$$\left(\frac{\partial s}{\partial p}\right)_v = \left(\frac{\partial v}{\partial T}\right)_s$$

Calling third Maxwell relation:

$$\left(\frac{\partial p}{\partial T}\right)_v = -\left(\frac{\partial s}{\partial v}\right)_T$$

Eq. (6.69) becomes,

$$\left(\frac{\partial s}{\partial T}\right)_v = -\left(\frac{\partial v}{\partial T}\right)_s \left(\frac{\partial s}{\partial v}\right)_T \qquad (6.70)$$

Substituting the values of $\left(\frac{\partial s}{\partial T}\right)_p$ from Eq. (6.68) and $\left(\frac{\partial s}{\partial T}\right)_v$ from (6.70) in Eq. (6.66), we get

$$\frac{c_p}{c_v} = \frac{-\left(\frac{\partial p}{\partial T}\right)_s \left(\frac{\partial s}{\partial p}\right)_T}{-\left(\frac{\partial v}{\partial T}\right)_s \left(\frac{\partial s}{\partial v}\right)_T} = \frac{-\left(\frac{\partial p}{\partial v}\right)_s}{-\left(\frac{\partial p}{\partial v}\right)_T} = \frac{1}{-\left(\frac{\partial v}{\partial p}\right)_s} \times \left(-\frac{\partial v}{\partial p}\right)_T$$

$$= \frac{-\frac{1}{v}\left(\frac{\partial v}{\partial p}\right)_T}{-\frac{1}{v}\left(\frac{\partial v}{\partial p}\right)_s}$$

$$\frac{c_p}{c_v} = \frac{K_T}{K_S}$$

Hence proved.

where

$$K_T = -\frac{1}{v}\left(\frac{\partial v}{\partial p}\right)_s, \quad \text{isothermal compressibility.}$$

and

$$K_S = -\frac{1}{v}\left(\frac{\partial v}{\partial p}\right)_s, \quad \text{adiabatic compressibility.}$$

also

$$\frac{c_p}{c_v} = \gamma \text{ , adiabatic index.}$$

$\therefore$

$$\gamma = \frac{c_p}{c_v} = \frac{K_T}{K_s}$$

6.13   Clapeyron Equation

The Clapeyron equation is used to determine the change in enthalpy when phase change takes place during a process, *i.e.*, enthalpy of vaporization h_{fg}.

Calling third Maxwell relation, Eq. (6.21) of Sect. 6.3,

$$\left(\frac{\partial p}{\partial T}\right)_v = \left(\frac{\partial s}{\partial v}\right)_T \tag{6.71}$$

When a phase change occurs, the saturated pressure p_s depends on saturated temperature T_s only and is independent of specific volume v.

That is,

$$p_s = f(T_s)$$

The partial derivative, $\left(\frac{\partial p}{\partial T}\right)_v$ can be written as a total derivative $\left(\frac{dp}{dT}\right)_{sat}$. The total derivative $\left(\frac{dp}{dT}\right)_{sat}$ is the slope on a saturated curve in p–T diagram at a saturated state, as shown in Fig. 6.1, is independent on specific volume v. Thus, the slope $\left(\frac{dp}{dT}\right)_{sat}$ is considered as a constant during the integration of Eq. (6.71) between two saturated liquid state f and saturated vapour state g.

Equation (6.71) is written as

$$\left(\frac{dp}{dT}\right)_{sat} dv = ds$$

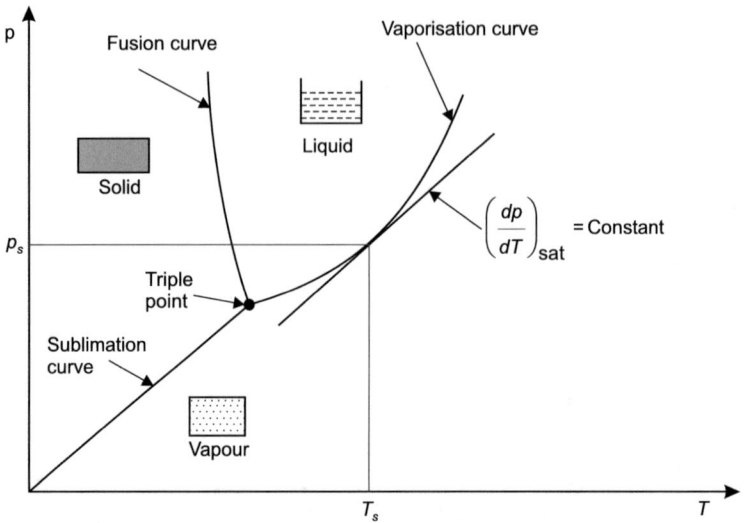

Fig. 6.1 Phase diagram on p–T planes

Integration between saturated liquid state f and saturated vapour state g, we get

$$\left(\frac{dp}{dT}\right)_{sat} \int_{v_f}^{v_g} dv = \int_{s_f}^{s_g} ds$$

$$\left(\frac{dp}{dT}\right)_{sat} \left(v_g - v_f\right) = s_g - s_f$$

$$\left(\frac{dp}{dT}\right)_{sat} v_{fg} = s_{fg}$$

$$\left(\frac{dp}{dT}\right)_{sat} = \frac{s_{fg}}{v_{fg}} \tag{6.72}$$

Calling Eq. (6.17) of Sect. 6.3,

$$dh = T\,ds + v\,dp$$

During the phase change, both temperature and pressure are constant, i.e.,

$$T = C, p = C$$

$\therefore$

$$dh = T ds$$

Integration between two saturated states, we get

$$\int_{h_f}^{h_g} dh = T \int_{s_f}^{s_g} ds$$

$$h_g - h_f = T\left(s_g - s_f\right)$$

$$h_{fg} = T s_{fg}$$

where
$h_{fg} = h_g - h_f$, specific enthalpy of vaporization
$s_{fg} = s_g - s_f$, change in specific entropy during a phase change occurs.

$\therefore$

$$s_{fg} = \frac{h_{fg}}{T}$$

Substituting the value of $s_{fg} = \frac{h_{fg}}{T}$ in Eq. (6.72), we get

$$\left(\frac{dp}{dT}\right)_{sat} = \frac{h_{fg}}{T v_{fg}} \tag{6.73}$$

Equation (6.73) is called the **Clapeyron equation**. This equation is valid only for any phase change at constant T and p.

Equation (6.73) can be written in a general form as:

$$\left(\frac{dp}{dT}\right)_{sat} = \frac{h_{12}}{T v_{12}}$$

$$\left(\frac{dp}{dT}\right)_{sat} = \frac{h_2 - h_1}{T(v_2 - v_1)}$$

where the subscripts 1 and 2 refer to the saturated states of the two phases.

For liquid–vapour and solid-vapour phase change process, the Clapeyron equation can be simplified by using some approximations.

At low pressure, $v_g > > v_f$ for liquid and solid

$\therefore$

$$v_{fg} = v_g \qquad \because v_f \text{ is neglected}$$

If vapour is considered as an ideal gas,

$\therefore$

$pv_g = RT$ from equation of state

or

$$v_g = \frac{RT}{p}$$

$\therefore$

$$v_{fg} = \frac{RT}{p}$$

Substituting the value of $v_{fg} = \frac{RT}{p}$ in Eq. (6.73), we get

$$\left(\frac{dp}{dT}\right)_{sat} = \frac{ph_{fg}}{RT^2}$$

or

$$\left(\frac{dp}{p}\right)_{sat} = \frac{h_{fg}}{R}\left(\frac{dT}{T^2}\right)_{sat} \tag{6.74}$$

For small temperature ranges, h_{fg} can be considered as constant at some mean value. Integration Eq. (6.74) between two saturated values, we get

$$\int_{p_1}^{p_2} \frac{dp}{p} = \frac{h_{fg}}{R}\int_{T_1}^{T_2} \frac{dT}{T}$$

$$\log_e\left(\frac{p_2}{p_1}\right)_{sat} = \frac{h_{fg}}{R}\left(\frac{1}{T_1} - \frac{1}{T_2}\right)_{sat}$$

$$\log_e\left(\frac{p_2}{p_1}\right)_{sat} = \frac{h_{fg}}{R}\left(\frac{T_2 - T_1}{T_1 T_2}\right)_{sat} \tag{6.75}$$

Equations (6.74) and (6.75) are called the Clapeyron–Clausius equations (*i.e.*, C–C equations). These are used to calculate the variation of saturated pressure with temperature.

For solid-vapour regions, h_{fg} has to be replaced by h_{sg} in Eq. (6.75), where h_{sg} is the specific enthalpy of sublimation.

$$\therefore \quad \log_e \left(\frac{p_2}{p_1} \right)_{sat} = \frac{h_{sg}}{R} \left(\frac{T_2 - T_1}{T_1 T_2} \right)_{sat} \quad \text{for solid-vapour phase change process}$$

Problem 6.4: Determine the specific enthalpy of vaporization of water at 250 °C, using the Clapeyron equation and compare it with the value obtained from the steam table.

Solution: Given data:
Saturated temperature:

$$T = 250 \ ^\circ C$$

We know that the Clapeyron equation

$$\left(\frac{dp}{dT} \right)_{sat} = \frac{h_{fg}}{T v_{fg}} \tag{6.76}$$

From saturated steam table on temperature based
At

$$T_s = 250 \ ^\circ C$$

T_{sat} °C	p bar	Specific volume	
		v_g m³/kg	m³/kg m³/kg
245	36.48	0.00125	0.05013
250	39.73		
255	34.19		

$\therefore$

$$v_{fg} = v_g - v_f \quad \text{at} \quad T = 250 \ ^\circ C$$
$$= 0.05013 - 0.00125$$
$$= 0.04888 \, \text{m}^3/\text{kg}$$

$$T = 250 \ ^\circ C = (273 + 250) K = 523 K,$$

and

$$\left(\frac{dp}{dT}\right)_{sat,250°C} = \frac{p_{sat,255°C} - p_{sat,245°C}}{255 - 245} = \frac{43.19 - 36.48}{255 - 245}$$

$$= 0.671 bar/°C = 67.1 kPa/°C$$

Substituting the value of T, v_{fg} and $\left(\frac{dp}{dT}\right)_{sat,250°C}$ in Eq. (6.76), we get

$$67.1 = \frac{h_{fg}}{523 \times 0.04888}$$

or

$$h_{fg} = 1715.36 kJ/kg.$$

Also, the specific enthalpy of vaporization h_{fg} obtained from the steam table at saturated temperature 250 °C is 1716.2 kJ/kg. The difference between the two values of h_{fg} is just because of the approximation used in finding the slope of the saturated curve at 250 °C.

Problem 6.5: Determine the latent heat of vaporization of NH_3 at 0.5 bar. Saturated property data for NH_3 is given below:

p bar	T °C	v_f m³/kg	v_g m³/kg
0.4	− 50.4	1.424×10^{-3}	2.680
0.5	− 46.5	1.433×10^{-3}	2.175
0.6	− 43.3	1.441×10^{-3}	1.835

Solution: Given data:

$$p = 0.5 \text{ bar}$$

Saturated temperature at $p = 0.5$ bar,

$$T = -46.5°C = (-46.5 + 273)K = 226.5 \text{ K}$$

Clapeyron equation,

$$\left(\frac{dp}{dT}\right)_{sat} = \frac{h_{fg}}{Tv_{fg}}$$

$$\left(\frac{dp}{dT}\right)_{sat,-46.5°C} = \frac{h_{fg}}{T\left(v_g - v_f\right)}$$

$$\frac{p_2 - p_1}{T_2 - T_1} = \frac{h_{fg}}{T\left(v_g - v_f\right)}$$

$$\frac{60 - 40}{-43.3 - (-50.4)} = \frac{h_{fg}}{226.5 \times (2.175 - 0.001433)}$$

$$\frac{20}{7.1} = \frac{h_{fg}}{492.31}$$

or

$$h_{fg} = 1386.78 \text{ kJ/kg}.$$

Summary

1. The relations are derived with the help of the first law, the second law of thermodynamics, and the mathematical concepts of partial differentiation, which are called thermodynamic property relations.
2. **Exact (or perfect) Differential:**
     Let

$$z = f(x, y)$$

According to the condition of exact differential

$$dz = \left(\frac{\partial z}{\partial x}\right)_y dx + \left(\frac{\partial z}{\partial y}\right)_x dy$$

3. **Cyclic Relation:**

$$\left(\frac{\partial x}{\partial y}\right)_z \left(\frac{\partial y}{\partial z}\right)_x \left(\frac{\partial z}{\partial x}\right)_y = -1$$

4. **Maxwell Relations:**

Ist Maxwell relation: $\left(\frac{\partial T}{\partial v}\right)_s = -\left(\frac{\partial p}{\partial s}\right)_v$

2nd Maxwell relation: $\left(\frac{\partial T}{\partial p}\right)_s = \left(\frac{\partial v}{\partial s}\right)_p$

3rd Maxwell relation: $\left(\frac{\partial p}{\partial T}\right)_v = \left(\frac{\partial s}{\partial v}\right)_T$

4th Maxwell relation: $\left(\frac{\partial v}{\partial T}\right)_p = -\left(\frac{\partial s}{\partial p}\right)_T$

5. **Change in Internal Energy:**

$$dU = mc_v dT + \left[T\left(\frac{\partial p}{\partial T}\right)_v - p\right]dV$$

Change in specific internal energy:

$$du = c_v dT + \left[T\left(\frac{\partial p}{\partial T}\right)_v - p\right]dv$$

6. **Change in Specific Enthalpy:**

$$dh = c_p dT + \left[v - T\left(\frac{\partial v}{\partial T}\right)_p\right]dp$$

7. **Change in Specific Entropy:**

$$ds = c_v \frac{dT}{T} + \left(\frac{\partial p}{\partial T}\right)_v dv$$
$$= c_p \frac{dT}{T} - \left(\frac{\partial v}{\partial T}\right)_p dp$$

8. **Volume Expansivity:** β. It is defined as the ratio of volumetric strain to an increase in temperature at constant pressure. Mathematically, it is defined as

Volume expansivity: $\beta = \frac{1}{V}\left(\frac{\partial V}{\partial T}\right)_p = \frac{1}{v}\left(\frac{\partial v}{\partial T}\right)_p$.

9. **Isothermal Compressibility:** K_T it is defined as the ratio of volumetric strain to an increase in pressure at constant temperature. Mathematically, it is defined as.

Isothermal compressibility: $K_T = -\frac{1}{V}\left(\frac{\partial V}{\partial p}\right)_T = -\frac{1}{v}\left(\frac{\partial v}{\partial p}\right)_T$.

10. **Adiabatic Compressibility:** K_s.

It is defined as the ratio of volumetric strain to an increase in pressure at constant entropy:

Mathematically, it is defined as

Adiabatic compressibility:

$$K_s = -\frac{1}{V}\left(\frac{\partial V}{\partial p}\right)_S = -\frac{1}{v}\left(\frac{\partial v}{\partial p}\right)_S$$

11. *Tds* **Relations:**

    (i) $Tds = c_v dT + \frac{\beta}{K_T} T dv$

    (ii) $Tds = c_p dT - v\beta T dp$

12. **Difference Between c_p and c_v:**

$$c_p - c_v = -T\left(\frac{\partial p}{\partial v}\right)_T \left(\frac{\partial v}{\partial T}\right)_p^2$$

$$= \frac{Tv\beta^2}{K_T}$$

13. **Clapeyron Equation:** It is used to determine the change in enthalpy when phase change takes place during a process.

$$\left(\frac{dp}{dT}\right)_{sat} = \frac{h_{fg}}{Tv_{fg}} \quad \text{Clapeyron equation.}$$

Assignment–1

1. State the Maxwell relations in the order they are known.
2. Define the following terms:

    (i)   Volume expansivity

    (ii)   Isothermal compressibility

    (iii)   Adiabatic compressibility

3. If $u = f(T, v)$ and $h = f(T, p)$, prove that

$$du = c_v dT + \left(\frac{T\beta}{K_T} - p\right)dv$$

and

$$dh = c_p dT + v(1 - \beta T)dp$$

4. Show that the internal energy of an ideal gas is function of temperature only.
5. Show that the change in specific entropy of an ideal gas is

(i) $ds = c_v \frac{dT}{T} + \left(\frac{\partial p}{\partial T}\right)_v dv$

(ii) $ds = c_p \frac{dT}{T} - \left(\frac{\partial v}{\partial T}\right)_p dp$

6. Using the Maxwell relation, derive the following Tds equation

$$Tds = c_p dT - T\left(\frac{\partial v}{\partial T}\right)_p dp$$

7. Using the Maxwell relations, show that for a pure substance
(i) $Tds = c_v dT + \frac{\beta T}{K_T} dv$
(ii) $Tds = c_p dT - v\beta T dp$

where

β is volume expansivity
K_T is isothermal compressibility
c_p and c_v are specific heats at constant pressure and constant volume, respectively.

8. Explain the terms

(i)   Coefficient of expansion, β
(ii) Isothermal compressibility, K_T

Hence, show that

$$\frac{\beta}{K_T} = \left(\frac{\partial p}{\partial T}\right)_v$$

9. Using Maxwell's and other equations, show that

$$c_p - c_v = -T\left(\frac{\partial p}{\partial v}\right)_T\left(\frac{\partial v}{\partial T}\right)_p^2$$

Hence, also show that

$$c_p - c_v = \frac{T v \beta^2}{K_T}$$

The symbols have their usual meaning.

10. Derive the following Clapeyron and Clapeyron–Clausius equations

$$\left(\frac{dp}{dT}\right)_{sat} = \frac{h_{fg}}{T v_{fg}}$$

and

$$\left(\frac{dp}{p}\right)_{sat} = \frac{h_{fg}}{R}\left(\frac{dT}{T^2}\right)_{sat}$$

Explain the physical significance of these equations.

Chapter 7
Combustion

Nomenclature

The nomenclature introduced in this chapter is as follows:

AF	—	Ratio of the mass of air to the mass of fuel
m_{air}	kg	Mass of air
m_{fuel}	kg	Mass of fuel
p	kPa	Pressure
n	kmol	Number of moles
R	kJ/kmol K	Universal gas constant
T	K	Absolute temperature
Q	kJ	Heat transfer
H_P	kJ	Enthalpy of the products
H_R	kJ	Enthalpy of the reactants
h_f^o	kJ/kg	Enthalpy of formation
HHV	—	Higher heating value
LHV	—	Lower heating value
W_S	kJ	Shaft work
c_P	kJ/kgK	Specific heat at constant pressure
U	kW/m^{2}K	Overall heat transfer coefficient
A	m^2	Surface area
U_P	kJ	Internal energy of products
U_Rt	kJ	Internal energy of reactan

© The Author(s) 2022
S. Kumar, *Thermal Engineering Volume 2*,
https://doi.org/10.1007/978-3-030-89216-6_7

7.1   Introduction

Combustion involves the burning of a fuel with oxygen or a gas containing oxygen such as atmospheric air. A chemical reaction equation relates the components before and after the chemical process takes place.

7.2   Combustion Equations of Fuel

(i)   When C burns with O_2, CO_2 is produced along with the release of a large amount of heat. The chemical reaction is represented by

$$C + O_2 \rightarrow CO_2 \tag{7.1}$$

$$1 \text{ mol} + 1 \text{ mol} = 1 \text{ mol by volume,}$$
$$12 \text{ kg} + 32 \text{ kg} = 44 \text{ kg,}$$
$$1 \text{ kg} + \frac{32}{12} \text{ kg} = \frac{44}{12} \text{ kg,}$$
$$1 \text{ kg} + \frac{8}{3} \text{ kg} = \frac{11}{3} \text{ kg.}$$

Note that the number of moles of the element on the left-hand side may not equal the number of moles on the right-hand side. However, the number of atoms of an element must remain the same before, after, and during a chemical reaction, this demands that the mass of each element be conserved during combustion.

1 kg of C requires $\frac{8}{3}$ kg of O_2 for its complete combustion and produces $\frac{11}{3}$ kg of CO_2.

For the combustion of carbon, we assumed that the process occurred in a pure oxygen environment. Actually, such a combustion process would normally occur in air. For our purpose, we assume that air consists of 21% O_2 and 79% N_2 by volume (ignore the small percentage of other gases), so that each mole of O_2 in a reaction we will have

$$\frac{V_{N_2}}{V_{O_2}} = \frac{79}{21}$$
$$= 3.76 \frac{\text{molN}_2}{\text{molO}_2}.$$

Thus, N_2 will not under go any chemical reaction, Eq. (7.1) becomes

$$C + (O_2 + 3.76N_2) \rightarrow CO_2 + 3.76N_2$$

The minimum amount of air that supplies sufficient O_2 for the complete combustion of the fuel is called stoichiometric air or theoretical air. When complete combustion is achieved with theoretical air, the products contain no O_2, as in the above equation. The amount of air supply more than the theoretical air is called excess air.

The parameter that relates to the amount of air used in a combustion process is the air–fuel ratio (AF), which is defined as the ratio of the mass of air to the mass of fuel.

Mathematically,

$$AF = \frac{m_{air}}{m_{fuel}}$$

Air–fuel ratio of carbon combustion with theoretical air as:

$$AF = \frac{m_{air}}{m_{fuel}} = \frac{O_2 + 3.76N_2}{C}$$
$$= \frac{32 + 3.76 \times 28}{12}$$
$$= 11.44 \text{ kg air/kg fuel.}$$

If, for the combustion of carbon, AF > 11.44, a lean mixture occurs.

If AF < 11.44, a rich mixture results.

Lean mixture: If the air–fuel ratio is more than theoretical air, is called lean mixture.

Rich mixture: If the air–fuel ratio is less than theoretical air, is called rich mixture.

(ii) When methane (CH_4) burns with O_2, CO_2, and H_2O are produced along with the release of a large amount of heat. The chemical reaction is represented by

$$CH_4 + 2O_2 = CO_2 + 2H_2O,$$

$$1 \text{ mol} + 2 \text{ mol} = 1 \text{ mol} + 2 \text{ mol},$$

$$16 \text{ kg} + 64 \text{ kg} = 44 \text{ kg} + 36 \text{ kg},$$

$$1 \text{ kg} + 4 \text{ kg} = \frac{11}{4} \text{ kg} + \frac{9}{4} \text{ kg.}$$

It means that 1 kg of methane requires 4 kg of oxygen and produces $\frac{11}{4}$ kg of carbon dioxide and $\frac{9}{4}$ kg of steam or water.

If the combustion process takes place in dry air, the above equation is written as

$$CH_4 + 2(O_2 + 3.76N_2) \rightarrow 4CO_2 + 2H_2O + 7.52N_2.$$

The air-fuel ratio is

$$AF = \frac{m_{\text{air}}}{m_{\text{fuel}}}$$

$$= \frac{2(32 + 3.76 \times 28)}{16}$$

$$= 17.16 \text{ kg air/kg fuel}.$$

(iii)   When butane (C_4H_{10}) burns with O_2, CO_2, and H_2O are produced along with a release of large amount of heat. The chemical reaction is represented by

$$C_4H_{10} + 6.5O_2 \rightarrow 4CO_2 + 5H_2O,$$

$$1 \text{ mol} + 6.5 \text{ mol} = 4 \text{ mol} + 5 \text{ mol},$$

$$58 \text{ kg} + 208 \text{ kg} = 176 \text{ kg} + 90 \text{ kg},$$

$$1 \text{ kg} + 3.58 \text{ kg} = 3.03 \text{ kg} + 1.55 \text{ kg}.$$

It means that 1 kg of butane requires 3.58 kg of oxygen and produces 3.03 kg of carbon dioxide and 1.55 kg of steam or water.

If the combustion process takes place in dry air, the above equation is written as

$$C_4H_{10} + 6.5(O_2 + 3.76N_2) \rightarrow 4CO_2 + 5H_2O + 24.44N_2.$$

The air-fuel ratio is

$$AF = \frac{m_{\text{air}}}{m_{\text{fuel}}}$$

$$= \frac{6.5(32 + 3.76 \times 28)}{58}$$

$$= 15.38 \text{ kg air/kg fuel}.$$

(iv)   When acetylene (C_2H_2) burns with O_2, CO_2, and H_2O are produced along with a release of large amount of heat. The chemical reaction is represented by

$$C_2H_2 + 2.5O_2 \rightarrow 2CO_2 + H_2O,$$

$$1 \text{ mol} + 2.5 \text{ mol} = 2 \text{ mol} + 1 \text{ mol},$$

$$26 \text{ kg} + 80 \text{ kg} = 88 \text{ kg} + 18 \text{ kg},$$

$$1 \text{ kg} + 3.07 \text{ kg} = 3.38 \text{ kg} + 0.69 \text{ kg}.$$

It means that 1 kg of acetylene requires 3.07 kg of oxygen and produces 3.38 kg of carbon dioxide and 0.69 kg of steam or water.

If the combustion process takes place in dry air, the above equation is written as

$$C_2H_2 + 2.5(O_2 + 3.76 \text{ N}_2) = 2CO_2 + H_2O + 9.4 \text{ N}_2$$

The air-fuel ratio is

$$\begin{aligned} \text{AF} &= \frac{m_{\text{air}}}{m_{\text{fuel}}} \\ &= \frac{2.5(32 + 3.76 \times 28)}{26} \\ &= 13.2 \text{ kg air/kg fuel.} \end{aligned}$$

Problem 7.1: Butane is burned with dry air at an air–fuel ratio of 20. Calculate the percentage of excess air, the volume percentage of CO_2 in the product, and the dew point temperature of the products.

Solution: The reaction equation for theoretical air is

$$C_4H_{10} + 6.5(O_2 + 3.76 \text{ N}_2) \rightarrow 4CO_2 + 5H_2O + 24.44 \text{ N}_2$$

The air–fuel ratio for theoretical air is

$$\begin{aligned} (\text{AF})_{\text{th}} &= \frac{m_{air}}{m_{fuel}} = \frac{6.5(0_2 + 3.76N_2)}{C_4N_{10}} \\ &= \frac{6.5(32 + 3.76 \times 28)}{58} \\ &= 15.38 \text{ kg air/kg fuel.} \end{aligned}$$

The actual air–fuel ratio is 20, i.e., $(\text{AF})_{\text{act}} = 20$

$$\begin{aligned} \%\text{excess air} &= \frac{(\text{AF})_{\text{act}} - (\text{AF})_{\text{th}}}{(\text{AF})_{\text{th}}} \times 100 \\ &= \frac{20 - 15.38}{15.38} \times 100 \\ &= \mathbf{30\%.} \end{aligned}$$

The reaction equation with 130% theoretical air is

$$C_4H_{10} + 6.5 \times 1.30(O_2 + 3.76 \text{ N}_2) \rightarrow 4CO_2 + 5H_2O + 1.95O_2 + 31.772 \text{ N}_2$$

The volume percentage is obtained using the total moles in the products of combustion.

For CO_2, we have

$$\% \, CO_2 = \frac{4}{4 + 5 + 1.95 + 31.772} \times 100$$
$$= \frac{4}{42.722} \times 100$$
$$= 9.36\%.$$

To find the dew point temperature of the product, we need the partial pressure p_v of water vapour, we know that equation of state,

$$p_v V = n_v \overline{R} T \quad \text{and} \quad p V = n \overline{R} T$$

where

$n_v =$ Number of moles of water vapour.
$n =$ Toal number of moles in product.

The ratio of the above two equations gives

$$\frac{p_v}{p} = \frac{n_v}{n}$$

or

$$p_v = \frac{n_v}{n} p.$$

where

$n_v =$ 5,
$n =$ $4 + 5 + 1.95 + 31.772 = 42.722,$
$p =$ 101.325 kPa, atmospheric pressure,

$\therefore$

$$p_v = \frac{5}{42.722} \times 101.325 = 11.85 \text{ kPa} = 0.1185 \text{bar}.$$

From saturated steam table (pressure based),
At

$$p_v = 0.1185 \text{ bar},$$

we get, dew point temperature $= 49 \, °C.$

Problem 7.2: Butane is burned with 90% theoretical air. Determine the volume percentage of CO in the products and the air–fuel ratio. Assume no hydrocarbons in the products.

Solution: For incomplete combustion results in products that contain CO.

$$C_4H_{10} + 0.90 \times 6.5(O_2 + 3.76\ N_2) \rightarrow aCO_2 + 5H_2O + 22N_2 + bCO$$

For atomic balances on the carbon, we get

$$4 = a + b \tag{7.2}$$

For atomic balances on the oxgyen, we get

$$11.7 = 2a + 5 + b$$

or

$$6.7 = 2a + b \tag{7.3}$$

Eq. (7.3)–Eq. (7.2), we get

$$6.7 = 2a + b$$
$$\underline{4 = a\ + b}$$
$$\overline{2.7 = a}$$

or

$$a = 2.7.$$

Substituting the value of $a = 2.7$ in Eq. (7.2), we get

$$4 = 2.7 + b$$

or

$$b = 1.3.$$

$$\therefore$$

$$C_4H_{10} + 5.85(O_2 + 3.76N_2) \rightarrow 2.7CO_2 + 5H_2O + 22N_2 + 1.3CO$$

The volume percentage is obtained using the total moles in the products of combustion.

For CO, we have

$$\% \, CO = \frac{1.3}{2.7 + 5 + 22 + 1.3} \times 100$$
$$= \frac{1.3}{31} \times 100 = \mathbf{4.19\%}.$$

The air–fuel ratio is

$$AF = \frac{m_{air}}{m_{fuel}}$$
$$= \frac{5.85(32 + 3.76 \times 28)}{58}$$
$$= 13.84 \, \mathbf{kg \, air/kg \, fuel}.$$

Problem 7.3: Butane is burned with dry air, and volumetric analysis of the products on a dry basis gives 11% CO_2, 1% CO, 3.5% O_2, and 84.5% N_2. The water vapour is not measured. Find the percent theoretical air.

Solution: The problem is solved assuming that there are 100 mol of dry product. The chemical equation is.

$$\therefore a C_4 H_{10} + b(O_2 + 3.76 N_2) \rightarrow 11 CO_2 + 1 CO_2 + 3.5 O_2 + 84.5 N_2 + c H_2 O$$

For atomic balances on the carbon, we get

$$4a = 11 + 1,$$

$$4a = 12,$$

or

$$a = \frac{12}{3} = 3.$$

For atomic balance on the hydrogen, we get

$$10a = 2c,$$
$$10 \times 3 = 2c,$$
$$30 = 2c,$$

or

$$c = \frac{30}{2} = 15.$$

For atomic balance on the oxygen, we get

$$2b = 22 + 1 + 7 + c,$$
$$2b = 30 + c,$$
$$2b = 30 + 15,$$
$$2b = 45,$$
$$b = \frac{45}{2},$$
$$b = 22.5.$$

Substituting the values of a, b and c in Eq. (7.2), we get

$$3C_4H_{10} + 22.5(O_2 + 3.75N_2) \rightarrow 11CO_2 + CO + 3.5O_2 + 84.5N_2 + 15H_2O \tag{7.4}$$

Chemical Eq. (7.4) dividing by 3 so that we have 1 mol fuel, we get

$$C_4H_{10} + 7.5(O_2 + 3.75N_2) \rightarrow 3.67CO_2 + 0.33CO + 1.17O_2 + 28.17N_2 + 5H_2O$$

$$\%\text{theoretical air} = \frac{\text{Amount of air required in above equation}}{\text{Amount of air required in theoretical equation}}$$
$$= \frac{7.5(O_2 + 3.75N_2)}{6.5(O_2 + 3.75N_2)} \times 100$$
$$= \frac{7.5}{6.5} \times 100 = \mathbf{115.38\%}.$$

Problem 7.4: Volumetric analysis of the products of combustion of an unknown hydrocarbon, measured on a dry basis, gives 10.4% CO_2, 1.2% CO, 2.8% O_2, and 85.6% N_2. Find the composition of the hydrocarbon and the percent excess air.

Solution: The problem is solved assuming that there are 100 mol of dry products. The chemical equation is

$$C_aH_b + c(O_2 + 3.76N_2) \rightarrow 10.4CO_2 + 1.2CO + 2.8O_2 + 85.6N_2 + dH_2O$$

Balanceing C, we get

$$a = 10.4 + 1.2$$

$$\boxed{a = 11.6}$$

Balancing H, we get

$$b = 2d \tag{7.5}$$

Balancing O, we get

$$2c = 20.8 + 1.2 + 5.6 + d$$

$$2c = 27.6 + d \tag{7.6}$$

Balancing N, we get

$$7.52c = 171.2$$

or

$$\boxed{c = 22.76}$$

Substituing $c = 22.76$ in Eq. (7.6), we get

$$2 \times 22.76 = 27.6 + d$$

$$45.52 = 27.6 + d$$

or

$$\boxed{d = 17.92}$$

Substituting $d = 17.92$ in Eq. (7.5), we get

$$b = 2 \times 17.92$$

$$\boxed{b = 35.84}$$

Substituting the values a, b, c, and d in the above chemical equation, we get

$$C_{11.6}H_{35.84} + 22.76(O_2 + 3.76N_2) \rightarrow 10.4CO_2 + 1.2CO + 2.8O_2 + 85.6N_2 + 17.92H_2O$$

The actual air–fuel ratio is

$$(AF)_{act} = \frac{m_{air}}{m_{fuel}} = \frac{22.76(32 + 3.76 \times 28)}{175.04}$$

$$= 17.85.$$

The chemical formula is $C_{11.6}H_{35.84}$.

To find the percent theoretical air, we must have the chemical equation using 100% theoretical air

$$C_{11.6}H_{35.84} + 20.56(O_2 + 3.76N_2) \rightarrow 11.6CO_2 + 17.92H_2O + 77.30N_2$$

The theoretical air–fuel ratio is

$$(AF)_{act} = \frac{m_{air}}{m_{fuel}} = \frac{20.56(32 + 3.76 \times 28)}{175.04} = 16.12$$

$$\% \text{ exass air} = \frac{[(AF)_{act} - (AF)_{th}]}{(AF)_{th}} \times 100$$

$$= \left(\frac{7.85 - 16.12}{16.12}\right) \times 100 = \mathbf{10.73\%}.$$

7.3   Enthalpy of Formation, Enthalpy of Combustion, and the First Law

Consider the combustion of C with O_2, formation of CO_2:

$$C + O_2 \rightarrow CO_2$$

If C and O_2 enter a combustion chamber at 25 °C and 1 atm and CO_2 leaves the combustion chamber at the same temperature and pressure, then the measured heat transfer will be -393.522 MJ for each kmol of CO_2 formed. The -ve sign on the heat transfer means that the heat energy left the control volume as shown in Fig. 7.1.

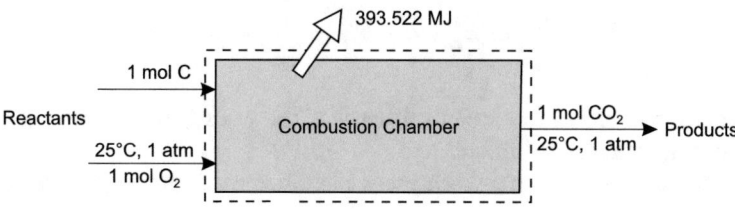

Fig. 7.1 The control volume used during combustion of carbon and oxygen

The first law applied to a combustion process in a control volume with no shaft work, neglecting the kinetic and potential energy change

$$Q = H_P - H_R$$

where

H_P = enthalpy of the products that leave the combustion chamber.
H_R = enthalpy of the reactant that enters the combustion chamber.

Enthalpy of formation. The energy absorbed or released when a compound is formed from its components at constant temperature and pressure is called the **enthalpy of formation**.

It is denoted by $h_f{}^o$. Note that some compound has a +ve $h_f{}^o$, indicating that they require energy to form (an endothermic reaction), other have a -ve $h_f{}^o$, indicating that they give off energy when they are formed (an exothermic reaction).

Endothermic reaction. A reaction that requires energy to form a compound is called an endothermic reaction.

Exothermic reaction. A reaction that releases energy when a compound is formed is called an exothermic reaction.

Enthalpy of combustion. The energy released when a compound undergoes complete combustion at constant temperature and pressure is called the enthalpy of combustion. For example, the enthalpy of formation of C is zero, yet when 1 mol C undergoes complete combustion to CO_2, it gives 393.522 MJ heat, the enthalpy of combustion of C is 393.522 MJ/kmol.

If the products contain liquid water, the enthalpy of combustion is the higher heating value (H HV), if the products contain water vapour, the enthalpy of combustion is the lower heating value (LHV). The difference between the higher heating value and the lower heating value is called the vaporization h_{fg} at standard conditions.

If the reactants and products consist of several components, the first law, allowing for the shaft and neglecting kinetic and potential energy changes

$$Q - W_s = \sum_{\text{Prod}} n_i \left(\bar{h}_f^o + \bar{h} - \bar{h}^o \right)_i - \sum_{\text{React}} n_i \left(\bar{h}_f^o + \bar{h} - \bar{h}^o \right)_i$$

where n_i = number of moles of substance i. The work is often zero, but not for example, in a combustion turbine.

If combustion occurs in a rigid chamber, like a bomb calorimeter, so that no work result, the first law is

$$Q = U_P - U_R = \sum_{\text{Prod}} n_i \left(\bar{h}_f^o + \bar{h} - \bar{h}^o - pv \right)_i - \sum_{\text{React}} n_i \left(\bar{h}_f^o + \bar{h} - \bar{h}^o - pv \right)$$

Since the volume of any liquid or solid is negligible compared to the volume of the gases.

$$Q = \sum_{Prod} n_i \left(\bar{h}_f^o + \bar{h} - \bar{h}^o - \bar{R}T \right) - \sum_{React} n_i \left(\bar{h}_f^o + \bar{h} - \bar{h}^o - \bar{R}T \right)_i$$

If $n_{Prod} = n_{React}$, then Q for the rigid volume is equal to Q for the control volume for the isothermal process.

In the above relation, we employ one of the following methods to find $(-^0)$:

- For a solid or liquid:

$$\bar{h} - \bar{h}^0 = c_p \Delta T$$

- For gases:

  Assume an ideal gas with constant specific heat so that

$$\bar{h} - \bar{h}^0 = c_p \Delta T.$$

7.4 Adiabatic Flame Temperature

A combustion process takes place adiabatically, with no shaft work, no change in kinetic and potential energy, then the temperature of the products is called the **adiabatic flame temperature**. The maximum adiabatic flame temperature that can be achieved occurs at theoretical air. This fact allows us to control the adiabatic flame temperature by the amount of excess air involved in the process. The greater the amount of excess air, the lower the adiabatic flame temperature. If the blades in a gas turbine can withstand a certain maximum temperature, we can determine the excess air needed so that the maximum allowable blade temperature is not exceeded.

If a significant amount of heat transfer does occur from the combustion chamber, we can account for it by including the following term in the energy equation:

$$Q = UA(T_P - T_E),$$

where

$U =$ overall heat transfer coefficient,
$T_P =$ temperature of products,
$T_E =$ temperature of environment,
$A =$ surface area of combustion chamber.

Summary

1. **Combustion.** It involves the burning of fuel with oxygen or a gas containing oxygen such as atmospheric air.
2. **Air–fuel ratio.** It is defined as the ratio of the mass of air to the mass of fuel. Mathematically,

$$\text{Air} - \text{fuel ratio} : AF = \frac{m_{air}}{m_{fuel}}$$

3. **Theoretical (or stoichiometric) air.** The minimum amount of air that supplies sufficient O_2 for the complete combustion of the fuel is called theoretical (or stoichiometric) air. The amount of air supply more than the theoretical air is called excess air.
4. **Lean mixture.** If the air–fuel ratio is more than theoretical air, is called lean mixture.
5. **Rich mixture.** If the air–fuel ratio is less than theoretical air, is called rich mixture.
6. **Reactants.** The components that enter a combustion chamber are called reactants.
7. **Products.** The components that leave a combustion chamber are called products.
8. **Endothermic reaction.** A reaction that requires energy to form a compound is called endothermic reaction.
9. **Exothermic reaction.** A reaction that releases energy when a compound is formed is called an exothermic reaction.
10. **Enthalpy of formation.** The energy absorbed or released when a compound is formed from its components at constant temperature and pressure is called the enthalpy of formation.
11. **Enthalpy of combustion.** The energy released when a compound undergoes complete combustion at constant temperature and pressure is called the enthalpy of combustion.
12. **Higher heating value (HHV).** If the products contain liquid water, the enthalpy of combustion is called higher heating value.
13. **Lower heating value (LHV).** If the products contain water vapour, the enthalpy of combustion is called lower heating value.
14. **Adiabatic flame temperature.** If a combustion process takes place adiabatically with no shaft work, no change in kinetic and potential energy, then the temperature of the products is called the adiabatic flame temperature.

Assignment–1

1. Determine the air–fuel ratio of the following fuels:

    (i)   Methane
    (ii)   Butane, and
    (iii) Acetylene

2. What do you understand by theoretical air and excess air in the context of combustion?
3. Define the following terms:

    (i)   Enthalpy of formation
    (ii)   Enthalpy of combustion

4. What is the difference between endothermic reaction and exothermic reaction?
5. What do you understand by adiabatic flame temperature?

Appendix

See Tables A.1, A.2, A.3, A.4, A.5, A.6, A.7, A.8, A.9, A.10, A.11, A.12, A.13, and A.14.

© The Author(s) 2022
S. Kumar, *Thermal Engineering Volume 2*,
https://doi.org/10.1007/978-3-030-89216-6

Table A.1 Properties of saturated steam (temperature based)

Sat. temp. °C T_{sat}	Sat. pressure kPa p_{sat}	Specific volume m³/kg		Specific internal energy kJ/kg		Specific enthalpy kJ/kg			Specific entropy kJ/kgK		
		Sat. liquid v_f	Sat. vapor v_g	Sat. liquid u_f	Sat. vapor u_g	Sat. liquid h_f	Evap. h_{fg}	Sat. vapor h_g	Sat. liquid s_f	Evap. s_{fg}	Sat. vapor s_g
0.01	0.6113	0.0010002	206.2	0.00	3237.3	0.01	2501.3	2501.4	0.000	9.156	9.156
1	0.6567	0.0010002	192.6	4.12	2376.7	4.2	2499.0	2503.2	0.015	9.115	9.130
2	0.6556	0.0010001	179.9	8.4	2378.1	8.4	2496.7	2505.0	0.031	9.073	9.104
3	0.7577	0.0010001	168.1	12.6	2379.5	12.6	2494.3	2506.9	0.046	9.032	9.077
4	0.8131	0.0010001	157.2	16.8	2380.9	16.8	2491.9	2508.7	0.061	8.990	9.051
5	0.8721	0.0010001	147.1	21.0	2382.3	21.0	2489.6	2510.6	0.076	8.950	9.026
6	0.9349	0.0010001	137.7	25.2	2383.6	25.2	2487.2	2512.4	0.091	8.909	9.000
7	1.002	0.0010002	129.0	29.4	2385.0	29.4	2484.8	2514.2	0.106	8.869	8.975
8	1.072	0.0010002	120.9	33.6	2386.4	33.6	2482.5	2516.1	0.121	8.829	8.950
9	1.148	0.0010003	113.4	37.8	2387.8	37.8	2480.1	2517.9	0.136	8.789	8.925
10	1.228	0.0010004	106.4	42.0	2389.2	42.0	2477.7	2519.7	0.151	8.750	8.901
11	1.312	0.0010004	99.86	46.2	2390.5	46.2	2475.4	2521.6	0.166	8.711	8.877
12	1.402	0.0010005	93.78	50.4	2391.9	50.4	2473.0	2523.4	0.181	8.672	8.852
13	1.497	0.001007	88.12	54.6	2393.3	54.6	2470.7	2525.3	0.195	8.632	8.828
14	1.598	0.0010008	82.85	58.8	2394.7	58.8	2468.3	2527.1	0.210	8.595	8.805
15	1.705	0.0010009	77.93	63.0	2396.1	63.0	2465.9	2528.9	0.224	8.5,57	8.781
16	1.818	0.001001	73.33	67.2	2397.4	67.2	2463.6	2530.8	0.239	8.519	8.758
17	1.938	0.001001	69.04	71.4	2398.8	71.4	2461.2	2532.6	0.253	8.482	8.735
18	2.064	0.001001	65.04	75.6	2400.2	75.6	2458.8	2534.4	0.268	8.444	8.712

(continued)

Table A.1 (continued)

Sat. temp. °C T_{sat}	Sat. pressure kPa p_{sat}	Specific volume m³/kg		Specific internal energy kJ/kg		Specific enthalpy kJ/kg			Specific entropy kJ/kgK		
		Sat. liquid v_f	Sat. vapor v_g	Sat. liquid u_f	Sat. vapor u_g	Sat. liquid h_f	Evap. h_{fg}	Sat. vapor h_g	Sat. liquid s_f	Evap. s_{fg}	Sat. vapor s_g
19	2.198	0.001002	61.29	79.8	2401.6	79.8	2456.5	2536.3	0.282	8.407	8.690
20	2.339	0.001002	57.79	84.0	2402.9	84.0	2454.1	2538.1	0.297	8.371	8.667
21	2.487	0.001002	54.51	88.1	2404.3	88.1	2451.8	2539.9	0.311	8.334	8.645
22	2.645	0.001002	61.45	92.3	2405.7	92.3	2449.4	2541.7	0.325	8.298	8.623
23	2.810	0.001002	48.57	96.5	2407.0	96.5	2447.0	2543.5	0.339	8.262	8.601
24	2.085	0.001003	45.88	100.7	2408.4	100.7	2444.7	2545.4	0.353	8.226	8.579
25	3.169	0.001003	43.36	104.9	2409.8	104.9	2442.3	2547.2	0.367	8.191	8.558
26	3.363	0.001003	40.99	109.1	2411.1	109.1	2439.9	2549.0	0.382	8.155	8.537
27	3.567	0.001004	38.77	113.2	2412.5	113.2	2437.6	2550.8	0.396	8.120	8.516
28	3.782	0.001004	36.69	117.4	2413.9	117.4	2435.2	2552.6	0.409	8.086	8.495
29	4.008	0.001004	34.73	121.6	2415.2	121.6	2432.8	2554.5	0.423	8.051	8.474
30	4.246	0.001004	32 89	125.8	2416.6	125.8	2430.5	2556.3	0.437	8.016	8.453
31	4.496	0.001005	31.17	130.0	2418.0	130.0	2428.1	2558.1	0.451	7.982	8.433
32	4.759	0.001005	29.54	134.2	2419.3	134.2	2425.7	2559.9	0.464	7.948	8.413
33	5.034	0.001005	28.01	138.3	2420.7	138.3	2423.4	2561.7	0.478	7.915	8.393
34	5.324	0.001006	26.57	142.5	2422.0	142.5	2421.0	2563.5	0.492	7.881	8.373
35	5.628	0.001006	25.22	146.7	2423.4	146.7	2418.6	2565.3	0.505	7.848	8.353
36	5.947	0.001006	23.94	150.9	2424.7	150.9	2416.2	2567.1	0.519	7.815	8.334

(continued)

Table A.1 (continued)

Sat. temp. °C	Sat. pressure kPa	Specific volume m³/kg		Specific internal energy kJ/kg			Specific enthalpy kJ/kg				Specific entropy kJ/kgK		
T_{sat}	p_{sat}	Sat. liquid v_f	Sat. vapor v_g	Sat. liquid u_f	Sat. vapor u_g		Sat. liquid h_f	Evap. h_{fg}	Sat. vapor h_g		Sat. liquid s_f	Evap. s_{fg}	Sat. vapor s_g
37	6.281	0.001007	22.74	155.0	2426.1		155.0	2413.9	2568.9		0.532	7.782	8.314
38	6.632	0.001007	21.60	159.2	2427.4		159.2	2411.5	2570.7		0.546	7.749	8.295
39	6.999	0.001007	20.53	163.4	2428.8		163.4	2409.1	2572.5		0.559	7.717	8.276
40	7.384	0.001008	19.52	167.6	2430.1		167.6	2406.7	2574.3		0.573	7.685	8.257
41	7.786	0.001008	18.57	171.7	2431.5		171.7	2404.3	2576.0		0.586	7.652	8.238
42	8.208	0.001009	17.67	175.9	2432.8		175.9	2401.9	2577.8		0.599	7.621	8.220
43	8.649	0.001009	16.82	180.1	2434.2		180.1	2399.5	2579.6		0.612	7.589	8.201
44	9.111	0.001010	16.02	184.3	2435.5		184.3	2397.2	2581.5		0.626	7.557	8.183
45	9.593	0.001010	15.26	188.4	2436.8		188.4	2394.8	2583.2		0.639	7.526	8.165
46	10.10	0.001010	14.54	192.6	2438.2		192.6	2392.4	2585.0		6.652	7.495	8.147
47	10.62	0.001011	13.86	196.8	2439.5		196.8	2390.0	2586.8		0.665	7.464	8.129
48	11.48	0.001011	13.22	201.0	2440.8		201.0	2387.6	2588.6		0.678	7.433	8.111
49	11.75	0.001012	12.61	205.1	2442.2		205.1	2385.2	2590.3		0.691	7.403	8.094
50	12.65	0.001012	12.03	209.3	2443.5		209.3	2382.7	2592.1		0.704	7.373	8.076
52	13.63	0.001013	10.97	217.7	2446.1		217.7	2377.9	2595.6		0.730	7.312	8.042
54	15.02	0.001014	10.01	226.0	2448.8		226.0	2373.1	2599.1		0.755	7.253	8.008
55	15.76	0.001015	9.568	230.2	2450.1		230.2	2370.7	2600.9		0.768	7.223	7.991
56	16.53	0.001015	9.149	234.4	2451.4		234.4	2368.2	2602.6		0.781	7.194	7.975

(continued)

Table A.1 (continued)

Sat. temp. °C T_{sat}	Sat. pressure kPa p_{sat}	Specific volume m³/kg		Specific internal energy kJ/kg		Specific enthalpy kJ/kg			Specific entropy kJ/kgK		
		Sat. liquid v_f	Sat. vapor v_g	Sat. liquid u_f	Sat. vapor u_g	Sat. liquid h_f	Evap. h_{fg}	Sat. vapor h_g	Sat. liquid s_f	Evap. s_{fg}	Sat. vapor s_g
58	18.17	0.001016	8.372	242.8	2454.0	242.8	2363.4	2606.2	0.806	7.136	7.942
60	19.94	0.001017	7.671	.251.1	2456.6	251.1	2358.5	2609.6	0.831	7.078	7.909
62	21.86	0.001018	7.037	259.5	2459.3	259.5	2353.6	2613.1	0.856	7.022	7.878
64	23.03	0.001019	6.463	267.9	2461.8	267.9	2348.7	2616.5	0.881	6.965	7.846
65	25.03	0.001020	6.197	272.0	2463.1	272.1	2346.2	2618.3	0.894	6.937	7.831
66	26.17	0.001020	5.943	276.2	2464.4	276.2	2343.7	2619.9	0.906	6.910	7.816
68	28.59	0.001022	5.471	284.6	2467.0	284.6	2338.8	2623.4	0.930	6.855	7.785
70	31.49	0.001023	5.042	293.0	2469.6	293.0	2333.8	2626.8	0.955	6.800	7.755
72	34.00	0.001024	4.650	301.4	2472.1	301.4	2329.8	2630.2	0.979	6.746	7.725
75	38.58	0.001026	4.131	313.9	2475.9	313.9	2321.4	2635.3	1.015	J.667	7.682
80	47.39	0.001029	3.407	334.9	2482.2	334.9	2308.8	2643.7	1.075	6.537	7.612
85	57.83	0.001033	2.828	355.9	2488.4	355.9	2296.0	2651.9	1.134	6.410	7.544
90	70.14	0.001036	2.361	376.9	2494.5	376.9	2283.2	2660.1	5.192	6.287	7.479
95	84.65	0.001040	1.982	397.9	2500.6	397.9	2270.2	2668.1	1.250	6.166	7.416
100	101.35	0.001044	1.673	418.9	2506.5	419.0	2257.0	2676.0	1.307	6.048	7.355
105	1.2082×10^2	0.001048	1.4194	440.02	2512.4	440.15	2243.7	2683.8	1.3630	5.9328	7.2958
110	1.4327×10^2	0.001052	1.2102	461.14	2518.1	461.30	2230.2	2691.5	1.4185	5.8202	7.2387
115	1.6906×10^2	0.001056	1.0366	482.30	2523.7	482.48	2216.5	2699.0	1.4734	5.7100	7.1833

(continued)

Table A.1 (continued)

T_{sat}	p_{sat}	Sat. liquid v_f	Sat. vapor v_g	Sat. liquid u_f	Sat. vapor u_g	Sat. liquid h_f	Evap. h_{fg}	Sat. vapor h_g	Sat. liquid s_f	Evap. s_{fg}	Sat. vapor s_g
		Specific volume m³/kg		Specific internal energy kJ/kg		Specific enthalpy kJ/kg			Specific entropy kJ/kgK		
Sat. temp. °C	Sat. pressure kPa										
120	1.9853×10^2	0.001060	0.8919	503.50	2529.3	503.71	2202.6	2706.3	1.5276	5.6020	7.1296
125	2.321×10^2	0.001065	0.7706	524.74	2534.6	524.99	2188.5	2713.5	1.5813	5.4962	7.0775
130	2.701×10^2	0.001070	0.6685	546.02	2539.9	546.31	2174.2	2720.5	1.6344	5.3925	7.0269
135	3.130×10^2	0.001075	0.5822	567.35	2545.0	567.69	2159.6	2727.3	1.6870	5.2907	6.9777
140	3.613×10^2	0.001080	0.5089	588.74	2550.0	589.13	2144.7	2733.9	1.7391	5.1908	6.9299
145	4.154×10^2	0.001085	0.4463	610.18	2554.9	610.63	2129.6	2740.3	1.7907	5.0926	6.8833
150	4.758×10^2	0.001091	0.3928	631.68	2559.5	632.20	2114.3	2746.5	1.8418	4.9960	6.8379
155	5.431×10^2	0.001096	0.3468	653.24	2564.1	653.84	2098.6	2752.4	1.8925	4.9010	6.7935
160	6.178×10^2	0.001102	0.3071	674.87	2568.4	675.55	2082.6	2758.1	1.9427	4.8075	6.7502
165	7.005×10^2	0.001108	0.2727	696.56	2572.5	697.34	2066.2	2763.5	1.9925	4.7153	6.7078
170	7.917×10^2	0.001114	0.2428	718.33	2576.5	719.21	2049.5	2768.7	2.0419	4.6244	6.6663
175	8.920×10^2	0.001121	0.2168	740.17	2580.2	741.17	2032.4	2773.6	2.0909	4.5347	6.6256
180	10.021×10^2	0.001127	0.19405	762.09	2583.7	763.22	2015.0	2778.2	2.1396	4.4461	6.5857
185	11.227×10^2	0.001134	0.17409	784.10	2587.0	785.37	1997.1	2782.4	2.1879	4.3586	6.5465
190	12.544×10^2	0.001141	0.15654	806.19	2590.0	807.62	1978.8	2786.4	2.2359	4.2720	6.5079
195	13.078×10^2	0.001149	0.14105	828.37	2592.8	829.98	1960.0	2790.0	2.2835	4.1863	6.4698
200	15.538×10^2	0.001157	0.12736	850.65	2595.3	852.45	1940.7	2793.2	2.3309	4.1014	6.4323
205	17.230×10^2	0.001164	0.11521	873.04	2597.5	875.04	1921.0	2796.0	2.3780	4.0172	6.3952

(continued)

Table A.1 (continued)

T_{sat}	p_{sat}	Sat. liquid v_f	Sat. vapor v_g	Sat. liquid u_f	Sat. vapor u_g	Sat. liquid h_f	Evap. h_{fg}	Sat. vapor h_g	Sat. liquid s_f	Evap. s_{fg}	Sat. vapor s_g
210	19.062×10^2	0.001173	0.10441	895.53	2599.5	897.76	1900.7	2798.5	2.4248	3.9337	6.3585
215	21.04×10^2	0.001181	0.09479	918.14	2601.1	920.62	1879.9	2800.5	2.4714	3.8507	6.3221
220	23.18×10^2	0.001190	0.08619	940.87	2602.4	943.62	1858.5	2802.1	2.5178	3.7683	6.2861
225	25.48×10^2	0.001199	0.07849	963.73	2603.3	966.78	1836.5	2803.3	2.5639	3.6863	6.2503
230	27.95×10^2	0.001209	0.07158	986.74	2603.9	990.12	1813.8	2804.0	2.6099	3.6047	6.2146
235	30.60×10^2	0.001219	0.06537	1009.89	2604.1	1013.62	1790.5	2804.2	2.6558	3.5233	6.1791
240	33.44×10^2	0.001229	0.05976	1033.21	2604.0	1037.32	1766.5	2803.8	2.7015	3.4422	6.1437
245	36.48×10^2	0.001240	0.05471	1056.71	2603.4	1061.23	1741.7	2803.0	2.7472	3.3612	6.1083
250	39.73×10^2	0.001251	0.05013	1080.39	2602.4	1085.36	1716.2	2801.5	2.7927	3.2802	6.0730
255	43.19×10^2	0.001263	0.04598	1104.28	2600.9	1109.73	1689.8	2799.5	2.8383	3.1992	6.0375
260	46.88×10^2	0.001276	0.04221	1128.39	2599.0	1134.37	1662.5	2796.9	2.8838	3.1181	6.0019
265	50.81×10^2	0.001289	0.03877	1152.74	2596.6	1159.28	1634.4	2793.6	2.9294	3.0368	5.9662
270	54.99×10^2	0.001302	0.03564	1177.36	2593.7	1184.51	1605.2	2789.7	2.9751	2.9551	5.9301
275	59.42×10^2	0.001317	0.03279	1202.25	2590.2	1210.07	1574.9	2785.0	3.0208	2.8730	5.8938
280	64.12×10^2	0.001332	0.03017	1227.46	2586.1	1235.99	1543.6	2779.6	3.0668	2.7903	5.8571
285	69.09×10^2	0.001348	0.02777	1253.00	2581.4	1262.31	1511.0	2773.3	3.1130	2.7070	5.8199
290	74.36×10^2	0.001366	0.02557	1278.92	2576.0	1289.07	1477.1	2766.2	3.1594	2.6227	5.7821
295	79.93×10^2	0.001384	0.02354	1305.2	2569.9	1316.3	1441.8	2758.1	3.2062	2.5375	5.7437

(continued)

Table A.1 (continued)

Sat. temp. °C	Sat. pressure kPa	Specific volume m³/kg		Specific internal energy kJ/kg		Specific enthalpy kJ/kg			Specific entropy kJ/kgK		
T_{sat}	p_{sat}	Sat. liquid v_f	Sat. vapor v_g	Sat. liquid u_f	Sat. vapor u_g	Sat. liquid h_f	Evap. h_{fg}	Sat. vapor h_g	Sat. liquid s_f	Evap. s_{fg}	Sat. vapor s_g
300	85.81×10^2	0.001404	0.02167	1332.0	2563.0	1344.0	1404.9	2749.0	3.2534	2.4511	5.7045
305	92.02×10^2	0.001425	0.019948	1359.3	2555.2	1372.4	1366.4	2738.7	3.3010	2.3633	5.6643
310	98.56×10^2	0.001447	0.018350	1387.1	2546.4	1401.3	1326.0	2727.3	3.3493	2.2737	5.6230

Table A.2 Properties of saturated steam (pressure based)

Sat. pressure bar p_{sat}	Sat. temp. °C T_{sat}	Specific volume m³/kg		Specific internal energy kJ/kg		Specific enthalpy kJ/kg			Specific entropy kJ/kgK		
		Sat. liquid v_f	Sat. vapor v_g	Sat. liquid u_f	Sat. vapor u_g	Sat. liquid h_f	Evap. h_{fg}	Sat. vapor h_g	Sat. liquid s_f	Evap. s_{fg}	Sat. vapor s_g
0.006113	0.01	0.0010002	206.140	0.00	2375.3	0.01	2501.3	2501.4	0.000	9.156	9.156
0.010	7.0	0.0010000	129.21	29.3	2385.0	29.3	2484.9	2514.2	0.106	8.870	8.976
0.105	13.0	0.0010007	87.98	54.7	2393.3	54.7	2470.6	2525.3	0.196	8.632	8.828
0.020	17.0	0.001001	67.00	73.5	2399.5	73.5	2460.0	2533.5-	0.261	8.463	8.724
0.025	21.1	0.001002	54.25	88.5	2404.4	88.5	2451.6	2540.1	0.312	8.331	8.643
0.030	24.1	0.001003	45.67	101.0	2408.5	101.0	2444.5	2545.5	0.355	8.223	8.578
0.035	26.7	0.001003	39.50	111.9	2412.1	111.9	2438.4	2550.3	0.391	8.132	8.523
0.040	29.0	0.001004	34.80	121.5	2415.2	121.5	2432.9	2554.4	0.423	8.052	8.475
0.045	31.0	0.001005	31.13	130.0	2417.9	130.0	2428.2	2558.2	0.451	7.982	8.433
0.050	32.9	0.001005	28.19	137.8	2420.5	137.8	2423.7	2561.5	0.476	7.919	8.395
0.055	34.6	0.001006	25.77	144.9	2422.8	144.9	2419.6	2565.5	0.500	7.861	8.361
0.060	36.2	0.001006	23.74	151.5	2425.0	151.5	2415.9	2567.4	0.521	7.809	8.330
0.065	37.6	0.001007	22.01	157.7	2426.9	157.7	2412.4	2570.1	0.541	7.761	8.302
0.070	39.0	0.001007	20.53	163.4	2428.8	163.4	2409.1	2572.5	0.559	7.717	8.276
0.075	40.3	0.001008	19.24	168.8	2430.5	168.8	2406.0	2574.8	0.576	7.675	8.251
0.080	41.5	0.001008	18.10	173.9	2432.2	173.9	2403.1	2577.0	0.593	7.636	8.229
0.085	42.7	0.001009	17.10	178.7	2433.7	178.7	2400.3	2579.0	0.608	7.599	8.207
0.090	43.8	0.001009	16.20	183.3	2435.2	183.3	2397.7	2581.0	0.622	7.565	8.187
0.095	44.8	0.001010	15.40	187.7	2436.6	187.7	2395.2	2582.9	0.636	7.532	8.168

(continued)

Table A.2 (continued)

Sat. pressure bar p_{sat}	Sat. temp. °C T_{sat}	Specific volume m³/kg		Specific internal energy kJ/kg		Specific enthalpy kJ/kg			Specific entropy kJ/kgK		
		Sat. liquid v_f	Sat. vapor v_g	Sat. liquid u_f	Sat. vapor u_g	Sat. liquid h_f	Evap. h_{fg}	Sat. vapor h_g	Sat. liquid s_f	Evap. s_{fg}	Sat. vapor s_g
0.10	45.8	0.001010	14.67	191.8	2437.9	191.8	2392.8	2584.7	0.649	7.501	8.150
0.11	47.7	0.001011	13.42	199.7	2440.4	199.7	2388.3	2588.0	0.674	7.453	8.117
0.12	49.4	0.001012	12.36	206.9	2442.7	206.9	2384.2	2591.1	0.696	7.390	8.086
0.13	51.0	0.001013	11.47	213.7	2.444.9	213.7	2380.2	2593.9	0.717	7.341	8.058
0.14	52.6	0.001013	10.69	220.0	2446.9	220.0	2376.6	2596.6	0.737	7.296	8.033
0.15	54.0	0.001014	10.02	225.9	2448.7	225.9	2373.2	2559.1	0.755	7.254	8.009
0.16	55.3	0.001015	9.43	231.6	2450.5	231.6	2369.9	2601.5	0.772	7.214	7.986
0.17	56.6	0.001015	8.91	236.9	2452.2	236.9	2366.8	2603.7	0.788	7.177	7.%5
0.18	57.8	0.001016	8.45	242.0	2453.8	242.0	2363.8	2505.8	0.804	7.14.1	7.945
0.19	59.0	0.001017	8.03	246.8	2455.3	246.8	2361.0	2607.8	0.818	7.108	7.926
0.20	60.1	0.001017	7.65	251.4	2456.7	251.4	2358.3	2609.7	0.832	7.077	7.909
0.22	62.1	0.001018	7.45	260.1	2459.4	260.1	2353.2	2613.3	0.858	7.018	7.876
0.24	64.1	0.001019	6.45	268.1	2461.9	268.1	2348.5	2616.6	0.882	6.964	7.846
0.25	65.0	0.001020	6.20	271.9	2463.1	271.9	2346.3	2618.2	0.893	6.938	7.831
0.26	65.9	0.001020	5.98	275.6	2464.2	275.6	2344.1	2619.7	0.904	6.914	7.818
0.28	67.5	0.001021	5.58	282.6	2466.4	282.6	2340.0	2622.6	0.925	6.868	7.793
0.30	69.1	0.001022	5.23	289.2	2468.4	289.2	2336.1	2625.3	0.944	6.825	7.768
0.32	70.6	0.001023	4.92	295.5	2470.3	295.5	2332.4	2627.9	0.962	6.784	7.746

(continued)

Table A.2 (continued)

Sat. pressure bar	Sat. temp. °C	Specific volume m³/kg		Specific internal energy kJ/kg		Specific enthalpy kJ/kg			Specific entropy kJ/kgK		
p_{sat}	T_{sat}	Sat. liquid v_f	Sat. vapor v_g	Sat. liquid u_f	Sat. vapor u_g	Sat. liquid h_f	Evap. h_{fg}	Sat. vapor h_g	Sat. liquid s_f	Evap. s_{fg}	Sat. vapor s_g
0.34	72.0	0.001024	4.65	301.4	2472.1	301.4	2328.8	2630.2	0.979	6.746	7.725
0.35	72.7	0.001024	4.53	304.2	2473.0	304.2	2327.2	2631.4	0.987	6.728	7.715
0.36	73.4	0.001025	4.41	307.0	2473.8	307.0	2325.5	2632.5	0.996	6.710	7.706
0.38	74.6	0.001026	4.19	312.4	2475.5	312.4	2322.3	2634.7	1.011	6.676	7.687
0.40	75.9	0.001027	3.99	317.6	2477.0	317.6	2319.2	2636.8	1.026	6.644	7.670
0.45	78.7	0.001028	3.58	329.6	2480.7	329.6	2312.0	2641.6	1.060	6.571	7.631
0.50	81.3	0.001030	3.24	340.5	2483.9	340.6	2305.4	2646.0	1.091	6.503	7.594
0.55	83.7	0.001032	2.96	350.5	2486.8	350.5	2299.3	2649.8	1.119	6.442	7.561
0.60	85.9	0.001033	2.73	359.8	2489.6	359.9	2293.6	2653.5	1.145	6.387	7.532
0.65	88.0	0.001035	2.53	368.5	2492.1	368.5	2288.3	2656.8	1.169	6.335	7.504
0.70	90.0	0.001036	2.37	376.7	2494.5	376.7	2283.3	2660.0	1.192	6.288	7.480
0.75	91.8	0.001037	2.22	384.3	2496.7	384.4	2278.6	2663.0	1.213	6.243	7.456
0.80	93.5	0.001039	2.087	391.6	2498.8	391.7	2274.1	2665.8	1.233	6.202	7.435
0.85	95.1	0.001040	1.972	398.5	2500.7	398.6	2269.8	2668.4	1.252	6.163	7.415
0.90	96.7	0.001041	1.869	405.1	2502.6	405.2	2265.7	2670.9	1.270	6.125	7.395
0.95	98.2	0.001042	1.777	411.3	25,044	411.4	2261.8	2673.2	1.287	6.090	7.377
1.00	99.6	0.001043	1.694	417.4	2506.1	417.5	2258.0	2675.5	1.303	6.057	7.360
1.0135	**100**	**0.001044**	**1.673**	**418.9**	**2506.5**	**419.0**	**2256.9**	**2676.0**	**1.307**	**6.048**	**7.355**

(continued)

Table A.2 (continued)

Sat. pressure bar	Sat. temp. °C	Specific volume m³/kg		Specific internal energy kJ/kg		Specific enthalpy kJ/kg			Specific entropy kJ/kgK		
p_{sat}	T_{sat}	Sat. liquid v_f	Sat. vapor v_g	Sat. liquid u_f	Sat. vapor u_g	Sat. liquid h_f	Evap. h_{fg}	Sat. vapor h_g	Sat. liquid s_f	Evap. s_{fg}	Sat. vapor s_g
1.1	102.3	0.001045	1.549	428.7	2509.2	428.8	2250.9	2679.7	1.333	5.994	7.327
1.2	104.8	0.001047	1.428	439.2	2512.1	439.3	2244.2	2683.5	1.361	5.937	7.298
1.3	107.1	0.001049	1.325	449.0	2514.8	449.1	2238.0	2687.1	1.387	5.884	7.271
1.4	109.3	0.001051	1.237	458.2	2517.3	458.4	2232.0	2690.4	1.411	5.835	7.246
1.5	111.4	0.001053	1.159	466.9	2519.7	467.1	2226.5	2693.6	1.434	5.789	7.223
1.6	113.3	0.001054	1.091	475.2	2521.9	475.4	2221.1	2696.5	1.455	5.747	7.202
1.7	115.2	0.001056	1.031	483.0	2523.9	483.2	2211.2	2699.2	1.475	5.706	7.181
1.8	116.9	0.001058	0.977	490.5	2525.9	490.7	2211.2	2701.8	1.494	5.668	7.162
1.9	118.6	0.001059	0.929	497.6	2527.7	497.8	2206.5	2704.3	1.513	5.631	7.144
2.0	120.2	0.001061	0.886	504.5	2529.5	504.7	2201.9	2706.7	1.530	5.597	7.127
2.1	121.8	0.001062	0.846	511.1	2531.2	511.3	2197.6	2708.9	1347	5.564	7.111
2.2	123.3	0.001063	0.810	517.4	2532.8	517.6	2193.4	2711.0	1.563	5.532	7.095
2.3	124.7	0.001065	0.777	523.5	2534.3	523.7	2189.3	2713.1	1378	5302	7.080
2.4	126.1	0.001066	0.747	529.4	2535.8	529.6	2185.4	2715.0	1.593	5.473	7.066
2.5	127.4	0.001067	0.719	535.1	2537.2	535.4	2118.5	2716.9	1.607	5.446	7.053
2.6	128.7	0.001069	0.693	540.6	2538.6	540.9	2177.8	2718.7	1.621	5.419	7.040
2.7	130.0	0.001070	0.669	546.0	2539.9	546.3	2174.2	2720.5	1.634	5.393	7.027
2.8	131.2	0.001071	0.646	551.2	2541.2	551.4	2170.7	2722.1	1.647	5.368	7.015

(continued)

Table A.2 (continued)

Sat. pressure bar	Sat. temp. °C	Specific volume m³/kg		Specific internal energy kJ/kg		Specific enthalpy kJ/kg			Specific entropy kJ/kgK		
p_{sat}	T_{sat}	Sat. liquid v_f	Sat. vapor v_g	Sat. liquid u_f	Sat. vapor u_g	Sat. liquid h_f	Evap. h_{fg}	Sat. vapor h_g	Sat. liquid s_f	Evap. s_{fg}	Sat. vapor s_g
2.9	132.4	0.001072	0.625	556.2	2542.4	556.5	2167.3	2723.8	1.660	5.343	7.003
3.0	133.5	0.001073	0.606	561.1	2543.6	561.6	2163.8	2725.3	1.672	5.320	6.992
3.1	134.7	0.001074	0.588	565.9	2544.7	566.3	2160.6	2726.8	1.684	5.297	6.981
3.2	135.8	0.001075	0.570	570.6	2545.8	571.0	2157.3	2728.3	1.695	5.275	6.970
3.3	136.8	0.001076	0.554	575.2	2546.9	575.5	2154.2	2729.7	1.706	5.254	6.960
3.4	137.9	0.001078	0.539	579.6	2547.9	580.0	2151.1	2731.1	1.717	5.233	6.950
3.5	138.9	0.001079	0.524	583.9	2548.9	584.3	2148.1	2732.4	1.728	5.213	6.941
3.6	140.0	0.001080	0.511	588.2	2549.9	588.6	2145.1	2733.7	1.738	5.193	6.931
3.7	140.8	0.001081	0.498	592.4	2550.9	592.8	2142.2	2735.0	1.748	5.174	6.922
3.8	141.8	0.001082	0.485	596.4	2551.8	596.8	2139.4	2736.2	1.758	5.155	6.913
3.9	142.7	0.001083	0.474	600.4	2552.7	600.8	2136.6	2737.4	1.767	5.137	6.904
4.0	143.6	0.001084	0.463	604.3	2553.6	604.7	2133.8	2738.5	1.777	5.119	6.896
4.1	144.5	0.001085	0.452	608.1	2554.4	608.6	2131.1	2739.7	1.786	5.102	6.888
4.2	145.4	0.001286	0.442	611.9	2555.3	612.4	2128.4	2740.8	1.795	5.085	6.880
4.3	146.3	0.001086	0.432	615.6	2556.1	616.1	2125.8	2741.9	1.804	5.068	6.872
4.4	147.1	0.001087	0.423	619.2	2456.9	619.7	2123.2	2742.9	1.812	5.052	6.684
4.5	147.9	0.001088	0.414	622.8	2557.6	623.3	2120.6	2743.9	1.821	5.036	6.857
4.6	148.7	0.001089	0.406	626.3	2558.4	626.8	2118.2	2744.9	1.829	5.020	6.849

(continued)

Table A.2 (continued)

Sat. pressure bar p_{sat}	Sat. temp. °C T_{sat}	Specific volume m³/kg		Specific internal energy kJ/kg		Specific enthalpy kJ/kg			Specific entropy kJ/kgK		
		Sat. liquid v_f	Sat. vapor v_g	Sat. liquid u_f	Sat. vapor u_g	Sat. liquid h_f	Evap. h_{fg}	Sat. vapor h_g	Sat. liquid s_f	Evap. s_{fg}	Sat. vapor s_g
4.7	149.5	0.001090	0.397	629.7	2559.1	63,012	2115.7	2745.9	1.837	5.005	6.842
4.8	150.3	0.001091	0.390	633.1	2559.8	633.6	2113.2	2746.8	1.845	4.990	6.835
4.9	151.1	0.001092	0.382	636.4	2560.6	636.9	2110.8	2747.8	1.853	4.975	6.828
5.0	151.9	0.001093	0.375	639.7	2561.2	640.2	2108.5	2748.7	1.861	4.961	6.821
5.2	153.3	0.001094	0.361	646.1	2562.6	646.7	2103.8	2750.5	1.876	4.932	6.808
5.4	154.8	0.001096	0.349	652.3	2563.9	652.9	2099.3	2752.1	1.890	4.905	6.795
5.5	155.5	0.001096	0.343	655.3	2564.5	655.9	2097.0	2752.9	1.897	4.892	6.789
5.6	156.2	0.001097	0.337	658.3	2565.1	658.9	2094.8	2753.8	1.904	4.879	6.783
5.8	157.5	0.001099	0.326	664.2	2566.3	664.8	2090.5	2755.3	1.918	4.853	6.771
6.0	158.9	0.001101	0.316	669.9	2567.4	670.6	2086.3	2756.8	1.931	4.829	6.760
6.2	160.1	0.001002	0.306	675.5	2568.5	676.2	2082.1	2758.3	1.944	4.805	6.749
6.4	161.4	0.001104	0.297	680.9	2569.6	681.6	2078.0	2759.6	1.956	4.782	6.738
6.5	162.0	0.001104	0.293	683.6	2570.1	684.3	2076.0	2760.3	1.963	4.770	6.733
6.6	162.6	0.001105	0.288	686.2	2570.6	686.9	2074.0	2761.0	1.969	4.759	6.728
6.8	163.8	0.001107	0.281	691.4	2571.6	692.1	2070.1	2762.2	1.981	4.737	6.718
7.0	165.0	0.001108	0.273	696.4	2572.5	697.2	2066.3	2763.5	1.992	4.716	6.708
7.2	166.1	0.001109	0.266	701.4	2573.4	702.2	2062.5	2764.7	2.004	4.695	6.699
7.4	167.2	0.001111	0.259	706.2	2574.3	707.1	2058.8	2765.9	2.014	4.675	6.689

(continued)

Table A.2 (continued)

Sat. pressure bar	Sat. temp. °C	Specific volume m³/kg		Specific internal energy kJ/kg		Specific enthalpy kJ/kg			Specific entropy kJ/kgK		
p_{sat}	T_{sat}	Sat. liquid v_f	Sat. vapor v_g	Sat. liquid u_f	Sat. vapor u_g	Sat. liquid h_f	Evap. h_{fg}	Sat. vapor h_g	Sat. liquid s_f	Evap. s_{fg}	Sat. vapor s_g
7.5	167.8	0.001112	0.256	708.6	2574.7	709.5	2057.0	2766.5	2.020	4.665	6.685
7.6	168.3	0.001112	0.252	711.0	2575.2	711.8	2055.2	2767.0	2.025	4.655	6.-680
7.8	169.4	0.001113	0.246	715.7	2576.0	716.5	2051.6	2768.1	2.036	4.635	6.671
8.0	170.4	0.001115	0.240	720.2	2576.8	721.1	2048.0	2769.1	2.046	4.617	6.663
8.2	171.5	0.001116	0.235	724.7	2577.6	725.6	2044.6	2770.2	2.056	4.598	6.654
8.4	172.5	0.001117	0.230	729.1	2578.3	730.1	2041.1	2771.2	2.066	4.580	6.646
8.5	173.0	0.001118	0.227	731.3	2578.7	732.2	2039.4	2771.6	2.071	4.571	6.642
8.6	173.5	0.001119	0.225	733.4	2579.1	734.4	2037.7	2772.1	2.076	4.562	6.638
8.8	174.4	0.001120	0.220	737.7	2579.8	738.6	2034.4	2773.0	2.085	4.545	6.630
9.0	175.4	0.001121	0.215	741.8	2580.5	742.8	2031.1	2773.9	2.095	4.528	6.623
9.2	176.3	0.001122	0.211	745.9	2581.1	747.0	2027.8	2774.8	2.104	4.511	6.615
9.4	177.2	0.001124	0.206	750.0	2581.8	751.0	2024.7	2775.7	2.113	4.495	6.608
9.5	177.7	0.001124	0.204	751.9	2582.1	753.0	2023.1	2776.1	2.117	4.487	6.604
9.6	178.1	0.001125	0.202	753.9	2582.4	755.0	2021.5	2776.5	2.122	4.479	6.601
9.8	179.0	0.001126	0.198	757.8	2583.1	758.9	2018.4	2777.3	2.130	4.463	6.593
10.0	179.9	0.001127	0.194	761.7	2583.6	762.8	2015.3	2778.1	2.139	4.448	6.587
10.5	182.0	0.001130	0.186	771.1	2585.1	772.2	2007.7	2779.9	2.159	4.411	6.570
11.0	184.1	0.001133	0.178	780.1	2586.4	781.3	2000.4	2781.7	2.179	4.374	6.553

(continued)

Table A.2 (continued)

Sat. pressure bar p_{sat}	Sat. temp. °C T_{sat}	Specific volume m³/kg		Specific internal energy kJ/kg		Specific enthalpy kJ/kg			Specific entropy kJ/kgK		
		Sat. liquid v_f	Sat. vapor v_g	Sat. liquid u_f	Sat. vapor u_g	Sat. liquid h_f	Evap. h_{fg}	Sat. vapor h_g	Sat. liquid s_f	Evap. s_{fg}	Sat. vapor s_g
11.5	186.1	0.001136	0.170	778.8	2587.6	790.1	1993.2	2783.3	2.198	4.340	6.538
12.0	188.0	0.001139	0163	797.3	2588.8	798.6	1986.2	2784.8	2.217	4.306	6.533
12.5	189.8	0.001141	0.157	805.4	2589.9	806.8	1979.4	2786.2	2.243	4.275	6.509
13.0	191.6	0.001144	0.151	813.4	2591.0	314.9	1972.7	2787.6	2.251	4.244	6.495
13.5	193.4	0.001146	0.146	821.2	2591.9	822.6	1966.2	2788.8	2.268	4.214	6.482
14.0	195.0	0.001149	0.141	828.7	2592.8	830.3	1959.7	2790.0	2.284	4,185	6.469
14.5	196.7	0.001151	0.136	836.0	2593.7	837.6	1953.5	27,911	2.300	4.157	6.457
15.0	198.3	0.001154	0.132	843.2	2594.5	844.9	1947.3	2792.2	2.315	4.130	6.445
15.5	199.9	0.001156	0.128	850.1	2595.3	851.9	1941.2	2793.1	2.330	4.103	6.433
16.0	201.4	0.001169	0.124	856.9	2596.0	858.8	1935.2	2794.0	2.344	4.078	6.422
16.5	202.9	0.001161	0.120	863.6	2596.6	865.5	1929.4	2794.9	2.358	4.053	6.411
17.0	204.3	0.001163	0.117	870.1	2597.3	872.1	1923.6	2795.7	2.372	4.028	6.400
17.5	205.8	0.001166	0.133	876.5	2597.8	878.5	1917.9	2796.4	2.385	4.005	6.390
18.0	207.2	0.001168	0.110	882.7	2598.4	884.8	1912.4	2797.2	2.398	3.981	6.379
18.5	208.5	0.001170	0.108	888.8	2598.9	891.0	1906.8	2797.8	2.411	3.958	6.369
19.0	209.8	0.001172	0.105	894.8	2599.4	897.0	1901.4	2798.4	2.423	3.936	3.359
19.5	211.1	0.001175	0.102	900.7	2599.9	903.0	1896.0	2799.0	2.435	3.915	6.350
20.0	212.4	0.001177	0.0996	906.4	2600.3	908.8	1890.7	2799.5	2.447	3.894	6.341

(continued)

Table A.2 (continued)

Sat. pressure bar p_{sat}	Sat. temp. °C T_{sat}	Specific volume m³/kg Sat. liquid v_f	Sat. vapor v_g	Specific internal energy kJ/kg Sat. liquid u_f	Sat. vapor u_g	Specific enthalpy kJ/kg Sat. liquid h_f	Evap. h_{fg}	Sat. vapor h_g	Specific entropy kJ/kgK Sat. liquid s_f	Evap. s_{fg}	Sat. vapor s_g
21.0	214.9	0.001181	0.0950	917.7	2601.0	920.2	1880.3	2800.5	2.470	3.852	6.323
22.0	217.3	0.001185	0.0907	928.5	2601.7	931.1	1870.2	2801.3	2.493	3.813	6.306
23.6	219.6	0.001189	0.0869	939.0	2602.3	941.8	1860.2	2802.0	2.514	3.775	6.289
24.0	221.8	0.001193	0.0833	949.2	2602.8	952.1	1850.5	2802.6	2.535	3.738	6.273
25.0	224.0	0.001197	0.0801	959.1	2603.1	962.1	1841.0	2803.1	2.555	3.703	6.258
26.0	226.1	0.001201	0.0769	968.7	2603.5	971.9	1831.6	2803.5	2.574	3.669	6.243
27.0	228.1	0.001205	0.0741	978.1	2603.7	981.3	1822.4	2803.8	2.593	3.635	6.228
28.0	230.1	0.001209	0.0715	987.2	2603.9	990.6	1813.4	2804.0	2.611	3.603	6.214
29.0	232.0	0.001213	0.0690	996.1	2604.0	999.6	1804.5	2804.1	2.628	3.572	6.200
30.0	233.9	0.001217	0.0667	1004.8	2604.1	1008.4	1795.7	2804.2	2.646	3.541	6.187
31.0	235.7	0.001220	00.645	1013.3	2604.1	1017.0	1787.1	2804.1	2.662	3.512	6.174
32.0	237.5	0.001224	0.0625	1021.6	2604.1	1025.5	1778.6	2804.1	2.679	3.483	6161
33.0	239.2	0.001227	0.0606	1029.7	2604.0	1033.7	1770.2	2803.9	2.695	3.454	6.149
34.0	240.9	0.001231	0.0588	1037.6	2603.9	1041.8	1761.9	2803.7	2.710	3.427	6.137
35.0	242.6	0.001235	0.0571	1045.4	2603.7	1049.7	1753.7	2803.4	2.725	3.400	6.125
36.0	244.2	0.001238	0.0555	1053.1	2603.5	1057.5	1745.6	2803.1	2.740	3.374	6.114
37.0	245.8	0.001242	0.0539	1060.6	2603.3	1065.2	1737.6	2802.8	2.755	3.348	6.103
38.0	247.4	0.001245	0.0523	1068.0	2603.0	1072.7	1729.7	2802.4	2.769	3.323	6.092

(continued)

Table A.2 (continued)

p_{sat}	T_{sat}	v_f	v_g	u_f	u_g	h_f	h_{fg}	h_g	s_f	s_{fg}	s_g
39.0	248.9	0.001249	0.0511	1075.2	2602.6	1080.1	1721.8	2801.9	2.783	3.298	6.081
40.0	250.4	0.001252	0.0498	1082.3	2602.3	1087.3	1714.1	2801.4	2.796	3.274	6.070
42.0	253.3	0.001259	0.0473	1096.2	2601.5	1101.5	1698.8	2800.3	2.823	3.227	6.050
44.0	256.1	0.001266	0.0451	1109.7	2600.6	1115.2	1638.8	2799.0	2.849	3.181	6.030
45.0	257.5	0.001269	0.0441	1116.2	2600.1	1121.9	1676.4	2798.3	2.861	3.159	6.020
46.0	258.8	0.001273	0.0431	1122.7	2599.5	1128.6	1669.0	2797.6	2.873	3.137	6.010
48.0	261.4	0.001279	0.0412	1135.4	2598.4	1141.5	1654.5	2796.0	2.897	3.095	5.992
50.0	264.0	0.001286	0.0394	1147.8	2597.1	1154.2	1640.1	2794.3	2.920	3.053	5.973
52.0	266.5	0.001293	0.0378	1159.9	2595.8	1166.6	1626.0	2792.6	2.943	3.013	5.956
54.0	268.8	0.001299	0.0363	1171.7	2594.4	1178.7	1612.0	2790.7	2.965	2.974	5.939
55.0	270.0	0.001302	0.0357	1177.4	2593.7	1184.5	1605.1	2789.6	2.975	2.955	5.930
56.0	271.2	0.001306	0.0350	1183.2	2592.9	1190.5	1598.2	2788.6	2.986	2.936	5.922
58.0	273.4	0.001312	0.0377	1194.4	2591.3	1202.0	1584.5	2786.5	3.006	2.899	5.905
60.0	275.6	0.001319	0.0324	1205.4	2589.7	1213.3	1571.0	2784.3	3.027	2.862	5.889
62.0	277.8	0.001325	0.0313	1216.3	2588.0	1224.5	1557.6	2782.1	3.046	2.827	5.873
64.0	279.9	0.001332	0.0302	1226.9	2586.2	1235.4	1544.3	2779.7	3.066	2.792	5.858
65.0	280.8	0.001335	0.0297	1232.0	2585.3	1240.7	1537.9	2778.6	3.075	2.775	5.850
66.0	281.9	0.001338	0.0292	1237.3	2584.4	1246.1	1531.1	2777.2	3.085	2.758	5.843

(continued)

Table A.2 (continued)

Sat. pressure bar	Sat. temp. °C	Specific volume m³/kg		Specific internal energy kJ/kg		Specific enthalpy kJ/kg			Specific entropy kJ/kgK		
p_{sat}	T_{sat}	Sat. liquid v_f	Sat. vapor v_g	Sat. liquid u_f	Sat. vapor u_g	Sat. liquid h_f	Evap. h_{fg}	Sat. vapor h_g	Sat. liquid s_f	Evap. s_{fg}	Sat. vapor s_g
68.0	283.9	0.001345	0.0283	1247.5	2582.5	1256.6	1518.1	2774.7	3.103	2.725	5.828
70.0	285.9	0.001351	0.0274	1257.5	2580.5	1267.0	1505.1	2772.1	3.121	2.692	5.813
72.0	287.8	0.001358	0.0265	1267.4	2578.5	1277.2	1492.2	2769.4	3.139	2.660	5.799
74.0	289.7	0.001364	0.0257	1277.2	2576.4	1287.3	1479.4	2766.7	3.156	2.629	5.785
75.0	290.6	0.001368	0.0253	1282.0	2575.3	1292.2	1473.1	2765.3	3.165	2.613	5.778
76.0	291.5	0.001371	0.0249	1286.8	2574.3	1297.2	1466.6	2763.8	3.174	2.597	5.771
78.0	293.3	0.001378	0.0242	1296.2	2572.1	1307.0	1453.9	2760.9	3.190	2.567	5.757
80.0	295.1	0.001384	0.0235	1305.6	2569.8	1316.6	1441.3	2758.0	3.207	2.536	5.743
85.0	299.3	0.001401	0.0219	1328.4	2564.0	1340.3	1410.0	2750.3	3.247	2.463	5.710
90.0	303.4	0.001418	0.0205	1350.5	2557.8	1363.2	1378.9	2742.1	3.286	2.391	5.677
95.0	307.3	0.001435	0.0192	1372.0	2551.2	1385.6	1348.0	2733.6	3.323	2.322	5.645
100.00	311.1	0.001452	0.0180	1393.0	2544.4	1407.6	1317.1	2724.7	3.360	2.254	5.614
105.0	314.7	0.001470	0.0170	1413.6	2537.7	1429.0	1286.4	2715.4	3.395	2.188	5.583
110.0	318.2	0.001489	0.0160	1433.7	2529.8	1450.1	1255.5	2705.6	3.430	2.123	5.553
115.0	321.5	0.001507	0.0151	1453.5	2522.2	1470.8	1224.6	2695.4	3.463	2.059	5.522
120.0	324.8	0.001527	0.0143	1473.0	2513.7	1491.3	1193.6	2684.9	3.496	1.996	5.492
125.0	327.9	0.001547	0.0135	1492.1	2505.1	1511.5	1162.2	2673.7	3.528	1.934	5.462
130.0	330.9	0.001567	0.0128	1511.1	2496.1	1531.5	1130.7	2662.2	3.561	1.871	5.432

(continued)

Table A.2 (continued)

Sat. pressure bar	Sat. temp. °C	Specific volume m³/kg		Specific internal energy kJ/kg		Specific enthalpy kJ/kg			Specific entropy kJ/kgK		
p_{sat}	T_{sat}	Sat. liquid v_f	Sat. vapor v_g	Sat. liquid u_f	Sat. vapor u_g	Sat. liquid h_f	Evap. h_{fg}	Sat. vapor h_g	Sat. liquid s_f	Evap. s_{fg}	Sat. vapor s_g
135.0	333.9	0.001588	0.0121	1529.9	2486.6	1551.4	1098.8	2650.2	3.592	1.810	5.402
140.0	336.8	0.001611	0.0115	1548.6	2476.8	1571.1	1066.5	2637.6	3.623	1.749	5.372
145.0	339.5	0.001634	0.0109	1567.1	2466.4	1590.9	1033.5	2624.4	3.654	1.687	5.341
150.0	342.2	0.001658	0.0103	1585.6	2455.5	1610.5	1000.0	2610.5	3.685	1.625	5.310
155.0	344.9	0.001684	0.00981	1604.1	2443.9	1630.3	965.7	2596.0	3.715	1.563	5.278
160.0	347.4	0.001711	0.00931	1622.7	2431.7	1650.1	930.6	7.580.6	3.746	1.499	5.245
165.0	349.9	0.001740	0.00883	1641.4	2418.8	1670.1	894.3	2564.4	3.777	1.435	5.212
170.0	352.4	0.001770	0.00836	1660.2	2405.0	1690.3	856.9	2547.2	3.808	1.370	5.178
175.0	354.7	0.001804	0.00793	1679.4	2390.2	1711.0	817.8	2528.8	3.839	1.302	5.141
180.0	357.1	0.001840	0.00749	1698.9	2374.3	1732.0	777.1	2509.1	3.871	1.233	5.104
185.0	359.3	0.001880	0.00708	1719.1	2357.0	1753.9	733.9	2487.8	3.905	1.160	5.065
190.0	361.5	0.001924	0.00666	1739.9	2338.1	1776.5	688.0	2464.5	3.939	1.084	5.023
195.0	363.7	0.001976	0.00625	1762.0	2316.9	1800.6	638.2	2438.8	3.975	1.002	4.977
200.0	365.8	0.002036	0.00583	1785.6	2293.0	1826.3	583.4	2409.7	4.014	0.913	4.927
205.0	367.9	0.002110	0.00541	1811.8	2265.0	1855.0	520.8	2375.8	4.057	0.812	4.869
210.0	369.9	0.002207	0.00495	1842.1	2230.6	1888.4	446.2	2334.6	4.107	0.694	4.801
215.0	371.9	0.002358	0.00442	1882.3	2183.0	1933.0	344.9	2277.9	4.175	0.535	4.710
220.0	373.8	0.002742	0.00357	1961.9	2087.1	2022.2	143.4	2165.6	4.311	0.222	4.533

(continued)

Table A.2 (continued)

Sat. pressure bar	Sat. temp. °C	Specific volume m³/kg		Specific internal energy kJ/kg		Specific enthalpy kJ/kg			Specific entropy kJ/kgK		
p_{sat}	T_{sat}	Sat. liquid v_f	Sat. vapor v_g	Sat. liquid u_f	Sat. vapor u_g	Sat. liquid h_f	Evap. h_{fg}	Sat. vapor h_g	Sat. liquid s_f	Evap. s_{fg}	Sat. vapor s_g
221.2	374.15	0.003155	0.003155	2029.6	2029.6	2099.3	0	2099.3	4.4298	0	4.4298

Table A.3 Superheated steam table

T °C	v m³/kg	u kJ/kg	h kJ/kg	s kJ/kgK	v m³/kg	u kJ/kg	h kJ/kg	s kJ/kgK	v m³/kg	u kJ/kg	h kJ/kg	s kJ/kgK
	p = 0.1 bar (45.8 °C)				p = 0.5 bar (81.3 °C)				p = 1 bar (99.6 °C)			
Sat	14.674	2437.9	2584.7	8.1502	3.240	2483.9	2645.9	7.5939	1.6940	2506.1	2675.5	7.3594
50	14.869	2483.9	2592.6	8.1749								
100	17.196	2515.5	2687.5	8.4479	3.418	2511.6	2682.5	7.6947	1.6958	2506.7	2676.2	7.3614
150	19.512	2587.9	2783.0	8.6882	3.889	2585.6	2780.1	7.9401	1.9364	2582.8	2776.4	7.6134
200	21.825	2661.3	2879.5	8.9038	4.356	2659.9	2877.7	8.1580	2.172	2658.1	2875.3	7.8343
250	24.136	2736.0	2977.3	9.1002	4.820	2735.0	2976.0	8.3556	2.406	2733.7	2974.3	8.0333
300	26.445	2812.1	3075.5	9.2813	5.284	2811.3	3075.5	8.5373	2.639	2810.4	3074.3	8.2158
400	31.063	2968.9	3279.6	9.6077	6.209	2968.5	3278.9	8.8642	3.103	2967.9	3278.2	8.5435
500	35.679	3132.3	3489.1	9.8978	7.134	3132.0	3488.7	9.1546	3.565	3131.6	3488.1	8.8342
600	40.295	3302.5	3705.4	10.1608	8.057	3302.2	3705.1	9.4178	4.028	3301.9	3704.4	9.0976
700	44.911	3479.6	3928.7	10.4028	8.981	3479.4	3928.5	9.6599	4.490	3479.2	3928.2	9.3398
800	49.526	3663.8	4159.0	10.6281	9.904	3663.6	4158.0	9.8852	4.952	3663.5	4158.6	9.5652
900	54.141	3855.0	4396.4	10.8396	10.828	3854.9	4396.3	10.0967	5.414	3854.8	4396.1	9.7767
1000	58.757	4053.0	4640.6	11.0393	11.751	4052.9	4640.5	10.2964	5.875	4052.8	4640.3	9.9764
1100	63.372	4257.5	4891.2	11.2287	12.674	4257.4	4891.1	10.4859	6.337	4257.3	4891.0	10.1659
1200	67.987	4467.9	5147.8	11.4091	13.597	4467.8	5147.7	10.6662	6.799	4467.7	5147.6	10.3463
1300	72.602	4683.7	5409.7	11.5811	14.521	4683.6	5409.6	10.8382	7.260	4683.5	5409.5	10.5183

Table A.4 Superheated steam table

T °C	v m³/kg	u kJ/kg	h kJ/kg	s kJ/kgK	v m³/kg	u kJ/kg	h kJ/kg	s kJ/kgK	v m³/kg	u kJ/kg	h kJ/kg	S kJ/kgK
	p = 2 bar (120.2 °C)				p = 3 bar (133.5 °C)				p = 4 bar (143.6 °C)			
Sat	0.8857	2529.5	2706.7	7.1272	0.6058	2543.6	2725.3	6.9919	0.4625	2553.6	2738.6	6.8959
150	0.9596	2576.9	2768.8	7.2795	0.6339	2570.8	2761.0	7.0778	0.4708	2564.5	2752.8	6.9299
200	1.0803	2654.4	2870.5	7.5066	0.7163	2650.7	2865.6	7.3115	0.5342	2646.8	2860.5	7.1706
250	1.1988	2731.2	2971.0	7.7086	0.7964	2728.7	2967.6	7.5166	0.5951	2726.1	2964.2	7.3789
300	1.3162	2808.6	3071.8	7.8926	0.8753	2806.7	3069.3	7.7022	0.6548	2804.8	3066.8	7.5662
400	1.5493	2966.7	3276.6	8.2218	1.0315	2965.6	3275.0	8.0330	0.7726	2964.4	3273.4	7.8985
500	1.7814	3130.8	3487.1	8.5133	1.1867	3130.0	3486.0	8.3251	0.8893	3129.2	3484.9	8.1913
600	2.013	3301.4	3704.0	8.7770	1.3414	3300.8	3703.2	8.5892	1.0055	3300.2	3702.4	8.4558
700	2.244	3478.8	3927.6	9.0194	1.4957	3478.4	3927.1	8.8319	1.1215	3477.9	3926.5	8.6987
800	2.475	3663.1	4158.2	9.2449	1.6499	3662.9	4157.8	9.0576	1.2312	3662.4	4157.3	8.9244
900	2.705	3854.5	4395.8	9.4566	1.8041	3854.2	4395.4	9.2692	1.3529	3853.9	4395.1	9.1362
1000	2.937	4052.5	4640.0	9.5663	1.9581	4052.3	4639.7	9.4690	1.4685	4052.0	4639.4	9.3360
1100	3.168	4257.0	4890.7	9.8458	2.1121	4256.8	4890.4	9.6585	1.5840	4256.5	4890.2	9.5256
1200	3.399	4467.5	5147.5	10.0262	2.2661	4467.2	5147.1	9.8389	1.6996	4467.0	5146.8	9.7060
1300	3.630	4683.2	5409.3	10.1982	2.4201	4683.0	5409.0	10.0110	1.8151	4682.8	5408.8	9.8780

Table A.5 Superheated steam table

T °C	v m³/kg	u kJ/kg	h kJ/kg	s kJ/kgK	v m³/kg	u kJ/kg	h kJ/kg	s kJ/kgK	v m³/kg	u kJ/kg	h kJ/kg	s kJ/kgK
	p = 5 bar (151.9 °C)				p = 6 bar (158.9 °C)				p = 8 bar (170.4 °C)			
Sat	0.3749	2561.2	2748.7	6.8213	0.3157	2567.4	2756.8	6.7600	0.2404	2576.8	2769.1	6.6628
200	0.4249	2642.9	2855.4	7.0592	0.3520	2638.9	2850.1	6.9665	0.2608	2630.6	2839.3	6.8158
250	0.4744	2723.5	2960.7	7.2709	0.3938	2720.9	2957.2	7.1816	0.2931	2715.5	2950.0	7.0384
300	0.5226	2802.9	3064.2	7.4599	0.4344	2801.0	3061.6	7.3724	0.3241	2797.2	3056.5	7.2328
350	0.5701	2882.6	3167.7	7.6329	0.4742	2881.2	3165.7	7.5464	0.3544	2878.2	3161.7	7.4089
400	0.6173	2963.2	3271.9	7.7938	0.5137	2962.1	3270.3	7.7079	0.3843	2959.7	3267.1	7.5716
500	0.7109	3128.4	3483.9	8.0873	0.5920	3127.6	3482.8	8.0021	0.4433	3126.0	3480.6	7.8673
600	0.8041	3299.6	3701.7	8.3522	0.6697	3299.1	3700.9	8.2674	0.5018	3297.9	3699.4	8.1333
700	0.8969	3477.5	3925.9	8.5952	0.7472	3477.0	3925.3	8.5107	0.5601	3476.2	3924.2	8.3770
800	0.9896	3662.1	4156.9	8.8211	0.8215	3661.8	4156.5	8.7367	0.6181	3661.1	4155.6	8.6033
900	1.0822	3853.6	4394.7	9.0329	0.9017	3853.4	4394.4	8.9486	0.6761	3852.8	4393.7	8.8153
1000	1.1747	4051.8	4639.1	9.2328	0.9788	4051.5	4638.8	9.1485	0.7340	4051.0	4638.2	9.0153
1100	1.2672	4256.3	4889.9	9.4224	1.0559	4256.1	4889.6	9.3381	0.7919	4255.6	4889.1	9.2050
1200	1.3596	4466.8	5146.6	9.6029	1.1330	4466.5	5146.3	9.5185	0.8497	4466.1	5145.9	9.3855
1300	1.4521	4682.5	5408.6	9.7749	1.2101	4682.3	5408.3	9.6906	0.9076	4681.8	5407.9	9.5575

Table A.6 Superheated steam table

T °C	v m³/kg	u kJ/kg	h kJ/kg	s kJ/kgK	v m³/kg	u kJ/kg	h kJ/kg	s kJ/kgK	v m³/kg	u kJ/kg	h kJ/kg	s kJ/kgK
	p = 10 bar (179.9 °C)				p = 12 bar (188 °C)				p = 14 bar (195 °C)			
Sat	0.19444	2583.6	2778.1	6.5865	0.16333	2588.8	2784.8	6.5233	0.14084	2592.8	2790.0	6.4693
200	0.2060	2621.9	2827.9	6.6940	0.16930	2612.8	2815.9	6.5898	0.14302	2603.1	2803.3	6.4975
250	0.2327	2709.9	2942.6	6.9247	0.19234	2704.2	2935.0	6.8294	0.16350	2698.3	2927.2	6.7467
300	0.2579	2793.2	3051.2	7.1229	0.2138	2789.2	3045.8	7.0317	0.18228	2785.2	3040.4	6.9534
350	0.2825	2875.2	3157.7	7.3011	0.2345	2872.2	3153.6	7.2121	0.2003	2869.2	3149.5	7.1370
400	0.3066	2957.3	3263.9	7.4651	0.2548	2954.9	3260.7	7.3774	0.2178	2952.5	3257.5	7.3026
500	0.3541	3124.4	3478.5	7.7622	0.2946	3122.8	3476.3	7.6759	0.2521	3121.1	3474.1	7.6027
600	0.4011	3296.8	3697.9	8.0290	0.3339	3295.6	3696.3	7.9435	0.2860	3294.4	3694.8	7.8710
700	0.4478	3475.3	3923.1	8.2731	0.3729	3474.4	3922.0	8.1881	0.3195	3473.6	3920.8	8.1160
800	0.4943	3660.4	4154.7	8.4996	0.4118	3659.7	4153.8	8.4148	0.3528	3659.0	4153.0	8.3431
900	0.5407	3852.2	4392.9	8.7118	0.4505	3851.6	4392.2	8.6272	0.3861	3851.1	4391.5	8.5556
1000	0.5871	4050.5	4637.6	8.9119	0.4892	4050.0	4637.0	8.8274	0.4192	4049.5	4636.4	8.7559
1100	0.6335	4255.1	4888.6	9.1017	0.5278	4254.6	4888.0	9.0172	0.4524	4254.1	4887.5	8.9457
1200	0.6798	4465.6	5145.4	9.2822	0.5665	4465.1	5144.9	9.1977	0.4855	4464.7	5144.4	9.1262
1300	0.7261	4681.3	5407.4	9.4543	0.6051	4680.9	5407.0	9.3698	0.5186	4680.4	5406.5	9.2984

Table A.7 Superheated steam table

T °C	p = 16 bar (201.4 °C)				p = 18 bar (207.2 °C)				p = 20 bar (212.4 °C)			
	v m³/kg	u kJ/kg	h kJ/kg	s kJ/kgK	v m³/kg	u kJ/kg	h kJ/kg	s kJ/kgK	v m³/kg	u kJ/kg	h kJ/kg	s kJ/kgK
Sat	0.12380	2596.0	2794.0	6.4218	0.11042	2598.4	2797.1	6.3794	0.09963	2600.3	2799.5	6.3409
225	0.13287	2644.7	2857.3	6.5518	0.11673	2636.6	2846.7	6.4808	0.10377	2628.3	2835.8	6.4147
250	0.14184	2692.3	2919.2	6.6732	0.12497	2686.0	2911.0	6.6066	0.11144	2679.6	2902.5	6.5453
300	0.15862	2781.1	3034.8	6.8844	0.14021	2776.9	3029.2	6.8226	0.12547	2772.6	3023.5	6.7664
350	0.17456	2866.1	3145.4	7.0694	0.15457	2863.0	3141.2	7.0100	0.13857	2859.8	3137.0	6.9563
400	0.19005	2950.1	3254.2	7.2374	0.16847	2947.7	3250.9	7.1794	0.15120	2945.2	3247.6	7.1271
500	0.2203	3119.5	3472.0	7.5390	0.19550	3117.9	3469.8	7.4825	0.17568	3116.2	3467.6	7.4317
600	0.2500	3293.3	3693.2	7.8080	0.2220	3292.1	3691.7	7.7523	0.19960	3290.9	3690.1	7.7024
700	0.2794	3472.7	3919.7	8.0535	0.2482	3471.8	3971.8	7.9983	0.2232	3470.9	3917.4	7.9487
800	0.3086	3658.3	4152.1	8.2808	0.2742	3657.6	4151.2	8.2258	0.2467	3657.0	4150.3	8.1765
900	0.3377	3850.5	4390.8	8.4935	0.3001	3849.9	4390.1	8.4386	0.2700	3849.3	4389.4	8.3895
1000	0.3668	4049.0	4635.8	8.6938	0.3260	4048.5	4635.2	8.6391	0.2933	4048.0	4634.6	8.5901
1100	0.3958	4253.7	4887.0	8.8837	0.3518	4253.2	4886.4	8.8290	0.3166	4252.7	4885.9	8.7800
1200	0.4248	4464.2	5143.9	9.0643	0.3776	4463.7	5143.4	9.0096	0.3398	4463.3	5142.9	8.9607
1300	0.4538	4679.9	5406.0	9.2364	0.4034	4679.5	5405.6	9.1818	0.3631	4679.0	5405.1	9.1329

Table A.8 Superheated steam table

T °C	p = 25 bar (224 °C)				p = 30 bar (233.9 °C)				p = 35 bar (242.6 °C)			
	v m³/kg	u kJ/kg	h kJ/kg	s kJ/kgK	v m³/kg	u kJ/kg	h kJ/kg	s kJ/kgK	v m³/kg	u kJ/kg	h kJ/kg	s kJ/kgK
Sat	0.07998	2603.1	2803.1	6.2575	0.06668	2604.1	2804.2	6.1869	0.05707	2603.7	2803.4	6.1253
225	0.08027	2605.6	2806.3	6.2639								
250	0.08700	2662.6	2880.1	6.4085	0.07058	2644.0	2855.8	6.2872	0.05872	2623.7	2829.2	6.1749
300	0.09890	2761.6	3008.8	6.6438	0.08114	2750.1	2993.5	6.5390	0.06842	2738.0	2977.5	6.4461
350	0.10976	2851.9	3126.3	6.8403	0.09053	2843.7	3115.3	6.7428	0.07678	2835.3	3104.0	6.6579
400	0.12010	2939.1	3239.3	7.0148	0.09936	2932.8	3230.9	6.9212	0.08453	2926.4	3222.3	6.8405
450	0.13014	3025.5	3350.8	7.1746	0.10787	3020.4	3344.0	7.0834	0.09196	3015.3	3337.2	7.0052
500	0.13993	3112.1	3462.1	7.3234	0.11619	3108.0	3456.5	7.2338	0.09918	3103.0	3450.9	7.1572
600	0.15930	3288.0	3686.3	7.5960	0.13243	3285.0	3682.3	7.5085	0.11324	3282.1	3678.4	7.4339
700	0.17832	3468.7	3914.5	7.8435	0.14838	3466.5	3911.7	7.7571	0.12699	3464.3	3908.8	7.6837
800	0.19716	3655.3	4148.2	8.0720	0.16414	3653.5	4145.9	7.9862	0.14056	3651.8	4143.7	7.9134
900	0.21590	3847.9	4387.6	8.2853	0.17980	3846.5	4385.9	8.1999	0.15402	3845.0	4384.1	8.1276
1000	0.2346	4046.7	4633.1	8.4861	0.19541	4045.4	4631.6	8.4009	0.16743	4044.1	4630.1	8.3288
1100	0.2532	4251.5	4884.6	8.6762	0.21098	4250.3	4883.3	8.5912	0.18080	4249.2	4881.9	8.5192
1200	0.2718	4462.1	5141.7	8.8569	0.22652	4460.9	5140.5	8.7720	0.19415	4459.8	5139.3	8.7000
1300	0.2905	4677.8	5404.0	9.0291	0.24206	4676.6	5402.8	8.9442	0.20749	4675.5	5401.7	8.8723

Table A.9 Superheated steam table

T °C	$v\,\text{m}^3/\text{kg}$	u kJ/kg	h kJ/kg	s kJ/kgK	$v\,\text{m}^3/\text{kg}$	u kJ/kg	h kJ/kg	s kJ/kgK	$v\,\text{m}^3/\text{kg}$	u kJ/kg	h kJ/kg	s kJ/kgK
	$p = 40$ bar (250.4 °C)				$p = 45$ bar (257.5 °C)				$p = 50$ bar (264 °C)			
Sat	0.04978	2602.3	2801.4	6.0701	0.04406	2600.1	2798.3	6.0198	0.03944	2597.1	2794.3	5.9734
275	0.05457	2667.9	2886.2	6.2285	0.04730	2650.3	2863.2	6.1401	0.04141	2631.3	2838.3	6.0544
300	0.05884	2725.3	2960.7	6.3615	0.05135	2712.0	2943.1	6.2828	0.04532	2698.0	2924.5	6.2084
350	0.06645	2826.7	3092.5	6.5821	0.05840	2817.8	3080.6	6.5131	0.05194	2808.7	3068.4	6.4493
400	0.07341	2919.9	3213.6	6.7690	0.06475	2913.3	3204.7	6.7047	0.05781	2906.6	3195.7	6.6459
450	0.08002	3010.2	3330.3	6.9363	0.07074	3005.0	3323.3	6.8746	0.06330	2999.7	3316.2	6.8186
500	0.08643	3099.5	3445.3	7.0901	0.07651	3095.3	3439.6	7.0301	0.06857	3091.0	3433.8	6.9759
600	0.09885	3279.1	3674.4	7.3688	0.08765	3276.0	3670.5	7.3110	0.07869	3273.0	3666.5	7.2589
700	0.11095	3462.1	3905.9	7.6198	0.09847	3459.9	3903.0	7.5631	0.08849	3457.6	3900.1	7.5122
800	0.12287	3650.0	4141.5	7.8502	0.10911	3648.3	4139.3	7.7942	0.09811	3646.6	4137.1	7.7440
900	0.13469	3843.6	4382.3	8.0647	0.11965	3842.2	4380.6	8.0091	0.10762	3840.7	4378.8	7.9593
1000	0.14645	4042.9	4628.7	8.2662	0.13013	4041.6	4627.2	8.2108	0.11707	4040.4	4625.7	8.1612
1100	0.15817	4248.0	4880.6	8.4567	0.14056	4246.8	4879.3	8.4015	0.12648	4245.6	4878.0	8.3520
1200	0.16987	4458.6	5138.1	8.6376	0.15098	4457.5	5136.9	8.5825	0.13587	4456.3	5135.7	8.5331
1300	0.18156	4674.3	5400.5	8.8100	0.16139	4673.1	5399.4	8.7549	0.14526	4672.0	5398.2	8.7055

Table A.10 Superheated steam table

T °C	v m³/kg	u kJ/kg	h kJ/kg	s kJ/kgK	v m³/kg	u kJ/kg	h kJ/kg	s kJ/kgK	v m³/kg	u kJ/kg	h kJ/kg	s kJ/kgK
	$p = 60$ bar (275.6 °C)				$p = 70$ bar (285.9 °C)				$p = 80$ bar (295.1 °C)			
Sat	0.03244	2589.7	2784.3	5.8892	0.02737	2580.5	2772.1	5.8133	0.02352	2569.0	2758.0	5.7432
300	0.03616	2667.2	2884.2	6.0674	0.02947	2632.2	2838.4	5.9305	0.02426	2590.9	2785.0	5.7906
350	0.04223	2789.6	3043.0	6.3335	0.03524	2769.4	3016.0	6.2283	0.02995	2747.7	2987.3	6.1301
400	0.04739	2892.0	3177.2'	6.5408'	0.03993	2878.6	3158.1	6.4478	0.03432	2863.8	3138.3	6.3634
450	0.05214	2988.9	3301.8	6.7193	0.04416	2978.0	3287.1	6.6327	0.03817	2966.7	3272.0	6.5551
500	0.05665	3082.2	3422.2	6.8803	0.04814	3073.4	3410.3	6.7980	0.04175	3064.3	3398.3	6.7240
550	0.06101	3174.6	3540.6	7.0288	0.05195	3167.2	3530.9	6.9486	0.04516	3159.8	3521.0	6.8778
600	0.06525	3266.9	3658.4	7.1677	0.05565	3260.7	3650.3	7.0894	0.04845	3254.4	3642.0	7.0206
700	0.07352	3453.1	3894.2	7.4234	0.06283	3448.5	3888.3	7.3476	0.05481	3443.9	3882.4	7.2812
800	0.08160	3643.1	4132.7	7.6566	0.06981	3639.5	4128.2	7.5822	0.06097	3636.0	4123.8	7.5173
900	0.08958	3837.8	4375.3	7.8727	0.07669	3835.0	4371.8	7.7991	0.06702	3832.1	4368.3	7.7351
1000	0.09749	4037.8	4622.7	8.0751	0.08350	4035.3	4619.8	8.0020	0.07301	4032.8	4616.9	7.9384
1100	0.10536	4243.3	4875.4	8.2661	0.09027	4240.9	4872.8	8.1933	0.07896	4238.6	4870.3	8.1300
1200	0.11321	4454.0	5133.3	8.4474	0.09703	4451.7	5130.9	8.3747	0.08489	4449.5	5128.5	8.3115
1300	0.12106	4669.6	5396.0	8.6199	0.10377	4667.3	5393.7	8.5475	0.09080	4665.0	5391.5	8.4842

Table A.11 Superheated steam table

T °C	p = 90 bar (303.4 °C)				p = 100 bar (311.1 °C)				p = 125 bar (327.9 °C)			
	v m³/kg	u kJ/kg	h kJ/kg	s kJ/kg K	v m³/kg	u kJ/kg	h kJ/kg	s kJ/kgK	v m³/kg	u kJ/kg	h kJ/kg	s kJ/kg K
Sat	0.02048	2557.8	2742.1	5.6112	0.018026	2544.4	2724.7	5.6141	0.013495	2505.1	2673.8	5.4624
325	0.02327	2646.6	2856.0	5.8712	0.019861	2610.4	2809.1	5.7568				
350	0.02580	2724.4	2956.6	6.0361	0.02242	2699.2	2923.4	5.9443	0.016126	2624.6	2826.2	5.7118
400	0.02993	2848.4	3117.8	6.2854	0.02641	2832.4	3096.5	6.2120	0.02000	2789.3	3039.3	6.0417
450	0.03350	2955.2	3256.6	6.4844	0.02975	2943.4	3240.9	6.4190	0.02299	2912.5	3199.8	6.2719
500	0.03677	3055.2	3386.1	6.6576	0.03279	3045.8	3373.7	6.5966	0.02560	3021.7	3341.8	6.4618
550	0.03987	3152.2	3511.0	6.8142	0.03564	3144.1	3500.9	6.7561	0.02801	3125.0	3475.2	6.6290
600	0.04285	3248.1	3633.7	6.9589	0.03837	3241.7	3625.3	6.9029	0.03029	3225.4	3604.0	6.7810
650	0.04574	3343.6	3755.3	7.0943	0.04101	3338.2	3748.2	7.0398	0.03248	3324.4	3730.4	6.9218
700	0.04857	3439.3	3876.5	7.2221	0.04358	3434.7	3870.5	7.1687	0.03460	3422.9	3855.3	7.0536
800	0.05409	3632.5	4119.3	7.4596	0.04859	3628.9	4114.8	7.4077	0.03869	3620.0	4103.6	7.2965
900	0.05950	3829.2	4364.8	7.6783	0.05349	3826.3	4361.2	7.6272	0.04267	3819.1	4352.5	7.5182
1000	0.06485	4030.3	4614.0	7.8821	0.05832	4027.8	4611.0	7.8315	0.04658	4021.6	4603.8	7.7237
1100	0.07016	4236.3	4867.7	8.0740	0.06312	4234.0	4865.1	8.0237	0.05045	4228.2	4858.8	7.9165
1200	0.07544	4447.2	5126.2	8.2556	0.06789	4444.9	5123.8	8.2055	0.05430	4439.3	5118.0	8.0937
1300	0.08072	4662.7	5389.2	8.4284	0.07265	4460.5	5387.0	8.3783	0.05813	4654.8	5381.4	8.2717

Table A.12 Superheated steam table

T °C	p = 150 bar (342.2 °C)				p = 175 bar (354.7 °C)				p = 200 bar (365.8 °C)			
	v m³/kg	u kJ/kg	h kJ/kg	s kJ/kg K	v m³/kg	u kJ/kg	h kJ/kg	s kJ/kg K	v m³/kg	u kJ/kg	h kJ/kg	s kJ/kg K
Sat	0.010337	2455.5	2610.5	5.3098	0.007920	2390.2	2528.8	5.1419	0.005834	2293.0	2409.7	4.9269
350	0.011470	2520.4	2692.4	5.4421								
400	0.015649	2740.7	2975.5	5.8811	0.012447	2685.0	2902.9	5.7213	0.009942	2619.3	2818.1	5.5540
450	0.018445	2879.5	3156.2	6.1404	0.015174	2844.2	3109.7	6.0184	0.012695	2806.2	3060.1	5.9017
500	0.02080	2996.6	3308.6	6.3443	0.017358	2970.3	3274.1	6.2383	0.014768	2942.9	3238.2	6.1401
550	0.02293	3104.7	3448.6	6.5199	0.019288	3083.9	3421.4	6.4230	0.016555	3062.4	3393.5	6.3348
600	0.02491	3208.6	3582.3	6.6776	0.02106	3191.5	3560.1	6.5866	0.018178	3174.0	3537.6	6.5048
650	0.02680	3310.3	3712.3	6.8224	0.02274	3296.0	3693.9	6.7357	0.019693	3281.4	3675.3	6.6582
700	0.02861	3410.9	3840.1	6.9572	0.02434	3324.6	3824.6	6.8736	0.02113	3386.4	3809.0	6.7993
800	0.03210	3610.9	4092.4	7.2040	0.02738	3601.8	4081.1	7.1244	0.02385	3592.7	4069.7	7.0544
900	0.03546	3811.9	4343.8	7.4279	0.03031	3804.7	4335.1	7.3507	0.02645	3797.5	4326.4	7.2830
1000	0.03875	4015.4	4596.6	7.6348	0.03316	4009.3	4589.5	7.5589	0.02897	4003.1	4582.5	7.4925
1100	0.04200	4222.6	4852.6	7.8283	0.03597	4216.9	4846.4	7.7531	0.03145	4211.3	4840.2	7.6874
1200	0.04523	4433.8	5112.3	8.0108	0.03876	4428.3	5106.6	7.9360	0.03391	4422.8	5101.0	7.8707
1300	0.04845	4649.1	5376.0	8.1840	0.04154	4643.5	5370.5	8.1093	0.03636	4638.0	5365.1	8.0442

Table A.13 Superheated steam table

T °C	v m³/kg	u kJ/kg	h kJ/kg	s kJ/kg K	v m³/kg	u kJ/kg	h kJ/kg	s kJ/kgK	v m³/kg	u kJ/kg	h kJ/kg	s kJ/kg K
	p = 250 bar				p = 300 bar				p = 350 bar			
375	0.001973	1798.7	1848.0	4.0320	0.0017892	1737.8	1791.5	3.9305	0.0017003	1702.9	1762.4	3.8722
400	0.006004	2430.1	2580.2	5.1418	0.002790	2067.4	2151.1	4.4728	0.002100	1914.1	1987.6	4.2126
425	0.007881	2609.2	2806.3	5.4723	0.005303	2455.1	2614.2	5.1504	0.003428	2253.4	2373.4	4.7747
450	0.009162	2720.7	2949.7	5.6744	0.006735	2619.3	2821.4	5.4424	0.004961	2498.7	2672.4	5.1962
500	0.011123	2884.3	3162.4	5.9592	0.008678	2820.7	3081.1	5.7905	0.006927	2751.9	2994.4	5.6282
550	0.012724	3017.5	3335.6	6.1765	0.010168	2970.3	3275.4	6.0342	0.008345	2921.0	3213.0	5.9026
600	0.014137	3137.9	3491.4	6.3602	0.011446	3100.5	3443.9	6.2331	0.009527	3062.0	3395.5	6.1179
650	0.015433	3251.6	3637.4	6.5229	0.012596	3221.0	3598.9	6.4058	0.010575	3189.8	3559.9	6.3010
700	0.016646	3361.3	3777.5	6.6707	0.013661	3335.8	3745.6	6.5606	0.011533	3309.8	3713.5	6.4631
800	0.018912	3574.3	4047.1	6.9345	0.015623	3555.5	4024.2	6.8332	0.013278	3536.7	4001.5	6.7450
900	0.021045	3783.0	4309.1	7.1680	0.017448	3768.5	4291.9	7.0718	0.014883	3754.0	4274.9	6.9386
1000	0.02310	3990.9	4568.5	7.3802	0.019196	3978.8	4554.7	7.2867	0.016410	3966.7	4541.1	7.2064
1100	0.02512	4200.2	4828.2	7.5765	0.020903	4189.2	4816.3	7.4845	0.017895	4178.3	4804.6	7.4037
1200	0.02711	4412.0	5089.9	7.7605	0.022589	4401.3	5079.0	7.6692	0.019360	4390.7	5068.3	7.5910
1300	0.02910	4626.9	5354.4	7.9342	0.024266	4616.0	5344.0	7.8432	0.020815	4605.1	5333.6	7.7653

Table A.14 Superheated steam table

T °C	p = 400 bar				p = 500 bar				p = 600 bar			
	v m³/kg	u kJ/kg	h kJ/kg	s kJ/kgK	v m³/kg	u kJ/kg	h kJ/kg	s kJ/kg K	v m³/kg	u kJ/kg	h kJ/kg	s kJ/kg K
375	0.0016407	1677.1	1742.8	3.8290	0.0015594	1638.6	1716.6	3.7639	0.0015028	1609.4	1699.5	3.7141
400	0.0019077	1854.6	1930.9	4.1135	0.0017309	1788.1	1874.6	4.0031	0.0016335	1745.4	1843.4	3.9318
425	0.002532	2096.9	2198.1	4.5029	0.002007	1959.7	2060.0	4.2734	0.0018165	1892.7	2001.7	4.1626
450	0.003693	2365.1	2512.8	4.9459	0.002486	2159.6	2284.0	4.5884	0.002085	2053.9	2179.0	4.4121
500	0.005622	2678.4	2903.3	5.4700	0.003892	2525.5	2720.1	5.1726	0.002956	2390.6	2567.9	4.9321
550	0.006984	2869.7	3149.1	5.7785	0.005118	2763.6	3019.5	5.5485	0.003956	2658.8	2896.2	5.3441
600	0.008094	3022.6	3346.4	6.0144	0.006112	2942.0	3247.6	5.8178	0.004834	2861.1	3151.2	5.6452
650	0.009063	3158.0	3520.6	6.2054	0.006966	3093.5	3441.8	6.0342	0.005595	3028.8	3364.5	5.8829
700	0.009941	3283.6	3681.2	6.3750	0.007727	3230.5	3616.8	6.2189	0.006272	3177.2	3553.5	6.0824
800	0.011523	3517.8	3978.7	6.6662	0.009076	3479.8	3933.6	6.5290	0.007459	3441.5	3889.1	6.4109
900	0.012962	3739.4	4257.9	6.9150	0.010283	3710.3	4224.4	6.7882	0.008508	3681.0	4191.5	6.6805
1000	0.014324	3954.6	4527.6	7.1356	0.011411	3930.5	4501.1	7.0146	0.009480	3906.4	4475.2	6.9121
1100	0.015642	4167.4	4793.1	7.3364	0.012496	4145.7	4770.5	7.2184	0.010409	4124.1	4748.6	7.1195
1200	0.016940	4380.1	5057.7	7.5224	0.013561	4359.1	5037.2	7.4058	0.011317	4338.2	5017.2	7.3083
1300	0.018229	4594.3	5323.5	7.6969	0.014616	4572.8	5303.6	7.5808	0.012215	4551.4	5284.3	7.4837

Bibliography

David R. Gaskell, Thermodynamics of material, 4th Edition, Taylor and Francis, New York.

V, Ganesan, Internal Combustion Engines, 3rd Edition, Tata McGraw Hill Publishing Company Ltd., New Delhi.

M. L. Mathur and R, P, Sharma, Internal Combustion Engines, 7th Edition, Dhanpat Rai Publications, New Delhi.

Mark Zemansky, Richard Dittman, Heat and Thermodynamics, 7th Edition, Tata McGraw Hill Publishing Company Ltd., New Delhi.

Michel A. Saad, Thermodynamics for Engineers, 2nd Edition, Prentice-Hall, New Delhi.

P.K. Nag, Engineering Thermodynamics, 4th Edition, Tata McGraw Hill Publishing Company Ltd., New Delhi.

M.C. Potter and E.P. Scott, Thermal SCIENCE, Cengage Learning India Pvt. Ltd. 2008.

E. Rathakrishnan, Engineering Thermodynamics, 2nd Edition, PHI Learning Pvt. Ltd., New Delhi.

Shiv Kumar, Fluid Mechanics, 1st Edition, Ane books Pvt. Ltd., New Delhi.

Sonntag, Borgnakke, Van Wylen, Fundamentals of Thermodynamics, 5th Edition, *Wiley India Private Limited*, New Delhi.

Yunus A. Cengel and Michael A. Boles, Thermodynamics, 5th Edition, Tata McGraw Hill Publishing Company Ltd., New Delhi.

© The Author(s) 2022
S. Kumar, *Thermal Engineering Volume 2*,
https://doi.org/10.1007/978-3-030-89216-6

Index

© The Author(s) 2022
S. Kumar, *Thermal Engineering Volume 2*,
https://doi.org/10.1007/978-3-030-89216-6

Printed in the United States
by Baker & Taylor Publisher Services